Atlas of American Agriculture

Atlas of American Agriculture
The American Cornucopia

Richard Pillsbury

John Florin

Macmillan Library Reference USA

Simon & Schuster Macmillan

New York

Prentice Hall International

London **Mexico City** **New Dehli** **Singapore** **Sydney** **Toronto**

Simon & Schuster Macmillan
1633 Broadway
New York, NY 10019

Library of Congress Catalog Card Number: 95–46039

Printed in the United States of America

printing number
1 2 3 4 5 6 7 8 9 10

Library of Congress Cataloging-in-Publication Data

Pillsbury, Richard.
    Atlas of American agriculture : the American cornucopia /
Richard Pillsbury, John Florin.
       p.   cm.
    Includes bibliographical references and index.
    ISBN 0–02–897333–X
    1. Agriculture—United States—Maps.   I. Florin. John William.
II. Title.
G1201.J1P5 1996 <G&M>
338.1 ' 0973 ' 022—dc20   95–46039
                                CIP
                                MAP

Contents

Contents

Maps

Contents

Charts

Tables

Contents

County Views

Acknowledgments

In many ways this book began on a fall weekend trip more than thirty years ago as two young graduate students left State College, Pennsylvania, with their professors heading for a professional meeting in Binghamton, New York. The students, the authors, were new to central Pennsylvania, indeed to the whole eastern United States, and their guides were two of the most articulate observers of the American landscape—Peirce Lewis and Wilbur Zelinsky. While the focus of the trip was on the historic material culture of the Northeast, this was an agricultural landscape, and the inexorable connection between the great Pennsylvania barns, herds of Holsteins, and corn ready for harvest was undeniable. The elaboration of artifacts and the scale of activities were mind wrenching for the students, newly arrived from the Great Plains and California's Central Valley. No classes, no experiences, no reading previously experienced could have prepared us for that trip across the Ridge and Valley into the dairy shed of New York and then southward to Lancaster County's burgeoning agrarian landscapes and finally home to central Pennsylvania. We were never the same again. We have traveled together many years since that trip, but no other journey could match the impact of that one on our intellectual growth. The last three years have provided the excuse and opportunity for an intensification of that tradition as we have interviewed hundreds of farmers while we peered into the agricultural landscapes of forty-four states in search for comprehension of American agriculture's changing face. The text, the photographs, and the maps reflect those trips and our hundreds of hours of musing about this changing scene—from whence it came and about where it might be going. Our vision of the whole has changed much since the beginning of the journey, and we hope that we have successfully conveyed those insights to the reader. Ultimately, however, the shortcomings are ours and the successes are as much due to those who have influenced us in our evolving educations as to our own selves.

A project such as this is only possible with the assistance of a long list of people who helped make it possible. Phil Friedman of Macmillan took our initial ideas scribbled on napkins while dining in a restaurant in the Salinas Valley and provided the encouragement and much of the funding to do this project. Both Georgia State University and the University of North Carolina, Chapel Hill, have provided significant assistance with released time from teaching, travel funds, and research assistance to develop the ideas presented here. Dean Ahmed Abdalal, College of Arts and Sciences at Georgia State, played an especially important role in providing one author with the time to make the project possible. Dean Stephen Birdsall, College of Arts and Sciences, University of North Carolina, Chapel Hill, provided similar assistance and especially encouragement to the other author. Without their aid we would never have gotten to this moment. The University Faculty Research Council and the Institute for Research in Social Science, both at the University of North Carolina at Chapel Hill, provided financial support in the early stages of the project.

Mention of our earlier years and interest is especially important in the final character of this work. For Richard Pillsbury, this meant the hundreds of thousands of miles of driving around the West as a child with parents who were knowledgeable and forthcoming about the passing agricultural landscape and how it came to be. In more recent years John Earl and Mary Pillsbury provided even more helpful insights into the psychology and reality of western ranching. Their understanding of the Central Valley and agriculture generally was crucial in the development of the foundations of this book. Pillsbury was especially fortunate to have attended Chico State College (now California State University, Chico) at a time when Arthur Karinen and David Lantis took him under their wing and provided the all-important intellectual foundations. Arthur Karinen's puckish humor and patience not only provided the author with his first academic paper, "Apples in Paradise," but a vision, which combined with David Lantis's intense energy, has continued to underlay much of Pillsbury's research to this day.

John Florin's evolution was quite different. A Middle Westerner, he was born and raised in Kansas City, Kansas, and Omaha, Nebraska. His father was in the meat packing business, and dinner conversation often centered on conditions in the industry. The city boy was introduced to farm life in summer visits to relatives' farms near

Acknowledgments

Moville, in northwestern Iowa. At the University of Kansas historian George Anderson almost literally lifted him out of the classroom and introduced him to the joy of research with the offer of a Carnegie Undergraduate Research Assistantship. Geographers David Simonett and Richard Smith, both filled with an obvious love of their discipline, then pulled this budding historian as a senior into geography. Still, it was the fortunate selection of the Pennsylvania State University for graduate school, a choice made almost on a whim to introduce this Grasslands boy to the foreign East, that opened him to the joy of landscape observation and interpretation, a joy that is rekindled with every trip through the American countryside.

A number of individuals provided important assistance during this project. Borden Dent and John Freeman have provided the authors with tidbits, important thoughts, and encouragement. Students assisting with data manipulation and endless work include Michael Weinburg, Julia Wharam, and Mark Rehder.

The project would have been impossible without the indefatigable efforts of Jane Andrassi at Macmillan who shepherded it through the system while patiently dealing with our seeming inability ever to learn the nuances of English grammar. We would especially like to reiterate our thanks to Phil Friedman, who presented the authors the rare opportunity to work directly with the book designer, Karen McMichael, and cartographer, Jeff McMichael. The McMichaels' patience and imagination were important parts in taking our ideas and concepts and shaping them into the book that evolved. It would have been impossible to have created this book in this manner without the assistance of these four individuals.

Finally, the project could never have been completed without the forbearance of our frequent absences, both mental and physical, and we must acknowledge the support and aid of Carolyn and Patricia, who have possibly invested more than anyone else in the successful completion of this work.

R. P., Sandy Springs, Georgia

J. F., Chapel Hill, North Carolina

I
The American Agricultural Scene

The replacement of the horse as the primary source of farm power revolutionized rural life in the early twentieth century. Note the heavy, black grassland soils being plowed on this Minnesota farm. (RP/JF)

The American agricultural cornucopia provides such a rich abundance of foodstuffs for the quarter of a billion people living in the United States that our country is able to export $42.4 billion of agricultural commodities to help feed the remainder of the world. A rich, natural agricultural potential and intense application of science and technology have guaranteed, unlike for so many other nations in the world, that the United States' primary agricultural problem will continue to be overproduction for the foreseeable future. American agriculture appears to be a moribund industry mired in tradition, yet even a casual survey, stripped of political and economic rhetoric, suggests that change has become an integral way of life. The crops grown and their geographies, markets, and competition, methods of distribution and sale, and even the farmer himself have all changed radically over the past fifty years. The following discussion will attempt to place this new American agriculture into perspective before more detailed explorations of the various regions and activities are presented.

Historical Evolution

Agriculture in America was well established at the beginning of the Colombian era in the Southwest and along the Atlantic seaboard. The southwestern United States was noted for its intensive irrigated cropping, though irrigation was not utilized in all areas by all farmers. Northern New Mexico had the largest area of canal-irrigated agriculture in pre-Colombian America, though the region's summer rain often made this water a supplement to cropping, rather than its foundation. Hundreds of miles of canals and thousands of acres of land were also irrigated in the Salt and Gila basins of Arizona beginning as early as 130 A.D., but primarily after about 1000 A.D.. Many of these canals were in continuous use for centuries, some for more than 700 years. Antonio de Espejo (1582) reported that their fields were planted with corn, beans, squash, and tobacco. Some southwestern farmers were using terraces and check dams to enhance available water supplies by 1000 A.D. These structures were built across the natural flow patterns to catch runoff and allow it to soak slowly into the ground behind the levees where crops were already planted. This system of water enhancement continued well into the twentieth century.

Eastern aboriginal agriculture is long believed to have been based on shifting (swidden) cultivation. Recent reinterpretations of the data, however, suggest that while fields were cleared by girdling and burning the trees, the fields were not regularly relocated. The difficulties of cutting trees with stone tools, the large acreages of many fields, and the lack of evidence of frequent field shifting all suggest permanent field occupancy. Accounts of ridged fields (upraised areas with long furrows for drainage) indicate that this labor-intensive technology was widely known from New England to the deep South. The continued rotation of beans and corn (or planting them together in tropical garden style) would renew the soil, making regular field relocation unnecessary. Small house gardens for cultivating vegetables, fruit, and other products were widespread in the East, rounding out a relatively complex Amerindian agricultural landscape at the beginning of the era of European occupation.

The arrival of Europeans in the New World immediately changed agriculture in both the New and Old Worlds. While much has been written about the impact of New World crops on the Old, a dramatic change also took place in Amerindian agriculture. The introduction of the horse may have been the most important of the European additions to American life. Though the actual point of exchange is debated, it is agreed that use of the horse was widespread in North America by 1650. Hogs entered the scene as early as the sixteenth century, though primarily as feral animals. Citrus trees grew so well in the Caribbean and Middle America that they were considered pests by the eighteenth century. Peach orchards had become an integral part of agrarian life throughout eastern North America by 1750, often long before the first credited white settler visited the area.

The development of local food production for the European colonists during the early seventeenth century brought increasing numbers of Old World crops to the

Amerindian Agriculture, 1492

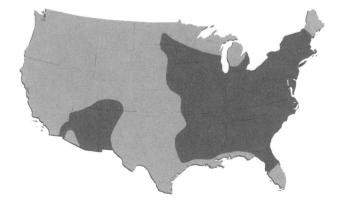

New World. Rice, sugar, indigo, and other commercial endeavors attracted colonists and opportunists who needed a reliable source of food. Wheat, oat, and barley fields, and apples, peaches, and other fruit orchards were all established to provide "proper" food for the European emigrants in North America; yams, rice, peanuts, and other crops were cultivated to feed the large numbers of residents from Africa. New World crops were also transferred from area to area within the New World in the search for wealth, as seen in John Rolfe's introduction of Caribbean tobaccos into Virginia for their sweeter flavors.

European-based American agriculture has always been characterized by its multiple personalities. Commercial plantation cultivation dominated much of the South; general farming geared to produce surpluses of grain and livestock predominated in the settled sections of the Middle Atlantic and New England; great livestock ranches spread across the Spanish-settled areas of the Far West; and subsistence agriculture filled the voids between these commercial agrarian endeavors. Agricultural improvement was the axiom of the colonial farmer. Parts of the Amerindian triad of corn, beans, and squash were adopted initially to provide a secure food base until traditional European crops could be integrated into the agricultural ecology. New foods and agricultural technologies were constantly being introduced from around the world to create an almost laboratory atmosphere of crop and technology experimentation. Potatoes, yams, rice, and a host of other foods new to the northwest European settlers were introduced to the nation's tables as the American diet began its long, distinctive evolution.

While agrarian life has long been viewed as static and slow to change, agricultural historians have begun to suggest that such a view might be simplistic. Atack's (1987) analysis of agrarian technological change in the nineteenth century demonstrated that the American farmer

Farm Size, 1992

Average Acres

- 751 to 84,757
- 401 to 750
- 251 to 400
- 151 to 250
- 0 to 150

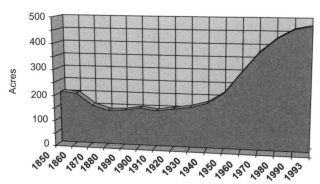

Average Farm Size: 1850-1993

was almost innovation driven in his desire to appear to be modern. Detailed cost/benefit analyses of new and old equipment illustrated that farmers often adopted new technology prior to its becoming economically beneficial. Mechanical reapers thus replaced hand scythes many years before they were cost effective. Combines replaced threshers prior to their perfection, and the tractor began replacing the horse while it was still little more than a curiosity.

The evolution of the American transportation system played a crucial role in the development of the nation's agriculture. Overland transportation of agricultural produce was so expensive before the development of good roads, canals, and railways that the value of goods was often exceeded by transportation costs over a relatively short distance. Grain imported from Europe thus was often less expensive in port cities than its American counterpart grown only a few miles inland during much of the seventeenth and eighteenth centuries. Though threatened by cheap imported grain for more than a century, the New England farmer was not mortally wounded until the opening of the Erie Canal in 1821. Inexpensive midwestern wheat flooded the market,

driving the higher-cost eastern producers into failure or product innovation. Thousands migrated westward, more moved to the burgeoning cities, and the survivors turned to animal husbandry and the production of increasingly exotic (and valuable) fruits and vegetables for urban markets.

The creation of a network of rail lines during the last half of the nineteenth century expanded the agricultural frontier westward into the Great Plains where wheat could be grown even less expensively. Mid-

western farming shifted toward livestock, which could be transported economically to eastern markets, and eastern farmers increasingly specialized in crops for the urban market. Southern plantation agriculture took on a new life as cotton production spread westward away from the coast and rivers into the broad interior of the Gulf South. More intense activities began to appear in the Far West as farmers and ranchers could finally economically ship goods to market.

Transportation costs continued to play

Number of Farms: 1850-1993

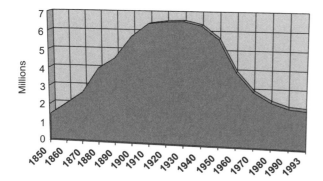

Declining rural populations have brought stagnation to thousands of American farm service centers. (JF/RP)

a declining role in agricultural cropping as now they are a relatively minor consideration for most agricultural activities. The American market is increasingly dominated by a wealth of goods produced in low-cost areas with environments better suited for their production. Coupled with the rise of global marketing, today's consumer is able to purchase almost any desired good at a reasonable price the year around. Peaches from Chile, apples from New Zealand, orange juice from Brazil, cheese from Finland, and onions from Peru have become as much a part of our shopping habits as American flour, apples, and grapefruit are in supermarkets elsewhere in the world.

The Industrial Revolution brought other kinds of technological change that aided in the restructuring of American agriculture. The refrigerated railcar, new wheat and barley varieties introduced by German and Russian immigrants, new plows, reapers, tractors, combines, and processing machinery all arrived on the scene in a matter of a few decades. Not to be outdone by the Carnegies, Rockefellers, and other industrialists of the day, agricultural entrepreneurs soon developed their own marketing cartels

Farm Value, 1992

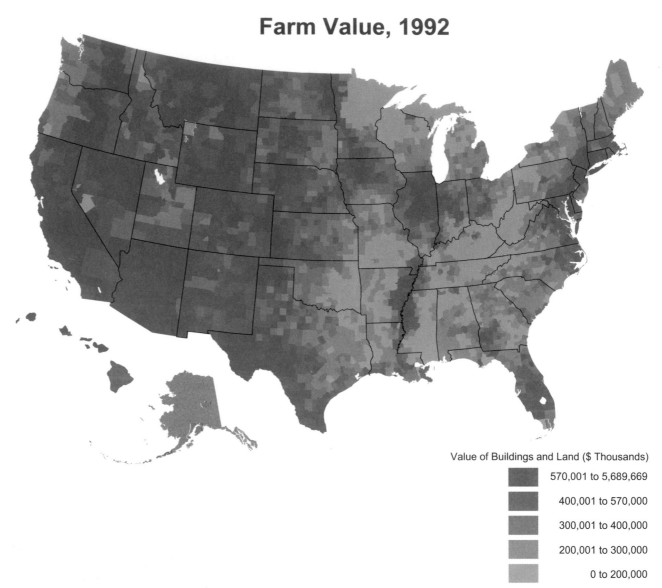

Value of Buildings and Land ($ Thousands)

570,001 to 5,689,669

400,001 to 570,000

300,001 to 400,000

200,001 to 300,000

0 to 200,000

Average Farm Value: 1850-1993

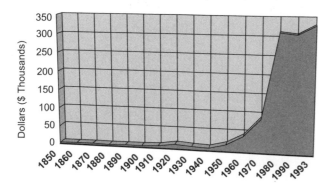

to concentrate agrarian profits in the hands of the processors, rather than the producers. Charles A. Pillsbury bought a half interest in the Minneapolis Flouring Mill in 1869. Thirteen years later he controlled the world's largest flour milling company. Philip Armour, Gustavus Swift, and other meat barons did much the same for the meat industries. Kraft soon dominated the processed dairy industry. Heinz, the

Stokley brothers, and others reshaped our consumption of fruits and vegetables. And possibly even more important, the great grocery chains began to form vast retail food empires, which further encouraged the consolidation of purchasing and ultimately production. Together these entrepreneurial giants lowered costs to consumers, increased the breadth of goods available to the public, and made vast profits while

Operator Age, 1992

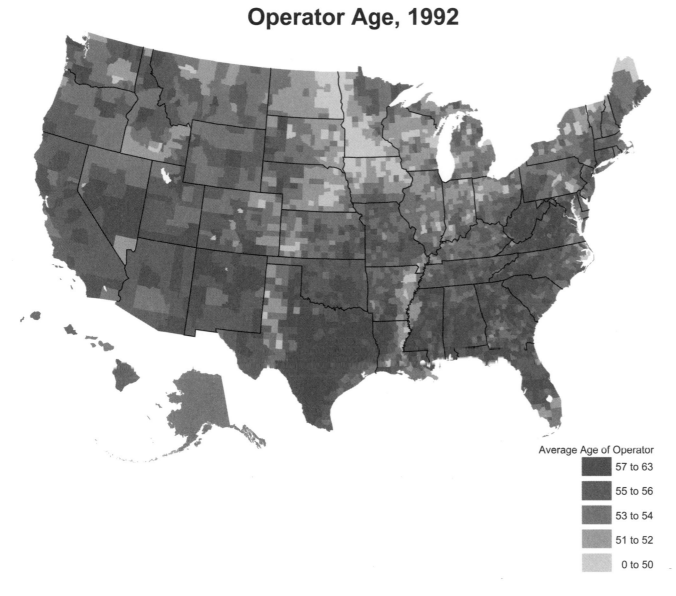

Average Age of Operator

■	57 to 63
■	55 to 56
■	53 to 54
■	51 to 52
□	0 to 50

of the global farm market is just beginning. More than a dozen themes of change within five general topic areas are sufficiently important to be explored before we examine the basic regional complexes and crop patterns individually.

The Larger and the Fewer

The most obvious recent change in American farming is the consolidation of land into fewer, larger units. Average farm size has increased from 134 acres in 1880 to 174 acres in 1940 and to 473 acres in 1993. The average size of working farms (those with sales in excess of $10,000) has increased to 782 acres. Possibly more important has been the rise in number of farms over 1,000 acres from 29,000 to 167,000 during the same period. Simultaneously, the number of farm units increased from slightly over 4 million in 1880 to a peak of 6.5 million during the 1920s and then declined to just over 2 million in 1993. The greatest declines in the numbers of farms have been concentrated in the eastern United States, especially the Atlantic and Gulf Coastal Plains, northern New England, and the northern Midwest. The

often crushing those individuals and companies with the temerity to stand in their way (for fictional treatments of this era see Frank Norris, *The Octopus,* and Upton Sinclair, *The Jungle*).

American Agriculture Today

A common vision of contemporary American agriculture is one of an industry in

financial trouble focusing more on past problems than on future growth. Nothing could be further from the truth. Agriculture in this country is changing at a breathtaking pace. Some changes, like the increasing size of farm units, are obvious; others, like the concentration of the food distribution network into fewer and fewer hands, are less visible. The disappearance of general farming is largely completed; the evolution

Farm Population: 1880-1992

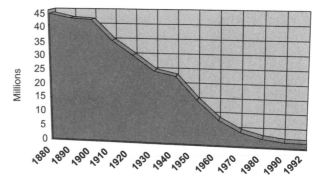

coastal Pacific Northwest has also sustained large declines. These decreases are usually accompanied by decreasing total cropped acreages as marginal farm producers have reverted to tree farming, pasture, and other low input activities. Many suburban counties also have had significant declines in farmed acreages, although the numbers of units may have actually increased.

Going against this tide are increasing numbers of small farms in the Intermontaine West and Rockies, southern Texas, central California, south Florida, and elsewhere. The Intermontaine West/Rockies have experienced the greatest increases, mostly through the establishment of minifarms carved from large livestock holdings. Increasingly intensive agriculture in south Florida, southern and central California, and parts of Texas has tended to create new smaller, irrigated specialty operations. An interesting phenomenon has been the increase in farm numbers in some exurban counties in the Northeast focusing on such specialty activities as horse training, Christmas trees, u-pick fruit and vegetables, herbs, and greenhouses.

Farm economics have been altered dramatically by this inexorable growth of farm size. Increased acreages have brought an explosion of expenditures on efficient equipment designed to handle the larger acreages with reduced labor costs. Less obvious have been the dramatic increases in the entry costs for new farmers. The average value of a farm unit peaked at $349,000 in 1984, and then declined to $319,000 (1992) per unit. These figures are a bit deceptive, however, in that average farm unit values have tended to increase in the most productive states, Florida ($1,315,000) and California ($680,000), while those in the least productive regions have slipped even more than the national averages suggest. Agriculture has clearly become big business, with large initial investments, high risks, and high potential rewards.

Increasing farm investments, in conjunction with the trend toward concentration of production, have focused farm income. Agricultural marketings in Fresno County, California, sales have now passed 2 billion annually, while four others exceed 1 billion, and fourteen more exceed one-half billion dollars. Possibly an even better index of the concentration of production are the nine counties with an average per farm income in excess of 1 million dollars annually.

Farm Marketings, 1992

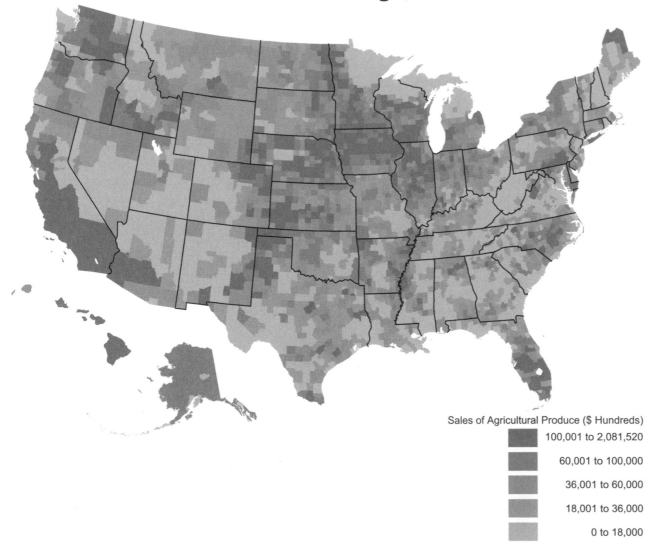

Sales of Agricultural Produce ($ Hundreds)

- 100,001 to 2,081,520
- 60,001 to 100,000
- 36,001 to 60,000
- 18,001 to 36,000
- 0 to 18,000

Changing Demographics

Possibly the most dramatic statistic of American agricultural life over the past century has been the decline of the farm population, both absolutely and as a percentage of the nation's population. Almost half of all Americans lived on a farm a century ago; today, a mere 1.9 percent reside there. Indeed, so few Americans are farmers that the Census Bureau has indicated that it is going to stop publishing that statistic.

The implications of this decline in farm population for day-to-day life in rural America have been disruptive and often devastating. The only labor available is often the farm family itself. If a crop cannot be planted, tended, and harvested by these few individuals, its production often ceases in that area. The purchase of more labor-efficient machinery can extend the ability to increase acreages in some crops, but for many others it does not represent a viable alternative for continuing production with minimal outside labor.

Immigrant farm labor remains available in many areas, though the nations of origin of these workers have changed dramatically over the past twenty years. Legal and illegal laborers from Mexico continue to be the largest source of foreign agriculture labor everywhere except in the northeastern United States, but increasing numbers of laborers from Central America, South America, and Asia are also increasingly seen in seemingly unlikely farm communities. The widening farm labor crisis has altered the balance of permanent and temporary labor as more and more farm operators restructure their operations to find ways to hire the best of their itinerant laborers as permanent farm help. Increasing

numbers of Mexican, Southeast Asian, and Caribbean emigrants are becoming permanent parts of the social and economic landscape of some of the nation's most isolated farm service centers.

Rural depopulation in most areas has brought repeated school consolidation, making the journey to education ever more difficult. Health care has become tenuous as rural facilities close, unable to continue with their declining potential markets. Urban reared and trained doctors and health workers are reluctant to leave the

perceived advantages of the city. Social and cultural amenities are frequently in short supply. Declines have also brought boarded-up Main Street shops and dwindling personal services that even the expansion of Wal-Mart has not been able to alleviate. These changes have in turn forced increasing numbers of the remaining farmers to move their residences into town where services are available. One third of all American farmers and 86 percent of farm laborers no longer live on the farm.

The aging of the American farm popu

Table I.1

Agricultural Marketings, 1992

Rank 1992	Rank 1982	County	Total ($Thousands)	Principal Activities
1	1	Fresno, CA	2,081,516	Fruit (grapes), cotton, cattle
2	3	Tulare, CA	1,386,744	Fruit (grapes), dairy, cattle
3	2	Kern, CA	1,336,886	Fruit (grapes), vegetables, cotton
4	6	Monterey, CA	1,202,715	Vegetables (lettuce), strawberries, nursery
5	4	Weld, CO	1,180,067	Feedlots (cattle and hogs), poultry
6	7	Merced, CA	907,600	Dairy, fruit (grapes), poultry
7	13	Stanislaus, CA	897,058	Dairy, fruit, poultry
8	14	Palm Beach, FL	891,196	Sugarcane, vegetables, nursery
9	9	Riverside, CA	846,932	Dairy, poultry, fruit (citrus)
10	10	San Joaquin, CA	785,050	Fruit (grapes), dairy, poultry
11	5	Imperial, CA	752,980	Cattle, off-season vegetables, hay
12	17	Yakima, WA	689,734	Fruit (apples), cattle, hops
13	11	Lancaster, PA	680,867	Poultry, dairy
14	18	Ventura, CA	667,826	Fruit (citrus), vegetables, nursery
15	12	Deaf Smith, TX	645,029	Feedlots (cattle)
16	16	Kings, CA	581,846	Cotton, dairy
17	15	San Bernadino, CA	567,783	Dairy
18	8	Maricopa, AZ	504,909	Cotton, dairy
19	27	Texas, OK	504,300	Feedlots (cattle)
20	19	San Diego, CA	496,508	Nursery (cut flowers), fruit (avocados)
21	32	Castro, TX	491,514	Feedlots (cattle)
22	24	Grant, WA	481,928	Cattle, fruit (apples and cherries)
23	28	Madera, CA	471,810	Fruit (grapes and oranges)
24	25	Parmer, TX	438,476	Feedlots (cattle)
25	31	Cuming, NE	434,696	Feedlots (cattle and hogs)

Source: *1992 Census of Agriculture*

Marketings per Farm, 1992

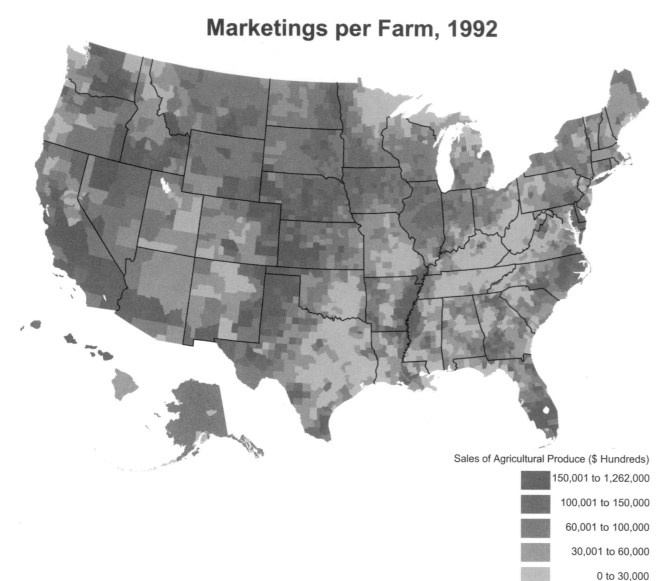

Sales of Agricultural Produce ($ Hundreds)

- 150,001 to 1,262,000
- 100,001 to 150,000
- 60,001 to 100,000
- 30,001 to 60,000
- 0 to 30,000

farm has largely disappeared. At first glance it would appear that this process is actually in its earliest stages with 86.7 percent (1987) of all farms still in individual or family ownership, 9.6 percent in partnerships, and only 3.2 percent in corporations. These statistics are misleading, however, when farm size and productivity are considered. The 3.2 percent of farms that are corporately owned accounted for $35 billion in sales (1987), or 25.6 percent of total farm product sales. The importance of corporate farming becomes much more

Table I.2

Agricultural Marketings per Farm, 1992

Rank	County	Average Sales ($)
1	Haskell, KS	1,261,995
2	Hartley, TX	1,207,121
3	Imperial, CA	1,207,121
4	Scott, KS	1,116,429
5	Grant, KS	1,115,271
6	Wichita, KS	1,099,811
7	Hansford, TX	1,081,689
8	Collier, FL	1,026,536
9	Deaf Smith, TX	1,011,018
10	Monterey, CA	966,036
11	Palm Beach, FL	964,497
12	Castro, TX	945,220
13	Elmore, ID	930,231
14	Moore, TX	908,051
15	Seward, KS	857,950
16	Sherman, TX	788,413
17	Dallas, TX	766,779
18	Yuma, AZ	761,719
19	Texas, OK	716,335
20	Kearny, KS	711,390
21	Hendry, FL	702,590
22	Finney, KS	696,603
23	Parmer, TX	679,808
24	Stevens, KS	678,970
25	Kern, CA	670,118

Source: *1992 Census of Agriculture*

lace is dramatic. The median age of farmers is 48, exceeded only by barbers among the thirty-five occupations with the longest tenure. Farmers' average occupational tenure of 21.8 years is also second only to barbers. More than a quarter of all farm operators are 65 years or older through vast areas of the nation, most notably in the South, Appalachia, southern Great Plains, and the mountainous West. The high entry costs for viable farming today means that yet another round of farm consolidations is on the horizon as these families retire and move to town.

Changing Ownership Patterns

Saving the family farm is one of the great axioms of American political life, yet for all intents and purposes the traditional family

apparent when hobby and nonviable units are not included in the statistics. Farm units with sales of $10,000 or more account for 5.3 percent (1987) of all farms, yet own 14 percent of the nation's farm acreage. Corporate farming is most common in south Florida and the Ranching and Oasis farm regions discussed below, and is least common in the Upland and Lowland South. Possibly more important than ownership are the increasing roles of contract farming and multinational farming enterprises.

Ultimately, the concern about corporate versus family farming has less to do with ownership than with the attitude and behavior of the farmer. Questions of stewardship and continuity are critical issues. Stewardship, the sense that the land is a part of continuing life for generation upon generation, has always been at the heart of the traditional American family farm. There is a gnawing concern in the American ethos that as farming becomes increasingly the province of faceless corporations, these entities will be more concerned with the quarterly return on profits than on the long-term ecology of the nation. Some family farmers may be good stewards and some may be bad, just as some corporations may be good or bad stewards. Yet, a continuing part of the American perception is the belief that there is something good, just, and *right* about the individual farmer tilling his fields that is lost forever when industrial agriculture takes his place.

Agricultural Subsidization

A continuing characteristic in this increasingly corporate and international agriculture is the intense explicit and implicit

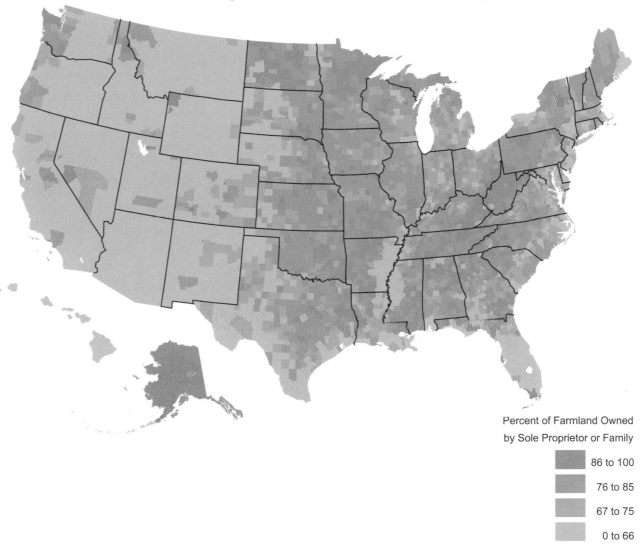

Sole Proprietor Farm Ownership, 1992

Percent of Farmland Owned
by Sole Proprietor or Family

- 86 to 100
- 76 to 85
- 67 to 75
- 0 to 66

governmental subsidization of the industry. Myriad federal agricultural policies, too numerous to explore fully here, have been designed to protect the farmer, the environment, and the nation's food supply. The largest federal programs include the Rural Electrification Administration, Crop Insurance, Farm Credit System, Farmers Home Administration, Commodity Credit Corporation, various marketing agreements and

programs for specific crops and areas, Conservation Reserve Program, Agricultural Conservation and Emergency Conservation Program, Water Bank Program, Forestry Incentives Program, Soil Conservation Service programs, the various livestock grazing programs of the United States Forest Service and Bureau of Land Management, various Department of Interior flood control and multiuse projects that provide irrigation water at below market or fully amortized production costs, and various Department of State programs based on shipping agricultural surpluses (often created because of other subsidy programs) as a part of the nation's foreign policy. Billions of dollars are spent each year implementing these and other programs, which often do not achieve their stated goals.

American farm programs may be broken into three basic types: attempts to guarantee sufficient food at reasonable prices, attempts to provide a better way of life for farmers and farm families, and attempts to reduce the negative impact of farming on the environment. The guarantee of continuing food supplies is the oldest area of government concern. Taxes and tariffs on imported foods such as sugar and grain

Commodity Credit Corporation Loans, 1991

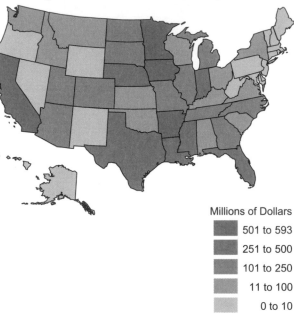

Millions of Dollars

■ 501 to 593
■ 251 to 500
■ 101 to 250
■ 11 to 100
□ 0 to 10

Total Payments: $6.4 billion

Government Payments, 1992

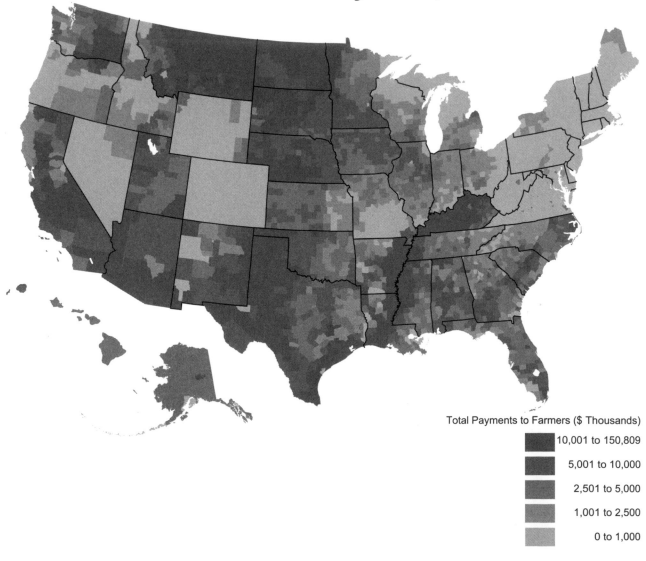

Total Payments to Farmers ($ Thousands)

■ 10,001 to 150,809
■ 5,001 to 10,000
■ 2,501 to 5,000
■ 1,001 to 2,500
□ 0 to 1,000

were the first attempts to ensure that American farmers would be able to market their produce without undue competition from low-cost foreign producers. These have been augmented by a variety of programs administered on the state level such as institutes devoted to locally important products like poultry and aquaculture, the deployment of county agents to provide current information, and the development of demonstration farms to test and illustrate new methods and crops.

Programs designed to control the volume of specified agricultural products in the marketplace, such as acreage allotments and crop price supports, have been enacted over the past half century to encourage continued production of key crops. Acreage and poundage limits established for a number of crops, such as tobacco and

Farm Real Estate Debt, 1991
Federal Credit System

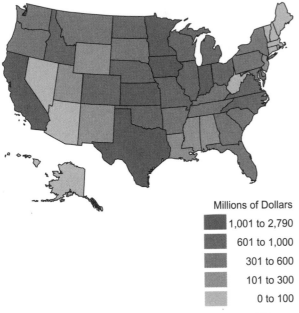

Millions of Dollars

1,001 to 2,790
601 to 1,000
301 to 600
101 to 300
0 to 100

Total Debt to Federal Credit System: $26.9 billion

Farmers of America). The Farm Home Administration and Farm Credit System were founded to make it easier for farmers to obtain funds for improvements of their work and home worlds. The development of Rural Free Delivery by the Post Office Department brought the farms closer to their catalog merchants.

The most recent trend stems from the recognition of the nation's farmers as potentially powerful change agents in the improvement of the nation's physical environment. The Food Security Act of 1985 is typical of this legislation, providing incentives for farmers to act as environmental stewards and penalties (in the form of loss of eligibility for other programs) if they later decide not to comply. The Food,

Agricultural Stabilization and Conservation Programs
Payments Received by States, 1991

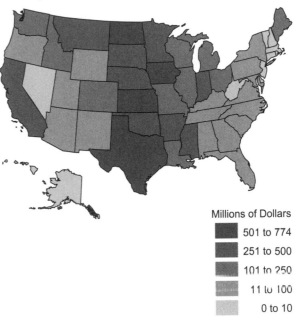

Millions of Dollars

501 to 774
251 to 500
101 to 250
11 to 100
0 to 10

Total Payments: $8.2 billion

Rural Electrification Administration
Long-Term Financing, 1935-1992

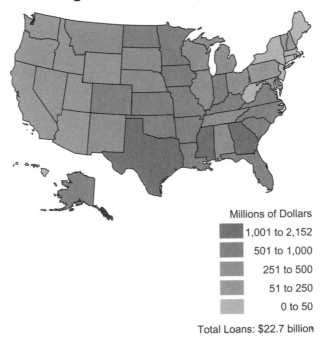

Millions of Dollars

1,001 to 2,152
501 to 1,000
251 to 500
51 to 250
0 to 50

Total Loans: $22.7 billion

peanuts, are designed to control the volume of available product to maintain adequate prices for producers. Federal programs to reduce the size of the nation's dairy herd work much the same. Parity programs, which establish an "appropriate" market price, used to guarantee prices for producers, with the federal government paying the difference if market prices fall below the parity-based price. Many of these price support programs, however, have been manipulated by large-scale producers to maximize their returns to the dismay of those who argue that these programs were designed to support the continuation of the smaller, *family* farms.

Programs to improve farm life have ranged from the REA (Rural Electrification Administration) to the sponsorship of organizations like the 4H and FFA (Future

Agriculture, Conservation, and Trade Act of 1990, for example, makes farmers who prepare wetlands for commodity crop production (even if they do not actually plant a crop) ineligible to receive USDA farm benefits. The Conservation Reserve Program component of this legislation further encourages farmers to create shelter belts, wildlife food plots, contour grass strips, wetland trees, and living snow fences. CRP enrollments have tended to be concentrated in the upper Midwest, Gulf South, and Northwest, though most states have received some benefits from the programs. CRP enrollments were used to assist farmers devastated by the 1993 Mississippi Valley floods with governmental purchase of flood-prone lands that provided farmers with capital while allowing them to purchase and till lands at less risk. As much as $400 million was spent in the Midwest in 1994 to implement this strategy.

The question of continuing agricultural subsidies is not necessarily one of right or wrong, but rather of propriety. As the share of agricultural commodity production from multinational corporations increases, and their ability to influence Congress continues unabated, concerned citizens become uneasy. Certainly disaster relief, the provision of food for the world's starving, research to keep American agriculture at the forefront of international production, and encouraging foreign trade to further reduce the continuing balance of payment shortcomings are admirable goals worthy of our national interest. But one is so often left wondering why the federal government is subsidizing the foreign advertising cam-

Conservation Reserve Program

Permanent Wildlife Habitat, 1986-1992

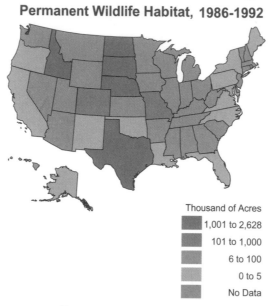

Thousand of Acres	
	1,001 to 2,628
	101 to 1,000
	6 to 100
	0 to 5
	No Data

Total Preserved Wildlife Habitat: 1.96 million acres

paigns of such companies as Grand Metropolitan (British), Beatrice (French), and Nestlé (Swiss), even if they are selling American products. One must wonder if it is appropriate to subsidize irrigation water to lands owned and/or operated by Fortune 500 companies. And possibly most of all, why is the government subsidizing the production and foreign sale of an agriculture product that carries a warning label in the United States indicating that its very use is hazardous to one's health?

Vertical and Horizontal Integration

Vertical and horizontal integration activities have long been admirable goals of American producers and processors. Horizontal integration, the expansion of the production unit to amortize overall costs across larger volumes of production, is the

most common form of expansion. An increasingly important form of horizontal integration is the development of production or marketing cooperatives such as Sunkist and Gold Kist rather than by simply increasing farm unit size. Farmers' cooperatives rose to popularity during the late nineteenth century as a way of aiding farmers in combating the monopolistic policies characteristic of the food processing giants of the period. Marketing cooperatives usually were organized to promote the orderly marketing and distribution of their members' products. Over the years many have added a host of services including product storage, processing, and the distribution of typically needed supplies to their membership. A final form of horizontal integration is the increasing use of contract farming. While some corporate farmers control 150,000 acres or more,

Conservation Reserve Program

Establishment of Permanent Native Grasses, 1986-1992

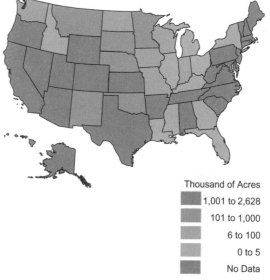

Thousand of Acres	
	1,001 to 2,628
	101 to 1,000
	6 to 100
	0 to 5
	No Data

Total Permanent Native Grasses Established: 8.3 million acres

Conservation Reserve Program

Total Enrollments, 1986-1992

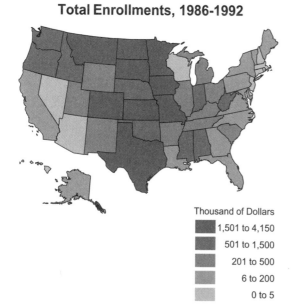

Thousand of Dollars	
	1,501 to 4,150
	501 to 1,500
	201 to 500
	6 to 200
	0 to 5

Total Debt to Federal Credit System: $36.4 million

Farmers' Cooperatives, 1991

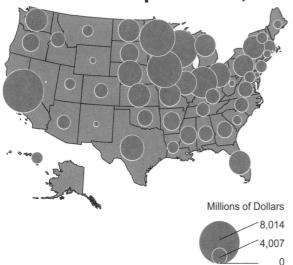

Millions of Dollars

8,014

4,007

0

Total Net Business: $76.6 billion

many distributors and processors have found it more efficient to leave actual production to specialists (farmers) and concentrate on processing, distribution, and marketing the finished products.

Vertical integration has long been seen as one of the most obvious ways to control costs, but the concept has taken on new meaning in recent years. Though the USDA estimates that less than 10 percent of the total agricultural output is directly associated with vertically integrated operators, the practice is becoming rapidly more important each year. Two examples exemplify the vertical integration process. Tenneco, originally a gas pipeline operator, moved into agricultural production when it purchased gas production lands in the southern San Joaquin Valley (California), which incidentally were also some of the nation's richest farmlands. Today the company owns or controls almost 200,000 acres of San Joaquin Valley farmlands. It also owns and operates its own fertilizer

factories, equipment manufacturing arm, farm product distributor, and chain of retail stores.

The J. R. Simplot Company of Boise, Idaho, is probably more typical of agricultural vertical integration. Simplot began in the 1930s with a small ranch in southern Idaho. He built a dehydration plant to maximize his returns on his potatoes in 1940 and moved into frozen food production after World War II. Perfecting the frozen French fried potato, he became the world's largest producer of frozen potatoes in the 1960s. He began beef and dairy feedlot operations to utilize the unused potato sugar beet materials; cheese manufacturing to use his excess milk production; pea, green bean, and other frozen vegetables to utilize unused time in his frozen-food processing plants; fertilizer mining, production, and sales to lower fertilizer costs; and trucking and barge operations to lower his transportation costs. While few companies have been as spectacularly successful as this entrepreneur, his use of vertical integration to control costs is typical of why increasing numbers of larger agricultural concerns engage in the practice.

Agricultural Cooperatives, 1913-1991: Net Business

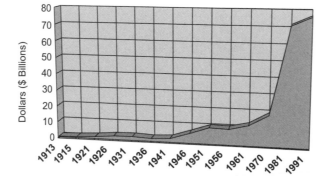

Meaningful statistics on vertical and horizontal integration will probably never be available because of the problems of delineating the boundaries of this phenomenon in its many manifestations. The areas of most rapid growth of vertical integration are among the processors and the largest of the producers, especially the cooperatives. Virtually all the largest cattle feedlot operatives practice varying degrees of vertical integration. The vast majority of all broilers are produced under contract to vertically integrated processor/distributors who in turn are increasingly absorbed with producing and marketing processed foods to retailers. Producers also are increasingly turning to vertical integration to reduce costs through farm cooperatives that provide myriad services to members. The Atlanta-based Gold Kist cooperative, for example, not only provides feeds to its members and processes and markets their output, but is engaged in basic research to develop better strains of poultry, hogs, and catfish, produces and distributes its own brand of fertilizers, and manufactures some of the most specialized farm machinery needed by its members. Indeed, even the increasingly common u-pick operations on the urban fringe are a form of vertical integration as operators lower costs by retailing their own products.

Niche Marketing

While the Doles, Simplots, and Tennecos of American industry continue to expand their operations into new product areas, others are finding it increasingly profitable to develop highly specialized production units. Though usually operating small production units with minimal invest ments, some have become quite large. The most obvious of these niche farmers are the increasing numbers of vegetable and fruit farmers in the urban fringe operating roadside stands and u-pick farms. Freshness, adventure, and nostalgia are the actual products of these retail operations as small Christmas tree growers exhort urban dwellers to bring their children to the country for the adventure of selecting their own tree, and vegetable growers lure the landless to put their hands in the dirt to pick strawberries, apples, or tomatoes in a pitch at capturing a nostalgia market.

Thousands of American farmers have also found that small acreages are sufficient to produce crops with market potentials too small for large producers to enter. The growing of high return crops (truffles), specialty versions of some traditional commodities ("purple" Andean potatoes), and products new to the market (bok choi) all provide niches where aggressive small farmers are able to find profitability. Some larger producers also have specialized in such narrow niches that competitors find it difficult to compete. Yoder Brothers, a nursery supply company headquartered in northern Ohio, produces chrysanthemum cuttings for some of the largest flower nurseries in the world. Operating out of comparatively small greenhouses (fewer than 150 acres) in Florida, California, South Carolina, and Kenya, this company produces more than 20 million cuttings for greenhouses in the United States and Europe with quality and cost controls that have made them the world leader in this tiny market.

The problem with niche production is that success breeds competition. Early California producers of kiwifruit, who largely popularized this product in the United States, created a market that eventually grew to the point where large producers were economically able to enter the market. Many of the smaller producers are now moving toward even more exotic fruit production. Similarly, the growing oriental vegetable market too will eventually attract the Simplots and Doles of the world, forcing the current producers to find other niches to exploit.

The Rise of the Multinational Food Corporation

The evolution of great fruit and vegetable processing companies, meat packers, sugar refiners, cooperative marketing associations, and other producers and distributors has guaranteed consumers a steady supply of reasonably priced food for almost 150 years. The continuing elaboration of the multinational corporation concept over the past decade, however, has begun to change

Direct sale of agricultural products has become an important source of farm income in most urban-fringe farming areas. (RP/JF)

the relationship between the producers, distributors, and retailers and the consumer. The growth of these entities has inexorably moved them from nationalistic companies doing international business to stateless entities. Though based in the United States, Japan, Europe, or elsewhere, these companies have such geographically diverse operations and management that the continued belief that they must always be good (or at least not detrimental) to the home community is no longer relevant in their decision-making process. The worldwide operations of Nestlé, Dole, Cargill, Archer-Daniels-Midland, and others insulate their

decision-makers so that planning is increasingly based on accounting principals, not traditions and long-term associations. A seemingly simple decision, such as producing winter vegetables for the U.S. market in Costa Rica and Chile, may seem relatively minor to a company like Dole, but its impact on the farmers of south Florida and the Imperial Valley (California) is potentially devastating.

The Rise of the Global Farm

A significant product of the stateless multi-national is the rise of the "global farm,"

Table I.3
U.S. Agricultural Foreign Trade, 1992

Selected Partners	($ Millions) Export	Import	Trade Balance
Australia	303	1,121	-818
Brazil	144	1,358	-1,214
Canada	4,812	3,930	882
China	691	369	322
Colombia	124	871	-747
Denmark	131	442	-311
Dominican Republic	258	242	16
France	618	833	-215
Germany	1,091	625	466
Guatemala	116	514	-398
Hong Kong	817	115	702
India	117	280	-163
Indonesia	353	789	-436
Italy	684	862	-178
Japan	8,383	256	8,127
Maylasia	154	339	-185
Mexico	3,676	2,286	1,390
Netherlands	1,813	795	1,018
Philippines	443	482	-39
Republic of Korea	2,200	59	2,141
Spain	951	416	535
Taiwan	1,916	137	1,779
Turkey	344	672	-328
United Kingdom	882	251	631

Source: *Agricultural Statistics*, 1993, Tables 674–75

Direct Farm Sales, 1992

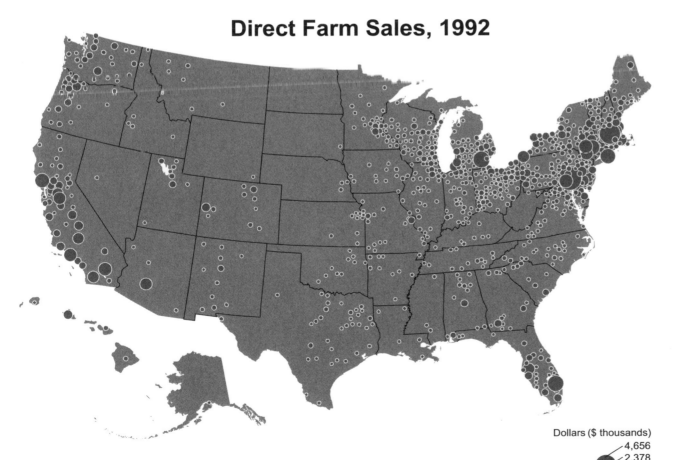

Dollars ($ thousands)
4,656
2,378
100

which markets not to the local region, or even its home nation, but to the world. The United States exported $38 billion worth of agricultural commodities and imported $23 billion in 1992. An analysis of some of these flows gives insight into the operation of this phenomenon. In 1992, for example, the United States exported $1,975 million of beef and veal and imported $1,702 million. Similar flows are seen in a variety of other commodities as the nation imports goods that match market demand and exports others (or the same products at

Table I.4
U.S. Foreign Trade, 1992

	($ Millions)	
	Exports	Imports
Beef, veal	1,975	1,702
Poultry and products	1,193	132
Hides and skins	1,245	126
Wheat	4,318	166
Corn	4,593	
Corn byproducts	870	
Rice		757
Grain sorghum	839	
Soybeans and products	5,597	
Cotton	2,197	10
Vegetables and products	2,790	2,125
Fruit and products	2,786	3,332
Nuts	1,155	432
Tobacco	1,568	1,299
Sugar and related products		114
Coffee		1,798
Cocoa (all products)		1,083
Rubber (all products)		651

Source: *Agricultural Statistics*, 1993, Tables 671–72

periods when they are not available locally to the importers) to regions where there is a demand. As the relative cost of transporting goods has plummeted and the reliability of the distribution system has improved, the importance of shipping distances has declined. Thus a walk through the produce aisle of an American market finds not only the obviously exotic gobee root from Japan and pineapple from Costa Rica, but large quantities of ordinary products such as onions, orange juice, and green peppers that have been imported as well. Similarly, American products are just as visible on the shelves of European, Japanese, and Latin American grocers.

Today's American farmer is thus not only at the mercy of local competition, the weather, and the distribution network, but also of competing agriculturalists around the world. Orange juice from Brazil, apple juice from Mexico, cucumbers from Costa Rica, and apples from New Zealand pour onto the world market in such plenty that all commercial agriculturalists are forced into ever-increasing complex decisions as to what should planted and where it should be marketed.

Consolidation of Production

Increasing market scale has also brought both rapid expansion of production and consolidation of processing units to amortize costs over larger and larger product volumes. These increases in output have emphasized the symbiotic relationship between production and processing to the point that neither can exist without an intense investment by the other. This circular process has processors replacing small plants with new, more efficient ones with vast appetites. Processors in turn demand that their suppliers increase output to allow these new plants to operate at optimal efficiency. Agriculturalists not in the areas served by these processing plants are ultimately forced into other endeavors as they find less and less demand for their output. Whether the end product is milk solids, processed vegetables, or hamburger, the scenario has been the same. Beef producers moved their packing plants from congested urban sites to meet environmental problems and to better serve the feedlots. The feedlots expanded to meet the enlarged needs of the packing plants. This brought an intense concentration of both packing and production while placing a de facto limitation on beef finishing in traditional locations poised to service the old processor sites.

Similarly, the dairy industry, especially solid milk products, has largely been spatially restructured over the past forty years. Wisconsin has fallen to second largest dairy state, as California dairy feedlots, often feeding cattle the waste from other com

Cheese Production

1909

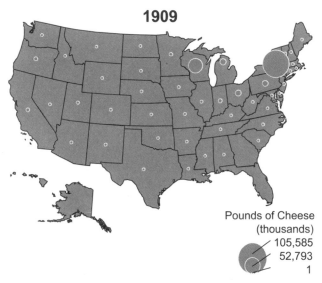

Pounds of Cheese
(thousands)
105,585
52,793
1

1992

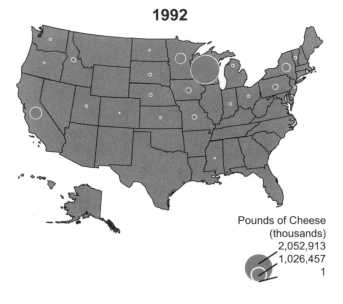

Pounds of Cheese
(thousands)
2,052,913
1,026,457
1

mercial agricultural operations (sugar beet pulp, almond hulls, cottonseed cakes, etc.), have expanded to service the large California market. Idaho has moved into seventh position with virtually no fluid milk market whatsoever. Feeding their dairy cattle sugar beet pulp, potato scraps, alfalfa, and other byproducts of the potato rotation cycle, Idaho milk processors have similarly focused production on cheese (primarily American) and dehydrated milk. Ultimately, large-scale, low-cost farmers will bring the concentration of milk, broilers,

and other products to the point that only a few areas will produce these once ubiquitous items, except for niche operators who focus on their specialty versions. Free-range chickens, for example, are a growing niche market that allows chicken operators outside the current big six production areas tied to their producers to break the consolidation process. But the niche can only exist so long as the market is too small for major broiler processors to consider entering.

The mechanization of contemporary agriculture includes everything from tracking commodity prices on home computers to spraying crops by helicopter. (RP)

Contract Farming

More than 30 percent of all agricultural crops are produced under some form of contract, and more than 90 percent of tomatoes, sugar beets, and broilers, among many others, are grown in this manner. Contract farming is simply the execution of a contract between producer and processor that states that the farmer will plant and deliver the output of a predetermined acreage to the processor on or about a scheduled date at a negotiated price usually based on the market price of the crop at that time. Numerous variations of this theme have evolved, ranging from the broiler industry in which the processor provides all the materials (chicks, feed, hormones, etc.) to the typical sugar beet contract in which a processor agrees to purchase the output of a stated acreage based on a proportion of the market price at the time of delivery. The advantage of contract farming for the processor is the certainty that sufficient product will appear at the processing plant when needed. The advantage to the farmer is that he will receive a known return for his labor and capital investment. Increasingly, contract farmers are also purchasing forward sales contracts on the commodity exchanges to protect themselves from wide price fluctuations after the contract is written.

Contract farming has come under attack by social critics because it transforms the farmer from independent entrepreneur to hired hand. There is little doubt in contract farming that the control of downside risks also removes the opportunity for high profits in the good years. Ultimately, what is lost, however, is the very individualistic spirit that originally attracted many to

mental degradation directly attributable to agriculture include water table drawdowns, most notably in Oglala aquifer (Great Plains) and the Everglades, but actually almost everywhere in the United States; increasing air pollution stemming from rising humidities associated with the irrigation of desert basins and smoke where stubble-burning is widely practiced; runoff carrying high levels of chemicals, especially around feedlots; the disposition of animal wastes; the introduction of numerous diseases and pests from overseas, including the whitefly in Hawaii, the Mediterranean fruit fly in California, and the water hyacinth in Florida. While all these problems are currently being addressed by federal, state, and local governments, as well as by the farmers themselves, the continued concentration of activities in fragile areas means that these problems are going to intensify, not abate.

More than 95 percent of all sugar beets are grown under contract to processors today. (RP)

farming. On a less philosophical note, there is also the very real possibility that the increasing domination of contract farming will ultimately allow processors to force inadequate terms on producers who will be given the alternatives of accepting the offered terms or watching the crop rot.

Increasing Environmental Costs

Concentrated, large-scale farming ultimately increases the potential damage to the environment as even relatively benign activities become so pervasive as to affect the entire local or regional ecosystem. Some of the most well-known environmental consequences of long-accepted farming practices include the ultimate banning of DDT in the United States with the discovery of its side effects when present in the large doses that were then being applied; the impact of phosphate fertilizers on groundwater supplies; and runoff of herbicides, growth inhibitors, other insecticides, and other chemicals from fields as concentrations of specialized field and orchard crops have increased dramatically during the past few years. Other areas of environ-

Water for Irrigation, 1990

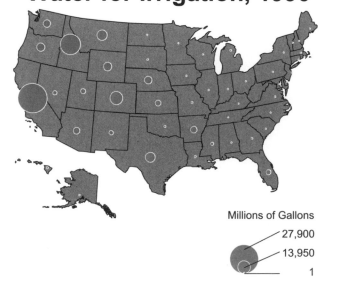

Millions of Gallons

27,900

13,950

1

Natural Farming

These continuing concerns about the effect of the chemicals and other additives used in food production have brought a boom in organic foods, which, in turn, has created a rebirth of "natural" farming. Research on natural pesticides, the increasing use of organic fertilizers, and the invention of a host of so-called natural farming strategies such as stubble-mulching and no-till cultivation spread slowly during the 1960s and 1970s as the small number of devotees honed their strategies. Increasing information about the impact of questionable practices has brought increasingly rapid acceptance by farmers throughout the nation. Several magazines devoted to the topic have appeared and are available to virtually all farmers. A part of the producer interest in natural farming is a rekindling of the issue of the role of environmental stewardship in the role of the farmer.

Lundberg Farms of Richvale, California, represents a classic example of a large farm that successfully adopted a whole new approach to its farming operations largely with the role of environmental stewardship as the goal, though profits have risen as well. Located at the north end of the Sacramento Valley Rice Belt, the Lundberg brothers practiced typical rice methodology for many years. Increasing levels of pollutants in the water, radically declining numbers of waterfowl in the valley, and the recognition of a market niche forced them to rethink their production system and "go organic." While other farmers in the region burn the rice stubble during the winter to clear the fields, the Lundbergs reflood their rice paddies to attract tens of thousands of the estimated 4.5 million waterfowl that winter in the region. The ducks and geese feast on the amazing crop of insects that make the rice fields their home, fertilize the paddies with the resulting excrement, and tromp down the rice stubble that was left after harvest. The Lundbergs are able to forego burning the rice stubble, fertilizing the paddies, and applying large quantities of pesticides for the bug population with little or no loss in productivity. Their strategy has been so successful that it is now being widely tested by other rice farmers to determine its utility on a larger scale.

Increasingly Symbiotic Relationship between Restaurants and Agriculture

The explosion of restaurant dining over the past forty years has reshaped not only the American diet, but American agriculture as well. Restaurant behemoths such as McDonalds Corporation, with more than 9,000 stores in the United States, require such vast quantities of food that a slight menu change can cause tremors in the marketplace. McDonalds and other major restaurant chains play an important role in shaping market demand and thereby the agriculture that fulfills the demand. A decision to add cherry pies, a pork barbecue sandwich, or to change the cheese on cheeseburgers from American to Swiss would immediately produce a demand exceeding supplies, as was demonstrated with the introduction of chicken filet sandwiches and bacon on some hamburgers by fast food companies in the 1970s. The recent expansion in bagel restaurants has driven up the price of durum wheat and incidentally increased national production.

This Wisconsin dairy farmer recycles a tractor trailer load of used newspaper as bedding straw every few weeks. (JF/RP)

Tillage Practices, 1992

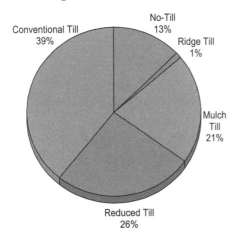

Conventional Till 39%

No-Till 13%

Ridge Till 1%

Mulch Till 21%

Reduced Till 26%

Restaurant chains now recognize the necessity of purchasing thousands or even millions of pounds of key ingredients prior to the first television advertisement of a new product to forestall the risk of lost revenues by being unable to meet demand.

The increasing role of restaurants in providing America's food supplies has created discernible changes in national per capita consumption of foods favored by restaurant chains. For example, the consumption of fresh potatoes has plummeted while frozen potato consumption (restaurant French fries are almost universally a frozen product) has increased. Fresh fruit and flour consumption has dropped, while soft drink, head lettuce, and tomato (ketchup) consumption has dramatically increased. Though these changes are not entirely a product of the restaurant revolution, they certainly are very much a part of it. While few companies currently supply their restaurant chains from their own farms, increasingly long-term contractual arrangements are appearing to ensure the availability of supplies. The Washington hop harvest, for example, is almost entirely produced under long-term contracts to American and European breweries. Most large aquaculture companies have similar long-term contracts with restaurant chains and suppliers to guarantee the availability of their products.

An Afterword

The result of these rapid and far-reaching trends is the transformation of a seemingly moribund industry into a dynamically changing entity. It has been said that American agriculture has undergone more change in the past forty years than in the previous two centuries. Though hard to document, this statement is probably not only true, but conservative. The rapid urbanization of our population, the increasingly multicultural society, and the constantly enlarging role of the global village have created a contemporary agriculture that would barely be recognized by our grandfathers.

II
Cornucopia's Regions

The distinctive character of this Corn Belt farm illustrates the enduring regional look of the American agricultural landscape. (RP/JF)

Agricultural Landscapes

On Division Number Three of the Los Mureitos ranch the wheat had already been cut, and S. Behrman...drove across the open expanse of stubble toward the southwest, his eyes searching the horizon for the feather of smoke that would mark the location of the steam harvester. However, he saw nothing. The stubble extended onward apparently to the very margin of the world.

Norris, *The Octopus*, 1901

One may instantly identify with the imagery of Frank Norris's Central Valley or sense the slowly changing continuity of Jane Smiley's Corn Belt, or the emptiness of Ole Rolvaag's Great Plains. While these are fictional writings, the places they describe are real. The identification and delineation of these and other agricultural landscapes are an important aspect of the development of a full understanding of American agriculture. The exploration of individual crops that proceeds this discussion is a valuable beginning, but can never be complete without placing the individual crops into their regional context. Corn, for example, is the most important crop by value in America today, yet a very important part of its role in American agriculture is its dominance of an entire region's agriculture for more than a century and its continued codominance of that region today with soybeans. One cannot speak of corn without evoking images of the Corn Belt: of the rolling low hills covered with alternating fields of corn and soybeans, of its thousands of compact farms that once produced pork and beef for the packing

houses in Chicago, Dubuque, and St. Louis, and in its own way the Corn Palace (Mitchell, South Dakota) covered with thousands of ears of corn woven into mosaics depicting the life beat of the region.

Three criteria are generally used to define agricultural regions: crops, technology and farming strategies, and environment. Crops are the most commonly used of these factors, but their use solely does not catch the essence of what makes an agricultural region distinctive. Certainly even a brief visit to California's Tulare

County and New York's Champlain lowland suggests that not all dairy farms are the same. Today's complex behavioral and agricultural decision-making environment reflects an intricate web of connections among the perceived goals of farmers (which are not necessarily to maximize net returns); regional attitudes about place, family, and environmental stewardship; and the relationship of the farmer to the marketplace, which increasingly controls crop distributions, utilized agricultural technologies, and ultimately the agricultural regions that reflect them.

Agricultural Provinces, 1909

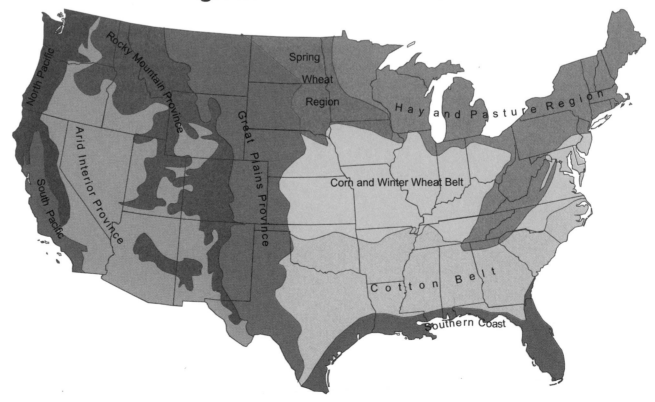

Historical Development of Region Building

The concept of agricultural regions predates attempts at mapping them. The term "Cotton Belt" was widely used by midnineteenth century, though the evolving region itself was apparent as early as the 1830s. The Corn, Wheat, and Dairy Belts are a bit more amorphous, and their identifying terminology didn't stabilize until the twentieth century. The first governmental attempt to create a comprehensive national agricultural regions map appeared in the *Yearbook of the Department of Agriculture, 1915* as an appendix titled "A Graphic Summary of American Agriculture." Graphic summaries have appeared periodically since that period, though only the early editions include agricultural regions maps.

The 1915 map, based on the 1909 census, recognized five eastern and five western provinces (regions) with the explanation that:

> The United States may be divided into an eastern and a western half, characterized, broadly speaking, one by a sufficient and the other by an insufficient amount of rainfall for the successful production of crops by ordinary farming methods.

Corn was the distinguishing crop in the East with about one third of farmed acreage, though the Corn Belt was incorporated into a larger Corn and Winter Wheat Belt that stretched from the Chesapeake Bay to eastern Kansas. The Cotton Belt was also identified, while the Eastern Dairy Belt and the eastern Upland South are combined

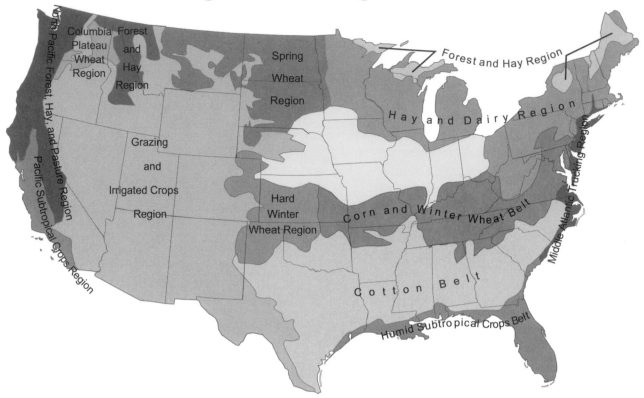

Agricultural Regions, 1925

with the Dairy Belt to form a Hay and Pasture Province. Although the foundations of the contemporary regionalization of the Northeast were present in 1909, they were too subtle for the authors to recognize with the data they had available.

The West's five provinces are surprisingly similar to the contemporary concept of the region, although the boundaries have evolved spatially with the expansion of irrigated agriculture. The Pacific Coast was subdivided into two regions quite similar to today's Pacific Industrial Region. The Arid Interior is possibly the most changed western province as irrigation has brought a blossoming to the lower Colorado, the central Columbia, and the upper Rio Grande and Snake River basins.

The Department of Agriculture's second attempt at mapping regions was little different from the first one. The 1931 version, however, shows increasing sophistication. Appearing as *Miscellaneous Publication 105*, this map is familiar to modern readers as the basis of most later textbook and atlas attempts to delineate agricultural regions. It broke the nation into fourteen regions. The Cotton, Corn, Wheat (in two parts), and subdivided Dairy Belts had clearly stabilized in the East. The beginnings of Megalopolis were identified as the "Trucking Region." The evolving domination of South Florida as the center of early- and off-season fresh vegetables was still in its infancy at this time, thus the southern production zone was designated as two independent subtropical specialty zones. The West was little different than today with the inclusion of the Rockies into the Grazing and Irrigated Crop Province that dominated the Intermontaine West. The Columbia Basin was clearly important enough to be recognized, though its eventual role in regional agronomics was not understood. The Pacific was unchanged.

The Depression and changing interests halted the development of these interpretive maps by the Department of Agriculture. Versions did continue to appear in textbooks and atlases, but a serious attempt to create a postwar regionalization did not occur until the 1954 Census of Agriculture's introduction of economic development regions. A new modernized regionalization quickly followed with the publication of Haystead and Fite's *The Agricultural Regions of the United States* (1955). Academic tastes had changed since 1925 and the concept of broad, interpretive regions had lost favor. Geography, agricultural economics, and other interested disciplines became obsessed with statistically derived solutions to problem solving, with declining interest in what has come to be called "exceptionalism." Haystead and Fite's map, as a result, has only nine categories (including a catchall general class in which no type exceeds 50 percent of activities) based on the primary types of cropping recognized. No attempt at generalization was made. The Census's use of smaller economic regions made this approach appear advantageous, and certainly some of the detail lost in broad interpretive regions is now visible. Haystead and Fite, for example, begin their book with a lengthy chapter titled "The nonexistent 'typical' farm" in which they state:

In short, our agriculture is of seemingly endless variety, with new crops constantly being bred at home or imported. The cost of the operation may run from a few hundred dollars capitalization to literally many millions, the size from a hydroponic tub to the nearly one million acres of the King Ranch. The outstanding fact about it is that *there is no typical American farm.*

While in detail this view is true, it is also true that the traditional ways of doing business for farmers continued, as did the agricultural regions that they supported.

Warren and colleagues' 1969 *A Regionalization of United States Farming* continued the largely quantitative approach to region building with a bit more sophistication.[1] The use of the smaller economic areas enabled the authors to delineate twenty-five agricultural subareas, which included many of the standard regions, but also many smaller "specialty" production areas. The creation of smaller regions with a minimum of variation in activity was a clear step forward. The deluge of detail, however, continued.

Haystead and Fite set the new pattern for regionalization. Writers and students since that time have either attempted to update the 1931 USDA on the basis of intuitive reasoning or to modify their statistically derived microregions to ever more complex depictions. Many authors of

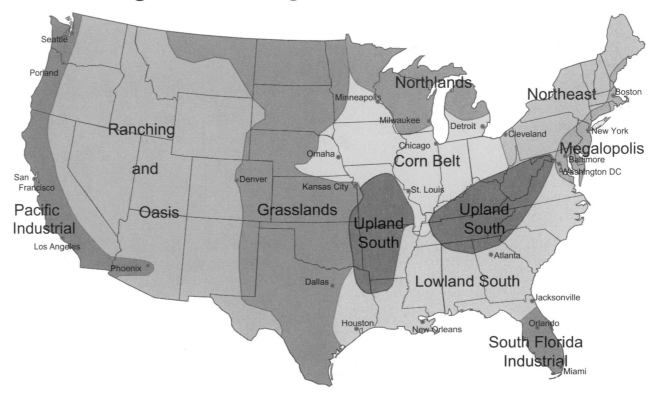

Agricultural Regions of the United States

regional geography texts, for example Fisher and DeBlij, have chosen to work with the earlier map. Others attempted to simplify the microregionalizations of Warren and his followers, most notably Patterson's most recent text on North America. Most, however, chose to ignore the problem. There has not been a major attempt to regionalize agriculture, in the manner of Haystead and Warren, for more than twenty-five years.

Contemporary Agricultural Regions

A region is a method of abstracting spatial associations to capture the similarity of places. Only ten agricultural regions are presented, with the belief that less is more. Two regions, Megalopolis and the Eastern Dairy Belt, are discussed together because of their symbiotic relationship, but it should be understood that each still stands independently.

The initial clustering of counties used here was based on factor scores generated by a principal components analysis of key crop and agricultural characteristic distributions. The final generalization was tweaked on the basis of detailed examinations of the variables and visits to every region and subregion.[2] This minimalist approach has been used because we believe that there is a commonalty of overall agricultural activity within each of the regions presented. Agriculture in the Northlands, for example, is dominated by the production of corn and hay silage to feed the region's thousands of dairy herds. Several important subareas, such as the eastern Michigan fruit and the Minnesota Valley Processed Vegetable Belts, however, are also very much a part of this

agrarian landscape. Both field and statistical examinations of this specialty area suggest that it would be simplistic to continue use of the traditional term "Dairy Belt" to describe this increasingly complex place. The following exploration of our nation's agricultural regions thus should be viewed as a symphony of ten parts, each with numerous, distinct countermelodies running through to elaborate the whole.

Notes

1. This work is adapted from an earlier regionalization presented by Morris E. Austin in USDA Agricultural Handbook 296.

2. A principal components analysis is a statistical routine that examines how the relative values of every observation (county in this instance) relate to a plane of "best fit." Values or scores are assigned on how well each observation (county) relates to that plane. In its simplest sense this routine could be described as a three-dimensional multiple regression analysis.

The Agricultural Northeast

Northeastern agriculture has been undergoing a rapid reorganization over the past fifty years that has transformed vast areas from traditional general farms to highly specialized holdings targeting niche markets. The region today must be seen as two interrelated agricultural zones. The eastern is fringe shaped and pummeled by the relentless urban sprawl of its cities, which forced the creation of an agriculture little seen or understood in the past. The western core continues much as before, but just as cognizant of the economic role of those urban centers lying a few miles to the East. Interrelated as these two regions remain, they must be discussed as separate entities as the core economic presumptions of each are so different. While Megalopolis has been transformed almost completely into the "city's countryside," the farmers of the adjoining Agricultural Northeast still fight to retain traditional agricultural ideals, crops, and livelihoods in the face of their inevitable demise.

Megalopolis

The term Megalopolis was introduced in 1955 by Jean Gottman in his pioneering study of the suburbanization of the eastern seaboard. Gottman's urbanized region stretched from Portland, Maine, to Richmond, Virginia, and permeated all activities, rural and urban alike. The agriculture of that region remains distinctive forty years later, though somewhat modified in appearance and geography. The Megalopolis

agriculture region is more than a statistical myth, it is a very real place.

The reach of the city into the countryside is pervasive. Its influence reaches more than fifty miles beyond the continuous

edge of suburbanization as farmers, speculators, and urban residents weigh the cost–benefit ratios of investment in the city's countryside. This process in turn reshapes each of the three basic elements of farm

Megalopolis and Northeast

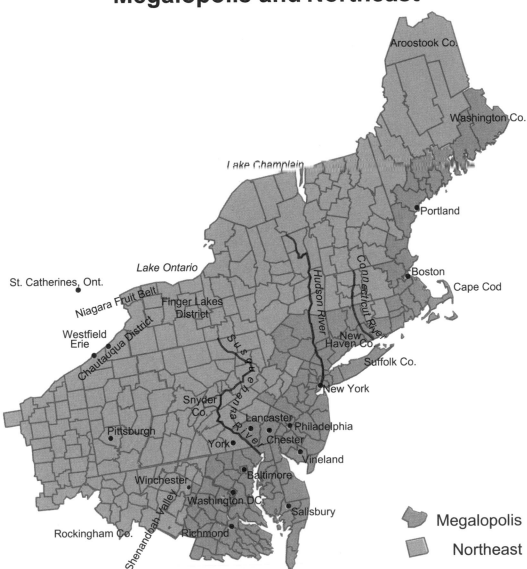

29

Agriculture in much of the Northeast has been little affected by the urban growth taking place just a few miles to the east. (JF/RP)

inputs—labor, capital, and land. Farmers note the specter of the oncoming urban monster and determine whether continuing to invest in permanent improvements is fiscally responsible. Speculators estimate the rate of spread and attempt to determine if purchasing land for future development would be economically beneficial. Residents seeking rural land for recreation weigh the problems of access and cost against the perceived benefits. The end product of these decisions is a process of agricultural change that gives farming in the urban

fringe a distinctive character. Although the following discussion focuses on the urbanized northeastern corridor, the same processes and changes are taking place in the urbanizing fringe of every city in the nation.

The cost, availability, and usage of labor change drastically as the growing urban fringe spreads into new areas. Traditional agricultural workers discover that the newly accessible urban jobs offer higher wages, less personal risk, long-term future employment opportunities, and health, retirement,

and other benefits. Their exodus leaves their employers in a quandary. There is no one to replace them. Farmers are forced to choose between automating their current activities and selecting crops that can be cultivated with little or no hired labor.

New Haven County, Connecticut

A classic exurban Megalopolis county at the northern edge of the New York City commuting zone. Straddling the Connecticut River Valley as it enters Long Island Sound, the area is a picturesque landscape of historic villages, a faltering industrial economy, and wealthy exurbanites. Nursery crops for the local market dominate the agricultural scene with lesser amounts of fruits and vegetables, often sold from hundreds of roadside stands that dot the county's rural roads.

	1992†	1982†
Farms (number)	381	440
Average size (acres)	68	68
Agricultural marketings	32,175	26,703
Direct sales	746	n/a
Nursery/greenhouse	20,946	13,480
Bedding plants	10,909	7,743
Cut flowers/foliage	3,235	n/a
Potted plants	3,147	n/a
Dairy sales	4,114	6,325
Vegetable sales	2,843	1,916
(corn, pumpkins, peppers)		
Fruit	2,054	2,024
(apples, peaches, strawberries)		
Poultry sales (eggs)	797	1,026
Beef cattle sales	623	716

†Figures in thousands of dollars unless otherwise noted.

Typically this has meant replacing large-scale commercial vegetables, fruit, and dairying with beef cattle or field crops. An alternative solution is to reduce acreages of fruits and vegetables while increasing net returns per acre through roadside retail sales and u-pick operations.

Capital factors shape agriculture by altering the balance of cost-effective and cost-ineffective investments. The arrival of suburban-style housing developments precludes increased investment in noxious activities, such as pig farms, and improvements that must be amortized over lengthy periods, such as permanent sprinkler systems. Rising land costs make it difficult to purchase additional land to amortize the increased costs of further mechanization. Renting land, however, becomes easier. Retired farmers, holding land for future sale, and speculators who have purchased land are encouraged by the real estate tax structure in many states that taxes active agricultural land at lower rates than other real estate.

The impact of these processes on the farming landscape is gradual, but ultimately overwhelming. The land on the fringe first takes on a frowzy look as increasing numbers of fields lie empty or are poorly tended. Neatly fenced fields of corn, timothy, beans, and potatoes increasingly are interspersed with abandoned land of waist-high weeds, grazing cattle, and a handful of horses under the trees in the corner. Individual, modern urban-style homes begin appearing behind moats of neatly tended lawn among the ancient farmhouses, unpainted barns, and rusted farm machinery. Hobby farms, estate farms, nurseries and greenhouses, roadside stands, and hand-painted signs exhorting the weekend wanderer to pick green beans and corn, select a Halloween pumpkin, or cut their own Christmas tree suddenly are no longer anachronisms, but have become common. Weathered taverns are resurrected to become chic bed and breakfasts or gourmet restaurants catering to clients who live miles away. Suddenly a subdivision of cheap, prefabricated homes on postage stamp lots surrounded by a picturesque fence with flags waving and signs extolling the virtues of country living appears along the curving country road. More cattle, more subdivisions, and, by osmosis, this agrarian landscape is transformed from farmland to the "city's countryside" and then into a part of the city itself.

Cattle, both beef and dairy, and the production of feed have been the most space-consuming agricultural activities in Megalopolis throughout the twentieth century. Interspersed within this livestock economy were specialty areas utilizing unique environments, market proximity, and local marketing genius to develop highly concentrated agricultural landscapes.

The Horse Farm Belt extending from Charlottesville, Virginia, northward into New Jersey is one of the most dramatic estate farm landscapes in the nation. (RP/JF)

Dairy and beef farms still remain, but production has dwindled. The specialty areas continue as well, but more often reflect the adaptive qualities of the modern farmer than traditional agricultural prowess. Nursery/greenhouse products are the leading agricultural group in most of the smaller states and a major force in the remainder. Direct sales, u-pick, and specialty production for local markets have become major marketing strategies, along with general adaptation traditional cropping to the new realities of production.

All the historic specialty cropping areas remain, but usually in somewhat abbreviated importance. Some specialty areas, such as the Cape Cod cranberry region, have been stable throughout the past fifty years, while others, such as Long Island, have virtually disappeared. Each must be viewed individually to understand why and how they have come to their current level of activity.

Henry Hall planted the nation's first commercial cranberries in a Cape Cod bog in 1810. The cranberry bogs of Cape Cod and mainland Plymouth County were soon joined by plantings in the Pine Barrens of central New Jersey and later by fields in the wetlands surrounding Lakes Wisconsin and Oshkosh (Wisconsin). Cranberry acreages expanded with the assistance of marketing cooperatives to peak at 27,640 nationally (1930–31). This cyclical industry bottomed nationally in 1963 (20,100 acres). Massachusetts acreage has been stable at between 11,000 and 12,000 acres since the turn of the century with only minor market-induced fluctuations. Total production for the state, about 90,000 tons in 1992, has continued to rise because of increasing yields per acre.

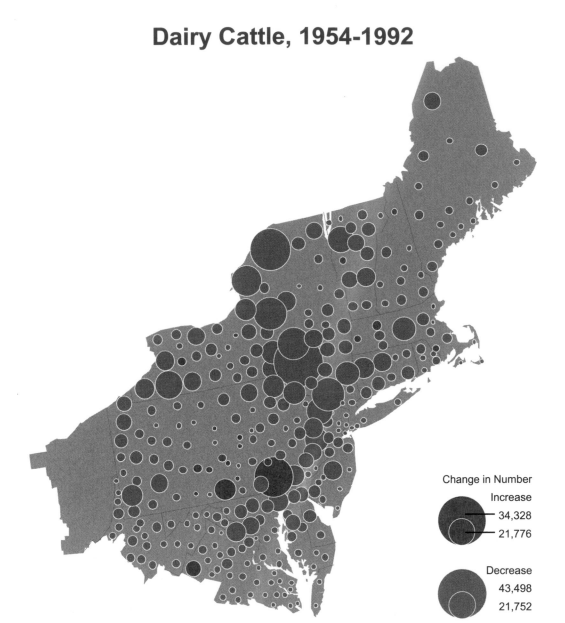

Dairy Cattle, 1954-1992

Change in Number

Increase

34,328

21,776

Decrease

43,498

21,752

The United Cape Cod Cranberry Company, later Ocean Spray Cranberries, Inc., introduced mass production of canned berries in the 1930s. Cape Cod growers average only about fourteen acres each and find it difficult to expand their operations because of high land costs. Unable to expand (as is true also in New Jersey), their relative position in the industry is declining. The industry in Massachusetts is protected to a degree by lower agricultural tax rates and the increasing use of farmland protection acts, but ultimately Cape Cod's role in the cranberry business will decline.

Wisconsin has risen to become the nation's second largest producing state with about 68,000 tons of berries harvested annually. New Jersey follows with about

18,000 tons, although Oregon's rapidly expanding cranberry bogs near Coos Bay have quadrupled production over the past decade to about 15,000 tons of harvested berries annually. Washington state farmers produce about half that amount.

The Connecticut River Valley cigar tobacco agricultural specialty area developed in the early nineteenth century. Though always small by national standards, Connecticut farmers harvested 471,000 pounds in 1839 and Massachusetts and Connecticut farmers, a total of 2,336,000 pounds in 1899. The region specializes in shade-grown (the plants are cultivated under a netting of specially designed cheesecloth), cigar wrapper, and to a lesser extent, filler tobaccos. Connecticut River Valley production peaked in 1919 with 62.4 million pounds cultivated on 40,000 acres. It still produced 40.7 million pounds as late as 1950. Tolerance of public cigar smoking, coupled with health concerns, has sharply reduced consumption. Production dropped to 8 million pounds (6,000 acres) in 1975. Only 2.9 million pounds of shade-grown tobacco were harvested in 1992, mostly for wrappers.

Long Island, for years an important part of the New York City truck garden area, was best known for ducks and new potatoes. The expansion of New York City's suburbs into these easily accessible agricultural lands after World War II ultimately forced space-consuming and noxious agriculture from the region. Long Island duck production slipped from 4.8 million (Suffolk County) in 1954 to about half that number in 1992. Suffolk County's 1954 potato production of 15.2 million bushels has virtually disappeared today. Continuing production of ducks at a reduced level, nursery/greenhouse products, grapes, and some fresh vegetable production, especially cabbage and cauliflower, remains, but the island's population growth from 500,000 (1950) to more than 2 million (1990) makes agriculture expensive.

The Hudson Valley developed into a major center of vegetable and fruit production for the New York City market in the nineteenth century. New York's wine industry began here in 1839 and some small wineries still exist. Apples, pears, and cherries were important, as well as a variety of fresh vegetables such as sweet corn and tomatoes. Apples and sweet corn are the most important survivors, although little local produce enters the national market.

New Jersey once was truly a garden state, producing vegetables, fruit, dairy products, and meat for nearby urban markets. Apples were the first important specialty fruit crop, often processed into cider and applejack. Essex County produced 300,000 gallons of applejack in 1810, while a local sparkling cider was often sold as champagne. Cranberries in the Pine Bar

Highly specialized crops dominate Megalopolis agriculture today. This New Jersey cranberry bog is harvested by flooding the field, loosening the fruit from the plants, and skimming the floating berries from the surface. (Burgess McSwain)

Megalopolis, Vegetable Production, 1992

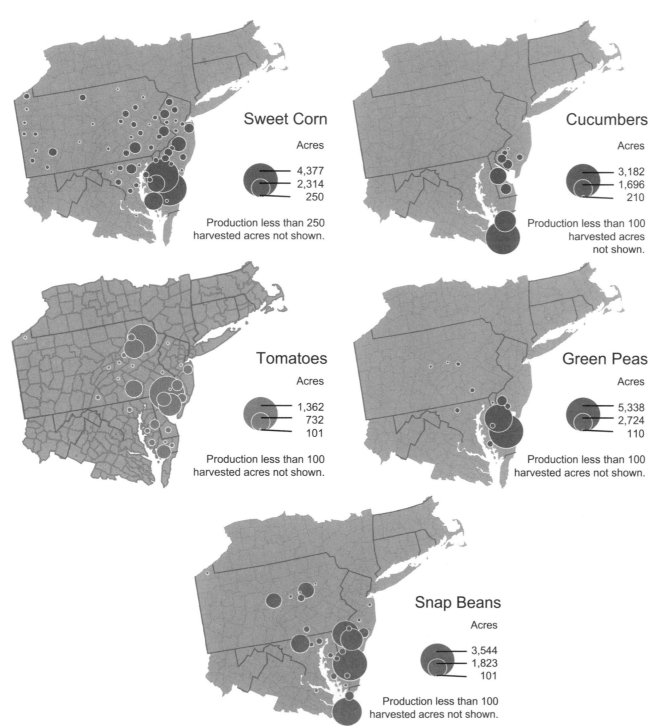

Sweet Corn

Acres

4,377
2,314
250

Production less than 250
harvested acres not shown.

Cucumbers

Acres

3,182
1,696
210

Production less than 100
harvested acres
not shown.

Tomatoes

Acres

1,362
732
101

Production less than 100
harvested acres not shown.

Green Peas

Acres

5,338
2,724
110

Production less than 100
harvested acres not shown.

Snap Beans

Acres

3,544
1,823
101

Production less than 100
harvested acres not shown.

rens in south Jersey soon followed. Viticulture began about 1860 in Egg Harbor, and several wineries soon followed. The Vineland community was created by a Philadelphia developer in 1861 as a future center for grape production, though wine was prohibited. Thomas Welch, a Wesleyan preacher, doctor, dentist and arch-prohibitionist, settled there in 1868. He developed a pasteurized grape juice based on Louis Pasteur's research, which he marketed widely by 1870. Losing his vineyards to black rot a few years later, Welch moved to Seneca Falls (New York) in 1896 and to Chautauqua County (New York) in 1897 where large acreages of Concord grapes were already widely grown.

Tomatoes were the most important vegetable, as well as asparagus, green beans, peas, and other cool weather crops. The state was an important early center for canned vegetables because of the role of the Campbell's and Franco-American packing companies.

The degradation of agriculture in the Garden State paralleled its transformation into residential suburbs after World War II. The rapid increase of population from 4.2 million in 1930 to 7.8 million in 1992 quickly eroded the stock of available farmland. The number of farms declined from 25,000 in the 1930s to 8,316 in 1964, where it has stabilized in number, if not output. New Jersey farm receipts have dropped to less than 0.4 percent of the nation's total, and the state ranks thirty-seventh in gross sales.

Nursery/greenhouse products are the state's most valuable ($182 million, 1992) crop complex today. Vegetable production is second with $56.6 million sales in 1991 and dairying is third with slightly less than

$50 million in sales. Cranberry production slowly continues to rise, but has become concentrated in Burlington County. Tame blueberries are similarly concentrated in adjacent Atlantic County. Gloucester County, in the southern section of the state, is the leading peach producing county in the Northeast with about 32 million pounds (1992). Apple production, in contrast, is declining rapidly from 92 million pounds in 1978 to 62 million in 1987 to only 39 million pounds in 1992. Gloucester is now the largest apple producing county in the state.

Dairying has long been the most space-consuming agricultural activity of southeastern Pennsylvania. Lancaster County's 92,000 dairy cows, more than twice the next largest county in the Northeast, reflect the traditional agrarian values held by this largely German, and sometimes Amish, farm population. Adjacent Chester County has long been the most important mushroom producing area with total production from Chester, Berks, and Lancaster counties passing $156 million per year. This area originally concentrated on processed product and did not participate in the rapid growth of fresh production in the 1960s with the rise of salad bars. Production continues despite increasing urban pressures because of the relatively small space needed for the crop. A small air-cured tobacco industry has also been found along the Maryland border for more than a century, but has declined to fewer than 13.7 million pounds (1991) per year.

Lancaster, Chester, Berks, and Lebanon county farmers have benefited significantly from the decline of New Jersey agriculture over the past decade or two. Lancaster, one of the richest agricultural counties in the

Broilers, 1954, 1992

1954

1992

Broilers Sold

194,185,728
97,097,864
10,000

Counties with sales less than 10,000 not indicated.

nation (Table I.1), has seen farm marketings increase from $575 to $681 million over the past decade, while Chester County marketings have risen 37 percent to $283 million, Lebanon County's have risen 47 percent to $132 million, and Berks County's 46 percent to $238 million. Animal husbandry has been the chief beneficiary of this growth with significant increases in dairy cattle, beef cattle sales, hogs, eggs, and broilers. State duck sales have increased to over 750,000 fowl per year.

The Delmarva peninsula, including Delaware and the eastern shore areas of Maryland and Virginia, has long been a center of vegetable production for northeastern markets. Maryland was the nation's largest producer of tomatoes in 1909 with 42,721 acres, followed by New Jersey with 26,552 acres. California had a mere 5,932 acres at that time. Other important vegetables included sweet corn, green peas, early potatoes, onions, watermelons, and muskmelons. Tomato production had risen to more than 50,000 acres by the 1930s, but was second to California. California continued to increase its dominance in the post–World War II era and Maryland production had slipped to 3,410 acres in 1978 and to only 2,700 acres in 1991. Similarly, fewer than 2,000 acres of potatoes remain and only about 7,000 acres of sweet corn are grown for processing.

Chickens are the success story for the Delmarva subregion. The growth of American chicken consumption in the post–World War II era was especially important as two of the nation's larger producers, Frank Purdue and Country Pride, developed there. From fewer than 2 million broilers in 1934, Maryland's production had reached 58 million by 1951, 221 million by 1978, and currently is more than 265 million annually. Sussex County, Delaware, and Wicomico, Worcester, and Somerset counties, Maryland, surrounding Salisbury, the state's broiler capital, produced 183 million birds in 1992.

Farmland Preservation Movement

Creeping suburbanization had become so pervasive by the 1960s that regional planners and environmentalists began to become concerned about the continuation of rural landscapes and the containment of the city. Although the greenbelt concept evolved in England in the 1890s, it never received widespread support in the United States because of the importance of individual property rights to the nation's self-image. The need to achieve the same effect in the exurban fringe, however, was present. Concerned groups began to purchase development rights to individual parcels in threatened areas of exceptional beauty such as southeastern Pennsylvania's Brandywine Valley and Long Island during the 1960s. State-financed programs began to evolve in the mid-1970s.

The Connecticut farmland preservation program, enacted in 1978, has preserved almost 24,000 acres on 157 farms (1994). The Connecticut farmland preservation program generally targets somewhat larger than average farms engaging in a wide range of farming activities. The program's goal is to preserve 85,000 acres of cropland to ensure that at least half the fluid milk and 70 percent of the in-season fruits and vegetables can continue to be produced locally. Increasing numbers of environmentally conscious farmers are now wholly or partially donating development rights to preserve their family farm's way of life for all time.

Northeast

America's first important dairy region, the Agricultural Northeast, includes most of northern Appalachia and New England. This area evolved from general farming in the early nineteenth century, to livestock in the midnineteenth century, and, with the advent of reliable rail transport, into providing fluid milk for the eastern industrial cities in the late nineteenth century. New York was the nation's leading dairy state in 1910 with 1.6 million dairy cows and $75 million of receipts (1909) from dairy product sales. The 1914 Census of Manufactures tabulated 995 cheese factories and 576 creameries in New York state alone. Only 768,000 dairy cattle remained on New York farms in 1990, and 1.8 million in the nine-state agricultural region. The parallel decline in supporting farm acreages of silage, hay, and other field crops has radically changed the agricultural landscape. Cropland has fallen below 30 percent of all land in more than half the region

Table II.1
New England Farmland Preservation Programs, 1994

State	Date of Inception	Total Farmland	Number Farms Preserved	Acres Preserved	Farmland (Percent) Preserved
MA	1977	680,000	314	29,460	4.30
CT	1978	420,000	154	23,535	5.60
NH	1979	480,000	33	2,787	0.60
RI	1982	66,000	30	2,428	3.70
VT	1987	1,510,000	84	27,422	1.80
ME	1990	1,142,000	1	307	0.03
Totals		4,298,000	616	85,939	2.00

Source: Joseph Dipple, director, Connecticut Farmland Preservation Program

and below 5 percent in much of New England.

Dairy production has become increasingly concentrated as the areas best situated to serve the large urban markets increase herd sizes and often total numbers of animals. Southeastern Pennsylvania, a short distance by interstate highway from Philadelphia, Baltimore, and Washington, DC, is prospering in this competitive environment with significant production increases. Though declining, central New York continues strongly, as does the Champlain Lowland in northern New York and Vermont. Declining milk production, however, is tied to increasing concerns about cholesterol, changing diets, and fewer numbers of young children. It is unlikely that the decline will abate, especially with the growing production of cheese and other processed products by low-cost western producers.

Five specialty agricultural zones are also located in the Northeast agricultural region. The American portion of the Niagara Fruit and Vegetable Belt along the southern shore of Lake Ontario is the largest. Utilizing the lake effect of moderating temperatures during the spring and fall, this area has specialized in vegetables and deciduous fruit since the late nineteenth century. More than 13 million bushels of apples were harvested in 1909, with smaller amounts of peaches (16,000 acres), plums (5000 acres), and cherries. Apples continue as the dominant fruit today (725 million pounds in 1991), with only small amounts of other fruit. Growers have tended to be slow in planting the newer, more popular varieties, and, as a result, have tended to receive lower prices for the less popular traditional nineteenth-century varieties such as McIn-

tosh and Northern Spy. About two thirds of the crop is processed almost equally into applesauce and juice/cider.

The region continues to be an important eastern center of frozen products partially because General Foods located many of their earliest plants there. Vegetables continue to be locally important,

especially cabbage, sweet corn, snap beans, onions, and green peas.

The Chautauqua fruit zone straddling the New York and Pennsylvania border south of Lake Erie has long been the largest center of grape production in the East. Catawba grapes were first grown in here in 1818, and the first winery was built in

Dairy Sales, 1992

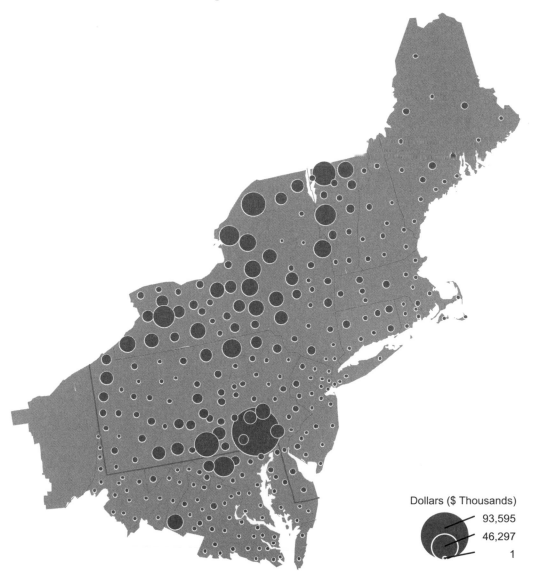

Dollars ($ Thousands)
93,595
46,297
1

Lakes Districts Fruit Belts, 1992

Apples

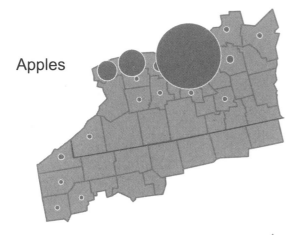

Cherries

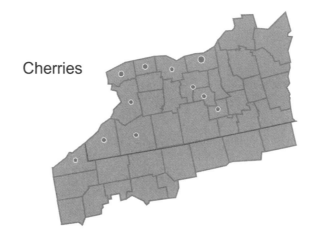

Grapes

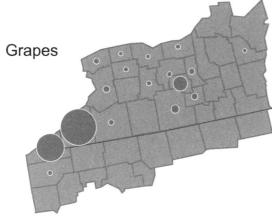

Peaches

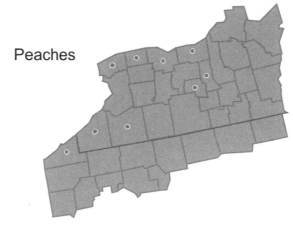

Plums

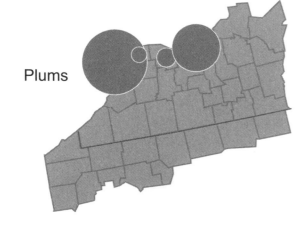

Pounds

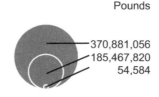

370,881,056
185,467,820
54,584

Production levels less than
50,000 pounds not shown.

1859. The introduction of the Concord grape in the 1870s and the arrival of Thomas Welch in 1897 changed the direction of future development. Welch built the world's first large-scale grape juice plant in Westfield and production soon hit 300,000 gallons. More than a million gallons were processed in 1909. Westfield remained the center of Welch's production, although plants were built in St. Catherines, Ontario; Lawton, Michigan; and Springdale, Arkansas, before 1920. About 200,000 tons of grapes are still produced annually with two thirds of production crushed for juice (Concord) and the remainder (primarily varietals) used in local wineries or sold to home producers.

Viticulture in the Finger Lakes region of New York began after 1829 when Catawba and Isabella shoots were brought from the Hudson Valley to Hammondsport. The industry today is dominated by five wineries, three owned by Seagram, though several dozen small wineries also dot the area. Traditionally the wineries depended on American grape varieties, especially Catawba, Isabella, and Delaware. French hybrids such as Aurora and several *Vitis Vinifera* varieties have been introduced in recent years with some interesting results. This area was previously the largest producer of American champagne, but recent advances have focused on light pinks and some very competitive varietals.

The limestone valley lands stretching from Winchester, Virginia, to the Susquehanna River in Pennsylvania provide a rich agricultural setting close to the nation's largest cities for specialty farming. Winchester has been a center of apple production since the nineteenth century. Most orchards are planted along the edges of ridges to take advantage of air drainage to produce some of the first apples of the season. Virginia farmers, like their New York counterparts, have been slow to switch to modern varieties and the fresh market has been eroding for many years. The vast majority of all fruit is now processed. Increasing suburbanization from Washington and the construction of government offices, residential subdivisions, and office parks westward along Interstate 70 into nearby West Virginia offer little incentive

Snyder County, Pennsylvania

Nestled among the alternating ridges and valleys of Pennsylvania, agriculture here was dominated by dairy farms raising cattle, corn, and hay for more than a century. Expanding market demand centered in Lancaster County, fifty miles to the south, has brought growth in broilers, eggs, cattle, and even the dairy industry as total agricultural marketings have increased by two thirds since 1982.

	1992†	1982†
Farms (number)	664	731
Average size (acres)	131	129
Agricultural marketings	55,582	37,066
Direct sales	341	n/a
Poultry sales	19,357	12,921
(Broiler sales, 5,267,075)		
(Turkey sales, 328,624)		
(Layer inventory, 341,422)		
Dairy sales	15,503	11,678
Hog sales	7,527	4,129
Cattle sales	6,205	2,662
(Cattle sold, 17,883)		
Corn sales	1,691	1,795

†Figures in thousands of dollars unless otherwise noted.

Lakes Districts Vegetable Belts, 1992

Cabbage
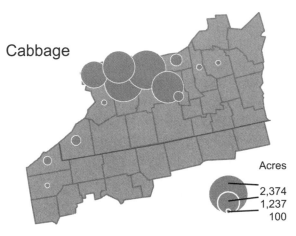

Acres
2,374
1,237
100

Production less than 100 harvested acres not shown.

Sweet Corn
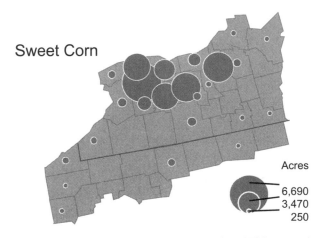

Acres
6,690
3,470
250

Production less than 250 harvested acres not shown.

Green Peas
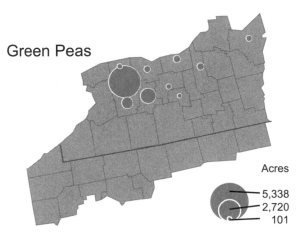

Acres
5,338
2,720
101

Production less than 100 harvested acres not shown.

Snap Beans
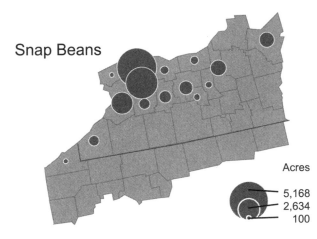

Acres
5,168
2,634
100

Production less than 100 harvested acres not shown.

for farmers to take the longer view of their activities.

The Cumberland Valley of Pennsylvania, the northern extension of the Shenandoah Valley, also has significant fruit orchards, especially apples, peaches, and cherries. Egg

production has increased rapidly in recent years to replace supplies that had previously been shipped from New Jersey. More than 5.5 billion eggs were produced in Pennsylvania in 1991.

Maine is the home of two of the

nation's more interesting relict specialty regions. The potato area along the Canadian border developed as a major potato area in the late nineteenth century after the development of rail links with southern New England. Aroostook potatoes soon became the premier source of seed potatoes in the nation, though never the largest area of overall production. The rise of frozen French fries and baked potato consumption hurt most eastern producers as it was discovered that the western russet potato produced a far superior product than the varieties suited for eastern cultivation. Production in New York, Pennsylvania, Maine, and other eastern states plummeted. The superior quality of Aroostook potatoes, however, has guaranteed that the region would continue, though at lower production levels. Aroostook County growers harvested more than 23 million hundredweight of potatoes in 1992, less than half its peak production level just prior to the fast food revolution, making it the largest single county of potato production in the nation. Idaho producers harvested almost five times more potatoes than Maine farmers in 1992.

The process for commercially canned blueberries was developed in Washington County, Maine, in 1866. This isolated area near the Canadian border has remained the dominant center of wild blueberry production in the nation since that time. The rise of tame blueberry production in Maine and elsewhere, however, has reduced the relative importance of this area in the nation's total blueberry production, although Washington County farmers harvested more than 70 million pounds of wild and 1.5 million pounds of tame blueberries in 1992.

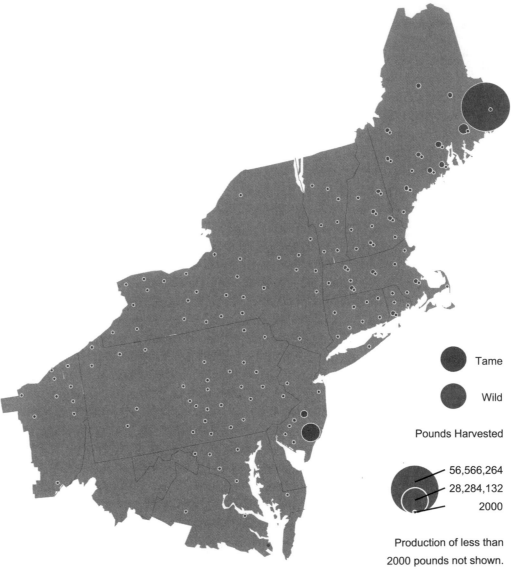

Blueberries, 1992

Tame

Wild

Pounds Harvested

56,566,264
28,284,132
2000

Production of less than
2000 pounds not shown.

The Corn Belt: America's Agricultural Heartland

The Corn Belt is the quintessential United States agricultural region. Ask most casual observers to describe the characteristic American agricultural scene and they will talk in terms of the family farm, operated by its owner with help from the children and just, perhaps, a hired hand or two. The imagined farmstead consists of a rectangular white farmhouse, a large barn to store harvested crops and house animals, and an assortment of smaller outbuildings. A large penned area near the farmstead is filled with hogs. The farm itself is fairly large to provide a decent income for a farmer and his family. The seemingly unending pattern of rectangular fields creates a patchwork of corn, small grain, open pasture grazed by cattle, and hay. The corn is grown to feed to the farm's hogs, while the hay feeds its cattle. Small towns, obviously agricultural service centers, dot the landscape in a remarkably consistent fashion, one about every ten miles, making them easily accessible to farmers marketing their crop and for shopping. What has been described is the American Corn Belt—or at least the classic image of this quintessential agricultural region.

Like so much of American agriculture, that image has a more solid basis in historic fact than in contemporary reality. The Corn Belt, so conservative and unchanging at first glance, is undergoing massive reorganization. Farms are becoming fewer and larger. Ever larger cultivated acreages are increasingly assembled through rental or sharecropping fields owned by retired farmers or from the unused land of aging neighbors. The upper Middle West leads the nation in the percentage of farmers who are part-owners. Unlike the stereotypical sharecropper or renter, however, these farms typically are larger and more profitable than those operated by neighboring full owners. More than one half (55 percent) of Iowa farmers rent land. The average farmer who is a part-owner cultivates 490 acres, compared with 165 acres for full owners. This expansion has created farms composed of noncontiguous fields scattered over tens of square miles, especially in the most productive areas. The sight of tractors, combines, tilling equipment, and spraying equipment lumbering down the backroads of the Corn Belt between scattered fields has become as common as test strips of corn and abandoned farmhouses in recent years.

The farmstead has been simplified, with many outbuildings discarded and destroyed. The standard agglomeration is now likely a farmhouse, a couple of low, rectan

Corn Belt

gular, metal equipment storage sheds, and possibly a barn now pressed into service as an additional equipment shed. The fields are larger; their patchwork irregularity simplified.

Today one can drive the secondary roads across the Iowa heart of the Corn Belt without seeing a single hog. Cattle and hayfields are less common than they were a few years ago. This once simple landscape has been made even more so with large swatches of corn interspersed with rolling fields of soybeans and little else. These large fields are commonly unfenced—why fence a field if there are no large farm animals to wander in and destroy the crop? The fences of decades past have largely been removed to accommodate the increasingly automated agribusiness that is conducted there. Only California's Central Valley among all of America's extensive agricultural landscapes is today more completely used for crop agriculture. Hogs and, increasingly, poultry are still important parts of the economy, but these are concentrated in great, flat barns holding hundreds of pigs and thousands of birds under a single roof, invisible to the casual traveler.

The farm community immortalized by Sinclair Lewis and others has dwindled in importance and size. A street or two of brick buildings, mostly derelict or boarded up, characterize these communities today. The co-op along the tracks, a hardware store, an antique shop, a pharmacy, a cafe, and a tavern or two remain as testaments to better times. Most of these places are too small or forlorn to attract a Wal-Mart or other discount store, and local farmers have become accustomed to driving to the next town to the consolidated high school or regional medical center.

Today's Corn Belt could probably more accurately be called the Corn-Soy Belt, or even the Corn/Soybeans/Hog Belt. Yet, it is corn that most clearly marks its location and defines its presence. This heartland region can be most simply described as that area where corn can most readily be grown without irrigation. Its northern margin marks the zone where the growing season is too short and cool for extensive corn production. Beyond its western boundary rainfall is too sparse to support its cultivation. The rugged terrain to the east does not support the large-scale activity needed to profit from growing grain. To the south it is not climatic barriers, but rather the opportunity to grow other crops with higher net returns that identifies the region's boundary.

Development of the Corn Belt

The origins of the Corn Belt are found in river valleys of southern Ohio, notably the Miami, which joins the Ohio River at Cincinnati. That city, founded in 1789,

The traditional Corn Belt hog lot is disappearing with the expansion of automated large-scale facilities. (RP)

provided an easy jumping-off point for settlers from Pennsylvania floating down the Ohio in search of agricultural opportunity. These early settlers brought with them their notions of mixed farming, crop rotation, and returning animal manure to the fields to maintain fertility from Pennsylvania. They, and most other early settlers across the breadth of the heartland, first turned to wheat as a major export crop. Wheat thrived in the better-drained, upland virgin soils of the frontier, was easily shipped, and maintained a ready market. The heavier, more nutritious wet prairie soils of the Grand Prairie of Illinois and the Des Moines glacial lobe in northcentral Iowa were generally ignored by the early wheat farmers. These soils produced a wheat crop that was tall and luxurious, but with very sparse seed heads.

Repeated wheat crops, however, created nitrogen deficiency and reduced yields. Thus, wheat production moved westward with the frontier. What emerged, first in southwest Ohio but eventually across the climatic margins of the Corn Belt, was a three-crop farming rotation based on corn (borrowed from southerners who had also drifted into southern Ohio), a small grain (typically either oats or winter wheat), and a hay (in early days clover, then in later years the more productive alfalfa). Both clover and alfalfa are legumes, capable of extracting nitrogen from the air and storing it underground. Corn was planted after hay to take advantage of this increased nitrogen. The small grain, less important than corn to the farm economy, survived on the remaining soil fertility after the corn crop.

This basic rotation dominated Corn Belt agriculture for over a century. The first modern U.S. Census of Agriculture in 1924 found that more than 95 percent of the harvested cropland of Iowa, Illinois, Indiana, and Ohio had been given over to corn, oats, wheat, and either clover or alfalfa. Oats were fed to the farm's draught animals, hay to its cattle, and corn to its cattle and hogs. Thus, the most important cash crops of the Corn Belt economy were beef and pork, not hay and corn. The relatively small wheat harvest was also sold as a cash crop, but never constituted a significant source of overall income.

Corn was the key crop in this rotation, and because hogs can convert corn to meat more efficiently than cattle, they became the dominant farm product for market. Cincinnati's first hog packing house was established in 1818. By the 1830s the city proudly called itself Porkopolis, boasting that "it was Cincinnati which originated and perfected the system which packs fifteen bushels of corn into a pig and packs that pig into a barrel, and sends him over the mountains and over the ocean to feed mankind" (Hudson, 1982). Chicago passed Cincinnati as the nation's leading center of hog slaughter around 1860, and by late century its own boast was that it was "Hog Butcher to the World." Today, the packing houses have moved even farther westward to smaller cities. The two largest producers now are the Iowa Beef Processors plant in Dennison, Iowa, and the Hormel packing house in Austin, Minnesota. Still, all of these processing centers were within the Corn Belt, taking advantage of the dominance of the corn/hog relationship in the agricultural region.

Twentieth-Century Modifications

By the 1930s the Corn Belt had pushed up against both its geographic and production limits. The westward and northward moving settlement frontier had carried the corn/hog economy with it, but always with some delay as wheat continued to be the first crop of choice. By 1880 the northern and western margins of the Corn Belt extended in a broad, undulating arc from southwestern Ohio across northcentral Indiana and northern Illinois and Iowa to southeastern Nebraska. It generally moved no more than fifty miles north or west during the next half century. While the corn/hog economy was usually more productive than the alternatives to the north and west, the climatic margins of corn had been met. Countless corn races were available, and new ones constantly produced, but all easily pollinated between the Flint and Dent types. Their production levels dropped drastically with shorter or drier growing seasons. Similarly, within the Corn Belt yield limits had been reached, and little expansion in per acre production was achieved in the fifty years prior to 1930.

Four massive changes have modified the nature of Heartland agriculture since the 1930s: hybridized corn, soybean substitution, farm mechanization, and restructuring of hog production. Two seed companies, Funks in central Illinois and the Pioneer Hi-Bred Corn Company in Des Moines, first advertised hybrid corn seeds for sale in 1926. In 1937 two centers of hybrid adoption, centered on the area of each of the two seed companies, had emerged. Within four years the majority of farmers across the Corn Belt were planting hybrid seeds.

Hybrid corn has had a dramatic impact on the Corn Belt. Utilizing heavy fertilization (40 percent of all agricultural fertilizer used in the United States is applied to corn) and closer planting (potentially doubling the number of plants per acre), the corn farmers have been able to sustain continuing, dramatic increases in yields since the 1940s. Per acre corn yields ranged from 120 to 130 bushels across the Corn Belt in 1990, a fivefold increase over the production in the early 1930s. Already the most valuable crop in the region, corn's role was intensified by hybridization. New hybrids that thrived under shorter growing seasons, with longer growing days pushed the Corn Belt northward into southern Michigan, a bit of southern Wisconsin, and across a broad area of southern and southwestern Minnesota. The current northern limit of the Belt was not established until the 1950s. Hybrids adapted to drier conditions and extended the Belt westward, especially along the Platte River Valley in Nebraska.

There were nearly four times as many farms as tractors in the Corn Belt in 1930. Twenty years later the numbers were roughly equal. The replacement of draft animals with tractors increased the potential acreage a farmer could cultivate, encouraging increases in farm size while reducing the amount of land that had previously been given over to producing feed for the draft animals. The development of better hybrid corns, which grew to uniform heights and had uniform maturation dates, encouraged the mechanization of the corn harvest. Larger producers most often had the capital (or credit) to purchase successively larger tractors and production equipment and were able to reduce unit costs and increase profits. This additional capital could then be reinvested in more land or purchase even more efficient equipment, which further increased the economic distance between the larger, highly mechanized farms and those that were less financially secure.

Soybeans were an ideal addition to corn in the region. Corn and soybeans thrive under similar soil and climate conditions. Their planting and harvest cycles are similar. The same farm machines can be modified to cultivate both crops. Soybeans have had few governmental planting restrictions. Global demand has remained generally strong, with relatively stable prices. Finally, soybeans are a legume, returning nitrogen to the soil and thus replacing clover and alfalfa in that role on many farms. As a result, the American soybean harvest increased from .5 million acres in 1924 to 11 million acres in 1949 to almost 58 million acres in 1991. Iowa and Illinois lead the nation with 9.2 million and 8.7 million acres planted, respectively, followed by Minnesota, Indiana, Ohio, and Missouri.

The last major change in the Corn Belt landscape has been an overwhelming shift in the structure of hog production. The rhythm of Corn Belt life on countless farms revolved for more than a century around the cultivation of corn to feed a small herd

Corn, 1992

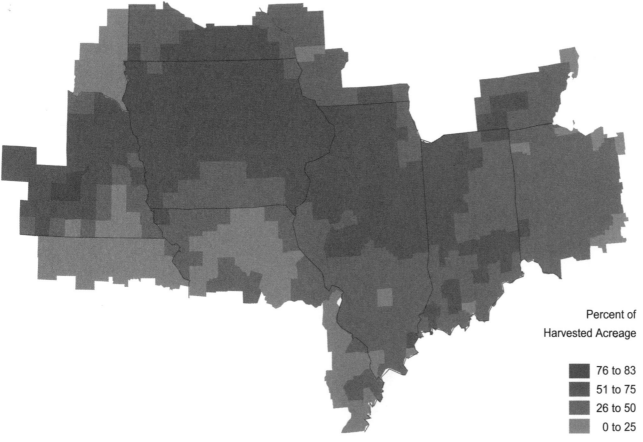

Percent of
Harvested Acreage

- 76 to 83
- 51 to 75
- 26 to 50
- 0 to 25

of hogs. The shift of meat packers to larger, more efficient, centralized slaughterhouses encouraged the evolution of large-scale hog production. The average sales of Iowa hog farmers climbed from 135 per farm in 1959 to 787 in 1992. Two thirds of 1992 Iowa hog sales were generated by farms with sales of 1,000 hogs or more, while eight counties averaged sales of more than 1,000 per hog farm.

Simultaneously the center of hog production moved westward during this period. Henry County, Illinois, was the leading hog producing county in 1964, with just over a half million hogs sold in that year. While production in Henry County remained almost static, phenomenal growth of production took place in western Iowa. Sioux County, Iowa, was the leading county in the nation in 1992 with 1.2 million hog and pig sales, and seven counties exceeded Henry County production, all in Iowa. Even further changes in hog production appear to be in the offing as North Carolina producers not only average almost twice the sales average as Iowa farmers, but currently are expanding production so rapidly that it is estimated it will become the largest producing state before the next agricultural census.

Diversity within the Corn Belt

The agricultural landscape of the Corn Belt is not as homogenous as the overwhelming importance of corn, hogs, and soybeans might suggest. That tripartite domination is most evident on the drained wet prairie soils of the Des Moines lob in northcentral and northwestern Iowa and southwestern Minnesota. This is an area of corn, soybeans, hogs, and not much else. The cli-

Soybeans, 1992

Percent of Harvested Acreage

- 51 to 75
- 26 to 50
- 0 to 25

mate and soil are ideally suited to the crop pair. Farmland prices, generally the highest in both states, reflect the intensity of use. Des Moines lob farmland in Iowa was generally valued at over $1,500 per acre in 1990. Farmland in the hilly, clay soiled, southcentral portion of the state, still a part of the Corn Belt, could be purchased for under $600 per acre.

The Grand Prairie of east central Illinois, ignored at first by settlers because of its very wet, heavy prairie soils, was the core of the development of a commercial grain economy in the region. Drainage, generally sponsored by large companies, gradually opened the area to settlement.

John Deere began his farm implement empire here by developing a self-scouring plow that could cut through these prairie soils. The farmland was so rich and productive that the three-part rotation including hay never became locally important. Farmers raised corn to sell and oats to feed to their draught animals. Soybeans replaced oats following the mechanization of farming in the region, and the resultant two-crop rotation still dominates the local agricultural economy.

Soybeans are the principal crop on the rich, flat alluvial lands of the Maumee Lake plain in northwestern Ohio and the Bootheel region of southeastern Missouri. Both

Hogs, 1964, 1992

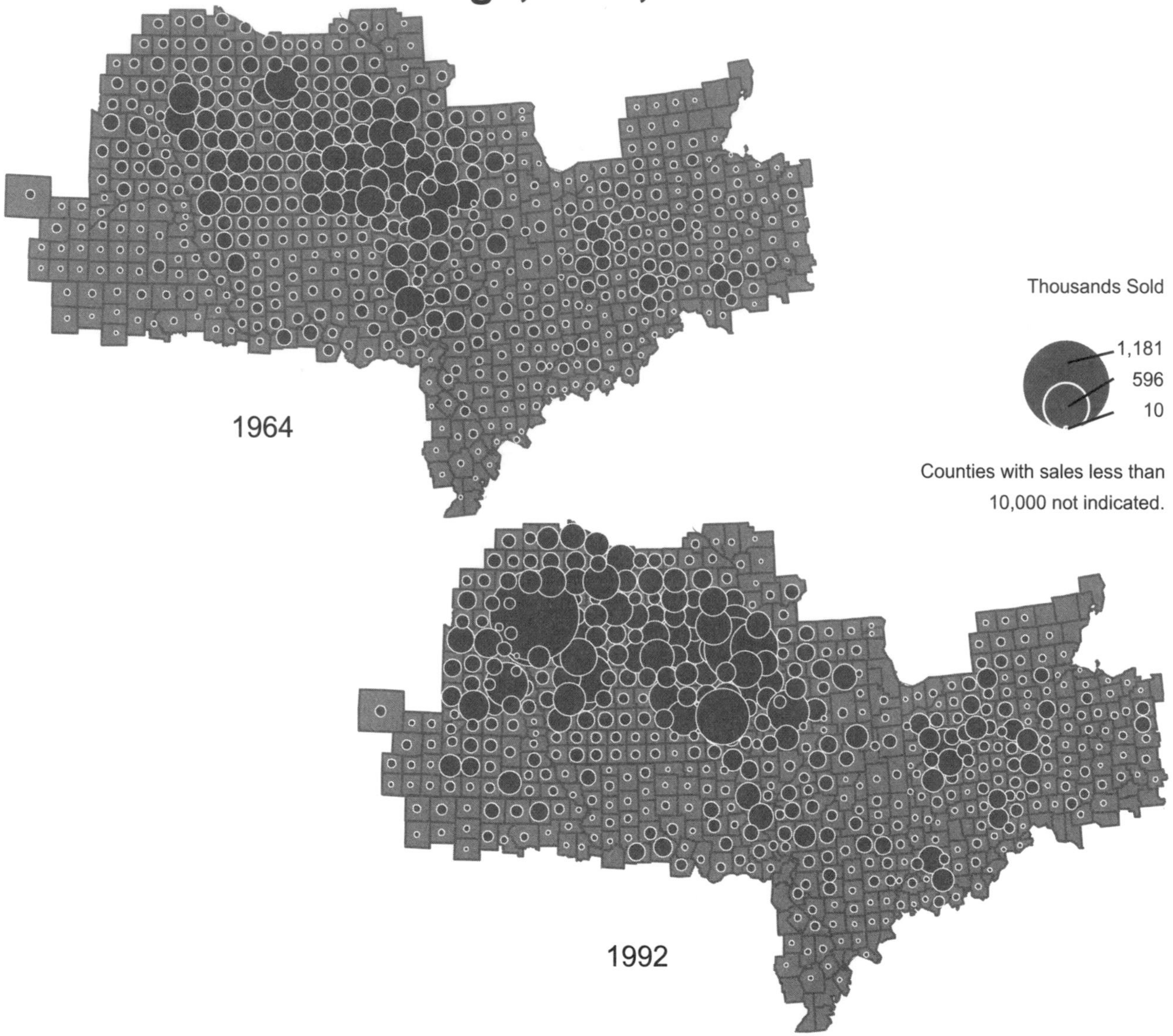

1964

Thousands Sold

1,181
596
10

Counties with sales less than 10,000 not indicated.

1992

product also providing feed for the regional feedlots.

The drier western portions of the Corn Belt generally have the largest beef cattle populations today. The regional mix of corn, soybeans, and hogs is joined by hay and beef cattle on the prairie lands along the Missouri River in southwestern Iowa, eastern Nebraska, and southeastern South Dakota. Most of the cattle are grain fed, although several communities in southeastern South Dakota vie for the title of national hay capital. The hilly country along

Butler County, Iowa

This classic Corn Belt agricultural community concentrates almost exclusively on corn and soybean cropping and the production of hogs and cattle. Typically, hay and oat production is declining.

A small poultry industry has been in the area for many years, but is being replaced by a growing turkey industry. Virtually no other significant agricultural activities are found there.

1992 Market Value of Agricultural Sales: $114,274,000

	1992	1982
Avg. farm size (acres)	275	229
Cattle and calves sold	20,777	30,817
Hogs sold	219,845	201,770
Corn (acres)	150,633	154,661
(59% of harvested acres 1992)		
Soybeans	88,938	87,484
(35% of harvested acres, 1992)		
Oats (acres)	4,571	11,090
Hay (tons)	28,472	55,864
Turkeys sold	157,802	90,211

regions were beyond the margins of the core of corn production, but are ideally suited to soybeans.

The westward extension of the Corn Belt along the Platte River in Nebraska has made Nebraska the nation's number three corn state after Iowa and Illinois. Much of that corn is grown under irrigation, and most is grown without rotation in a corn-only system. Nebraska growers have a substantial market advantage in the Asian export market because of lower costs, and thus dominate that trade. The Platte Valley has become a major center for cattle feedlots, and the beef industry and the remainder of the local output is used by feedlots. Southeastern Nebraska is a national leader in sorghum production, with much of the

the Missouri-Iowa border is largely cattle and hay country.

Some parts of the region also specialize in crops that would seem to have less to do with the Corn Belt complex. The middle Minnesota River Valley and surrounding areas are national leaders in sweet corn, peas, and shelled beans. Minnesota is the number two turkey producing state in the nation, with most raised along the northern edge of the Corn Belt in that state. Indiana ranks fourth in chicken population and Ohio seventh, mostly for eggs. Most of these fowl are found in a belt from Columbus to southeast of Chicago. Farmers in the largely Amish area surrounding Nappanee, Indiana, have created a distinctive agricultural complex to meet its specialized needs. Dairy remains strong with some expansion. This is also the center of the largest duck raising area in the nation. There are also growing numbers of chicken layer operations. The duck population is expanding in several areas around the southern edge of Lake Michigan. There are also several areas of commercial processed vegetable production in northern Indiana and nearby Ohio, relics of the beginnings of the American processed food industry.

Corn and Soybeans, 1992

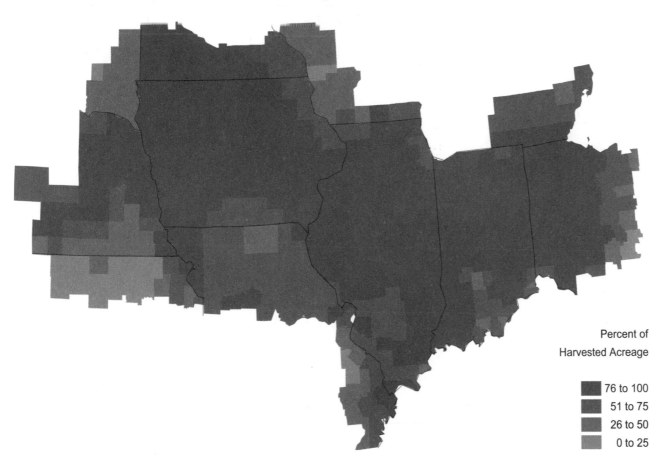

Percent of
Harvested Acreage

76 to 100
51 to 75
26 to 50
0 to 25

Hay and corn silage dominate the farm landscape throughout most of the Northlands. (JF/RP)

The Northlands

The southern margin of the Northlands is a transition between the older glaciated landscape of the Corn Belt and the more dramatic rolling topography, frequent swamps, and marshlands of the Northlands. The soils are more shallow and less fertile than the Corn Belt's. The growing season is shorter—farmers in northern Illinois can expect five months between killing frosts in spring and fall, while those in central Wisconsin must make do with only four. The physical border is marked further by the transition from the natural prairie to the south to the coniferous northern forest.

Northland agriculture is markedly different from that of the Corn Belt. The corn/soybeans/hog triumvirate is replaced by a diversified cropping pattern stemming from the search for suitable crops in a difficult environment. This diversity is characteristic of all peripheral agriculture regions in the nation today, notably the Upland South and Northeast, where aggressive farmers search for ways of making good returns from their marginal situations. Typical of the trend is the cultivation of ginseng by a handful of farmers in the central part of Wisconsin. This exotic crop is today the state's most important (by value) foreign export crop.

Agriculture in the Northlands is largely concentrated in a fairly narrow band extending through the core of Michigan's Lower Peninsula, the southern two thirds of Wisconsin, and westward to south central Minnesota between the Minnesota and Mississippi Rivers. The northern margin of significant agriculture is sometimes quite abrupt. It generally is defined by an intensification of the physical features characteristic of the region—a short growing season and thin, immature, often acidic woodland soils. This northern Paul Bunyon country is often called the "Cut-Over District" because of the intensity of the logging that ripped through this area between 1875 and 1910. Agriculture moved into this area during the late nineteenth century, but left just as quickly as the loggers moved west ward, taking with them whatever local market that had existed in this place. Forest has been slow to return to this cold region of barren soils. Tourism and recreation now dominate the local economy—fishing at the lake in summer, deer hunting in the fall, and a modest involvement with ice fishing, snowmobiles, and cross-country skiing in the winter.

The southern boundary has become a transition zone as plant geneticists have

The Northlands

decreased the maturation time of new hybrid corns. Dry bean farmers in the Saginaw Valley of eastern Minnesota have expanded their repertoire to include soybeans. In both cases farmers have moved toward the opportunity of increased income with less effort as they have blurred the boundaries between the two regions.

Wheat Frontier

The early development of Northlands farming was characteristic of frontier farm settlement across the northern tier of the country. Grain farming, most especially wheat, led the advance of crop agriculture. Virgin soils, even those of the Northlands, produced large wheat crops. It was comparatively easy to grow and ship, an ideal situation for a crop grown in marginal regions. The first major trans-Appalachian wheat producing area had been the Genesee County south of Rochester, New York, followed by central Ohio on the highlands between the Lake Erie and Ohio River drainage basins. Wheat farming began on a large scale in southern Michigan after 1840, then marched inexorably across the region for the next half century. Wheat became especially important in Wisconsin. In 1860, Wisconsin wheat's "Golden Year," state farmers harvested 17 percent of the nation's crop. Wisconsin was at least temporarily known as the nation's breadbasket. The area of greatest wheat acreage tended to be on the western margins of production of that particular season. New areas to the west offered better soil and thus greater potential profits. Rock County, in southeastern Wisconsin, was the leading wheat producer in 1849, Green Lake, in the center of the state, followed as the leader in 1859, and St. Croix in north central Wisconsin led in 1869.

Wheat monoculture inevitably brought its own destruction. Declines in soil fertility and increases in pests soon began inhibiting production in each new area. Wheat dominated the agricultural scene for only a few decades before a general precipitous decline in both yields and acreage. By the end of the nineteenth century wheat farming had all but disappeared from the region.

The Emergence of Dairyland

Wheat's decline brought the beginning of a far more diversified farming economy. Initially these farmers adopted a modified three-field crop rotation to control pests and maintain soil fertility. They turned to a rotation of corn, oats, and clover for soil improvement. The potential for killing frosts encouraged corn farmers to harvest the crop as silage for animal fodder. Cattle were the obvious choice as they could consume the crops generally under cultivation, along with the hay harvested from

Dairying continues to dominate our image of Northlands agriculture, although its relative importance to the region is declining. (RP)

poorly drained lands. They could also graze those slopes too steep for plowing. The result was the emergence of a dairy industry focused on the earlier wheat areas, just as had happened in upstate New York and middle Ohio. Wisconsin was in the process of shifting from America's breadbasket in its dairyland.

The shift to dairying during the last half of the nineteenth century was encouraged by the replacement of the departing wheat farmers by westward migrating farmers from New York who brought the latest concepts of dairy cattle breeding and modern dairy operations. Most of the vacated wheat lands, however, were purchased by German and Scandinavian immigrants who were comfortable with the labor-intensive dairy farm routine. The rise of Chicago, Milwaukee, and other smaller industrial cities just beyond the southern margins of the region provided a market for fluid milk products, while the development of the refrigerated railcar allowed the shipment of large quantities of solid product to eastern markets. Wisconsin cheese and butter expanded rapidly and were acclaimed as second only to New York's at the 1876 Centennial Exposition in Philadelphia.

The emergence of the Northland dairy region was a gradual process. Wisconsin production climbed to become the eighth most important state by 1870, while Michigan was tenth. Iowa replaced New York by 1890 with 1.5 million dairy cattle. Wisconsin remained eighth, but the total number of dairy cattle had risen to 800,000, while Minnesota took over tenth position with 600,000 cattle. Wisconsin became the national leader in the 1919 census with 1.8 million cattle, while Min-

nesota had risen to third with 1.2 million. The triangular core of this dairy region, encompassing much of southern Wisconsin and parts of eastern Minnesota remained the nation's dominant dairy area for the following 75 years, although California edged Wisconsin as the top producer of milk in 1993. Both states have about the same milk cow populations as in 1920, but

yield per cow has increased dramatically to almost 15,000 pounds per cow for Wisconsin and almost 19,000 pounds per cow for California.

The greatest difference between the two states, however, is in average herd size. Wisconsin's 1.5 million dairy cows were scattered among more than 30,000 dairy farms in 1992, while California had only

Dairy Cows, 1964, 1992

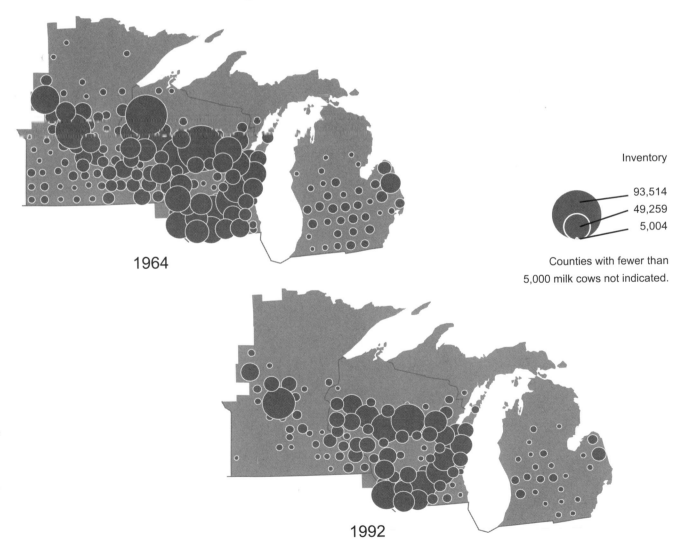

Inventory

93,514
49,259
5,004

Counties with fewer than
5,000 milk cows not indicated.

1964

1992

Fond du Lac County, Wisconsin

This east central Wisconsin county illustrates the plight of many Northland farmers. The number of dairy farms and dairy cows is declining while average herd size and gross income are increasing. Vegetable production has remained a stable supplement to dairy income, while some other specialties, such as nursery stock, have been increasing in importance.

1992 Agricultural Sales: $151,097,000

	1992	1982
Farm size (acre)	227	191
Dairy cows	45,481	53,312
Dairy sales ($ thousands)	92,978	83,757
Avg. herd size	58	48
Hogs sold	57,337	71,975
Corn (acres)	21,846	16,684
Soybeans (acres)	21,013	9,323
All hay (tons)	173,753	308,037
Sweet corn (acres)	21,846	16,684

3,124 dairy farms in that year. Wisconsin's average herd size of 50 places these farmers at a great economic disadvantage to the much larger dairy operations found in California. Almost three-quarters of all dairy cattle in California are in herds of 500 or more, while less than 1 percent of those in Wisconsin are located in such large herds.

Milk production in the Northlands Dairy Belt quickly exceeded regional demand, and much of the area's output has long been converted to solid product. Production of Grade A milk (designated for fresh consumption) is concentrated near the urban markets in southeastern Wisconsin and near Minneapolis/St. Paul. About half of all Wisconsin dairy farms are classified as Grade B milk producers, milk used in the manufacture of solid dairy products. The percentage of Grade B producers is even higher in Minnesota. Much of the Grade A milk production of the region is also converted into solid product because of overproduction. In 1991, for example, Wisconsin supplied 26 percent of the nation's butter, 32 percent of its cheese, and over 90 percent of its condensed milk. Minnesota provided an additional 10 percent of the 6.5 billion pounds of cheese manufactured. Only about 20 percent of the region's milk is sold fresh today.

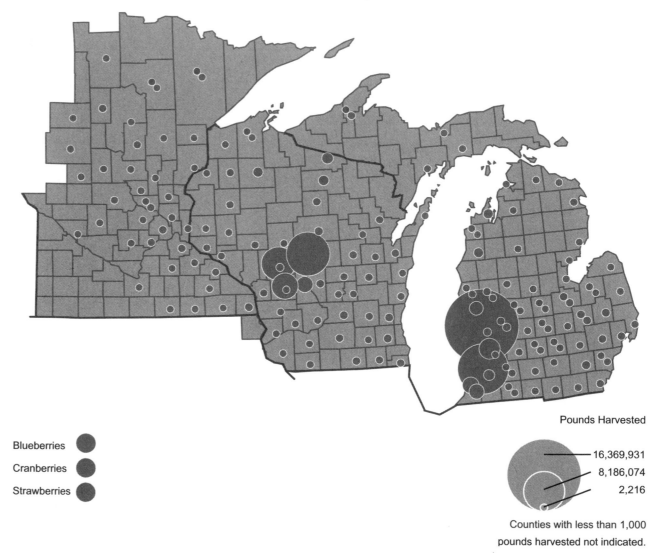

Berries, 1992

Blueberries ●
Cranberries ●
Strawberries ●

Pounds Harvested

16,369,931
8,186,074
2,216

Counties with less than 1,000 pounds harvested not indicated.

Cheese retains about 10 percent of the bulk of milk, while butter utilizes only about 5 percent of its bulk. This difference explains the geography of fresh milk, cheese, and butter production in Wisconsin. Fresh milk is produced closest to market; butter farthest from those markets. Overall production in the state has a geography that resembles a doughnut, with little dairying in the Central Sand Plain, an area of loose, sandy soils that were largely ignored during the period of dairy expansion. Most Grade A production is found in the southeastern section of the state along the margin of Lake Michigan. About two thirds of Wisconsin milk is consumed outside the state, primarily in metropolitan Chicago. Cheese production is concentrated in three regions around the Central Sand Plain, but far from Chicago: the Southwestern Cheese Regions of the Driftless area, the Lakeshore Cheese Region north of Milwaukee, and the Central Cheese Region north of the Sand Plain. At one time every Wisconsin community seemed to have its own cheese from its own cheese factory. Changes in technology have led to fewer, larger creameries. The state's creameries slipped from about 1,000 in 1960 to fewer than 800 at the start of the 1980s, to 176 in 1991. The number continues to dwindle each year. Nevertheless, the basic location of the Cheese Belt has not changed in a century. Butter manufacture is concentrated in western Wisconsin, even farther from large urban centers.

The demands of the dairy industry created a distinctive, attractive rural landscape. Regional dairy farms are fairly small, usually around 200 acres, and most have fairly small herds, nearly always under 100 head. Dairy cattle demand a complex diet during lactation to maintain high production. Farmers attempt to grow most of the needed foodstuffs to reduce production costs. Corn cut green and ground into silage provides much of the needed energy, while hay, especially alfalfa, also cut green and stored as haylage, provides needed protein. Large glass-lined metal silos, used to store both silage and haylage, loom over the usually well-maintained farmstead. Massive barns with huge lofts for hay storage are giving way to lower, smaller units for the storage of equipment. Hay is stored in the silos these days. Herds once free to roam pastures are now kept penned, freeing additional farmland for hay and corn production. As a result most fences have been removed as unnecessary hindrances to effective plowing and harvesting. Short season corn hybrids with improved yields now enable many farmers to produce grain surpluses that are sold. Only about 20 percent of the state's corn is ground into silage.

Contributions to Diversity

The common conception of the Northlands as Dairyland misses much of the substantial agricultural diversity that, in fact, characterizes the region. Michigan leads the country in the production of navy beans, tart cherries (historically providing about 75 percent of the national crop), tame blueberries, and cucumbers for pickles; it is second in all dry beans, and third in asparagus, snap beans for processing, celery, carrots, prunes, and plums. In addition to milk products, Wisconsin leads the nation in snap beans, sweet corn, and green peas for processing, and is second in cranberries. Minnesota is second in turkeys raised and sweet corn for processing, and third in snap beans for processing. While the total area given over to any one specialty crop is generally modest, collectively they have a substantial economic impact.

The Michigan Fruit Belt, a narrow band along the western side of the state's Lower Peninsula, is home for most of its $200 million fruit crop. Cold air passing over the unfrozen lake in winter is warmed, while hot air passing over in the summer is cooled. This "lake effect" helps provide a substantial snow cover to insulate delicate root systems during the worst of the winter, while the cooler temperatures of the spring delay fruit tree bloom a week or two and lessen the possibility of freeze damage when the crop is most vulnerable.

Apples are most important in the southern lakeshore area around Benton Harbor, while grapes are even more concentrated just north of the Indiana border. Michigan was one of the leading peach states many years ago with several important varieties developed there, but is much less important today. The surviving peach orchards tend to be located about midway up the western shore of Lake Michigan, especially from Holland to Muskegon, while cherries are found even farther north. The vast majority of all this fruit is processed—grapes into juice, apples into cider and sauce, and cherries frozen or canned. Similarly, the large berry farms of this area also concentrate primarily on production for processing.

Early Dutch settlers along the western Lake Michigan shore introduced a variety of horticultural activities. Celery has been an important local crop for more than a century, although production has been declining in recent years. The glaciated soils are also well adapted to berries, and the area is one of the nation's largest blueberry production areas. Finally, the most famous Dutch products of all, floriculture specialties, are also locally important, especially in the production of tulips and other bulbs. Dozens of commercial wholesale nurseries are scattered among the vineyards, berry patches, and apple orchards of southwestern Michigan along the backroads of the region as reminders of the early Dutch influences found here.

The fertile soils of the Saginaw Valley and "thumb" of east central Michigan have been an important specialty agricultural area for more than a century. This is the largest production center for navy beans and the fourth largest producer of sugar beets. Soybean farming has been increasing in importance in recent years.

Tobacco is produced in two small regions in Wisconsin, one south of Madison and the other south of La Crosse. Both areas historically produced cigar binder, but have largely switched to chewing tobacco. The large, slatted air curing barns of these areas add a distinctive touch to the landscape that is easily recognizable even to the casual visitor. Commercial tobacco production started in the state in the 1850s and peaked in 1903. Production peaked with about a 70-million-pound annual crop, but had declined to about 25 million pounds by 1980 and to fewer than 15 million pounds today.

Wisconsin farmers produce about 35 percent of the nation's cranberry crop, second only to Massachusetts. The two states account for almost 80 percent of total U.S. output. Cranberry farming began in 1850 at Berlin on the bed of former Lake Oshkosh. Production shifted to the lakebed of former Lake Wisconsin near Marshfield after 1870, a flat, poorly drained area with large beds of peat, muck, and sand—ideal for cranberries. Production had increased to 1,200 acres by 1900 in Wood County. Most of Wisconsin's cranberry acreage remains in Wood and neighboring Jackson, Juneau, and Monroe counties. Total cranberry acreage in the state has continued to expand, and the yield per acre usually exceeds that of Massachusetts.

The dairy country of southeastern Wisconsin is also one of the nation's oldest and largest centers of cool weather vegetables for processing (canning). In addition to sweet corn, snap beans, and green peas, the state is a national leader in the cultivation of cabbage for sauerkraut, pickles for cucumbers, and canned carrots. Traditionally, these crops have been produced under contract by dairy farmers attempting to diversify their income. The canneries themselves are concentrated in a

The adaptability of silage to outside storage has played an important role in its growing popularity among Northland farmers. (RP/JF)

band extending from Madison to Fond du Lac. Production, especially of green peas and potatoes, expanded westward after the introduction of central pivot irrigation to the Central Sand Plain.

Minnesota ranks second to Wisconsin in the growth of sweet corn for processing (each accounts for about one-quarter of the total crop), and is nearly tied with Wisconsin in green peas harvested for processing, although that state's Le Seur peas and golden nibblets are actually more famous. Production is concentrated in the Minnesota River Valley south of Minnesota. While Le Seur may be synonymous with peas, the center of production is farther south in Renville.

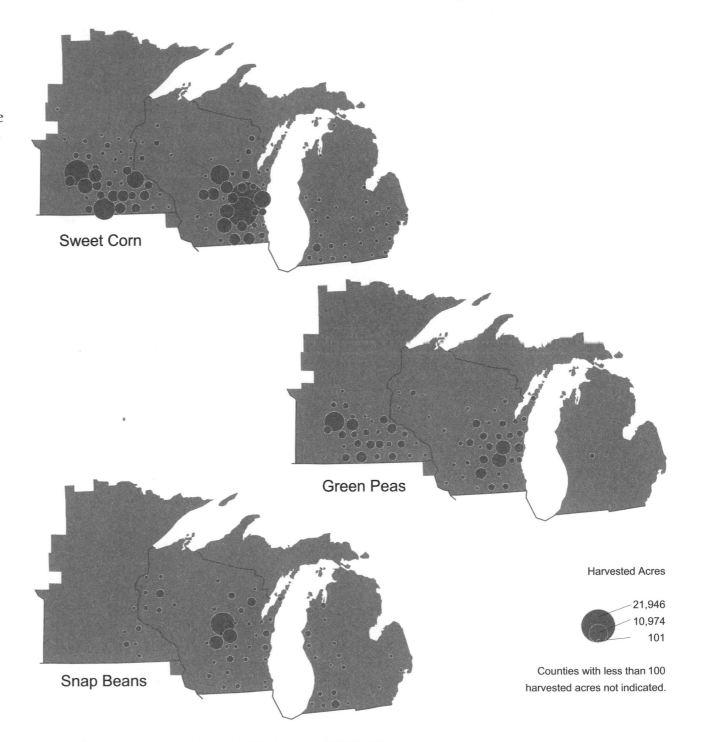

Sweet Corn, Green Peas, and Snap Beans, 1992

Sweet Corn

Green Peas

Snap Beans

Harvested Acres

21,946
10,974
101

Counties with less than 100 harvested acres not indicated.

The Homestead Act and the General Land Office survey system created a regular landscape of one mile squares throughout the Grasslands and Corn Belt. (RP)

The Grasslands

A broad area east of the Rocky Mountains, characterized by sparse and irregular rainfall, was once covered by a veritable sea of grasses. This broad swath of territory, stretching westward from roughly the ninety-eighth meridian to the mountains and extending from the Rio Grande to the Canadian border, is broken only by trees in narrow gallery forests along its rivers flowing eastward out of the mountains, a scrub forest on the rolling Edwards Plateau of south central Texas, a brushy thicket south of San Antonio, and the true pine forest of the Black Hills of South Dakota and northeastern Wyoming. Everywhere else there was grass.

American agriculture, bred in northwest Europe and honed in the humid environment of the eastern United States, at first struggled to adapt to this drier environment. The general lack of trees meant that there was no easy way to build the many wooden structures characteristic of the farmsteads of the period. The earliest housing was often of sod. Wooden fencing, the only kind then known, was crucial in keeping the many large grazing animals, whether buffalo or cattle, separated from cropland. Live fences of Osage Orange and other flora were attempted for a time, but never gained wide popularity because of their high maintenance costs.

Standard farming techniques, too, were poorly suited for this arid land. Most of the commonly grown crops of eastern farmers did poorly in the grassland's short wet season. The relatively small farm of the early period, generally under 100 acres, was

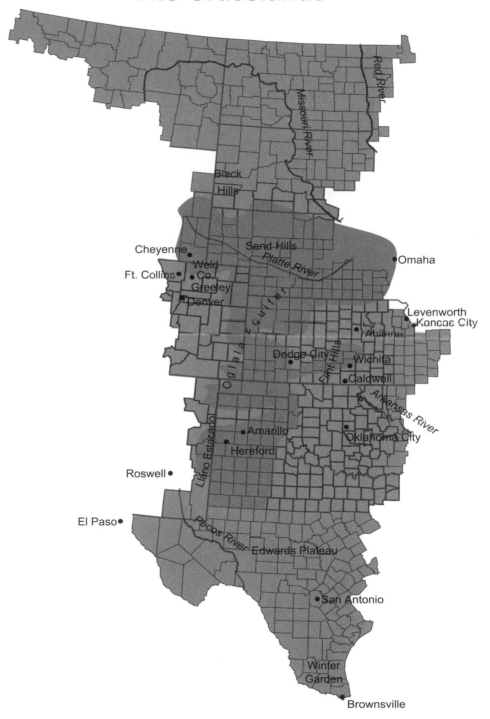

too small to produce a livable return with the crops of the time. The earliest settlers quickly claimed the land nearer the few perennial streams. Later arrivals were thus left with no access to water in this arid land, farms too small to provide a living for a man's family, little or no wood for cooking, heating, building houses, and constructing fences, and few crops that could survive in this harsh land. These dry grasslands were for the American farmer of 1850 truly a Great American Desert—dry and unusable.

The Cattle Kingdom

While the farming frontier hesitantly pushed onto the Grasslands, a cattle economy exploded across the Grasslands northward out of south Texas after the Civil War to reach the Canadian border and beyond within a decade. The Spanish introduced cattle to their American colonies with Columbus's second voyage to the New World in 1493. Hernando Cortez stocked his Mexican ranch, Cuernavaca (Cow Horn), in 1521, only two years after he had begun his conquest of the Aztec Empire. Introduced to America from Spain, these lanky, tough, long-horned animals originally evolved on the even more arid shores of northern Africa. Capable of surviving on minimal browse, they were carried northward with Spanish settlement into the brush country of south Texas. There the Spanish established the northern limits of their fenceless, open range ranching system. Animals were rounded up annually and calves given the brand of their mother.

This ranching system had been in place for a century when American settlers began entering Texas after Mexican independence in 1821. These new arrivals, drawn mostly from the western margins of the American South, were themselves used to an open range animal economy, though with some differences. While the Spanish ranchers rounded up their cattle and calves each spring and gave them the brands of their mothers, the Southern woodlands farmer relied on ear notches and other methods of identifying their animals. While cattle were periodically rounded up from their piney woods retreats, hogs generally were allowed to forage until they were needed for sale or slaughter. The new American arrivals easily adapted to the Texas ranching system. There were differences of course—hogs were absent, branding was used to identify ownership, the animals were allowed to roam over a much wider area, and, most important, the drover (the cowboy of American legend) rode a horse instead of walking. Walter Prescott Webb, our most noted historian of the Grassland, noted in *The Great Plains* that "It was the use of the horse that primarily distinguished ranching in the West from stock farming in the East."

The Civil War disrupted the annual cycle of roundup and branding in the south Texas brush country. Many of the cowboys went east to the war. Roundups were few and most calves went unbranded. The south Texan soldiers discovered several million longhorn cattle wandering the countryside upon their return home— unbranded and thus essentially unowned. Worth only a few dollars in the devastated economy of Texas, middle-western buyers might pay $20 a head for these animals in the North. Railroad lines snaking westward from Kansas City offered possible access to this lucrative market. A young Texan named Joseph McCoy (responsible for the phrase "the real McCoy") bought a few hundred acres of land near the Kansas railroad town of Abilene, built cattle pens with lumber imported from the East, and in 1867 induced the Union Pacific railroad to build a siding at Abilene and cattle transfer and feed yards in Leavenworth. The pathway to the East was opened. The first cattle trail to these railheads, the Chisholm Trail, named for the trader who first traveled and marked the route, extended from near Brownsville in south Texas northward across Indian Territory to central Kansas.

Thirty thousand cattle reached Abilene from Texas in 1867. The number jumped to 75,000 in 1868, 150,000 in 1869, and over 300,000 in 1870. Kansas City meat packers began rivaling those in St. Louis and Chicago. The construction of new railroads, the westward creep of settlement, and a westward moving series of quarantine lines banning tick-infested Texas cattle from areas settled with vulnerable eastern cattle pushed the "cow town" westward to Ellsworth, Wichita, Dodge City, and finally Caldwell, Kansas.

The American Grassland became a great highway during this early bovine invasion. This role was further supplemented as cattle began to be driven northward to stock Grassland ranches in Colorado, Wyoming, and beyond. Cattle ranches were established across the Grasslands almost everywhere farming was absent between the late 1860s and early 1880s. Most of the grazing land actually remained publicly owned through this period as ranchers controlled vast swaths of grassland through ownership of key water sites. Cattleman John W. Iliff, for example, actually only owned 105 parcels totaling 16,000 acres in far western Kansas, but through ownership of virtually all the land surrounding the surface water of the region, he was able to control the use of almost 6,000 square miles of rangeland. Individual cattlemen and cattle companies, often foreign owned, similarly claimed range rights over four, five, and even six million acres across the western Grassland.

The introduction of barbed wire and an inexpensive, reliable windmill that could be placed untended where cattle could graze if they had water completed the requirements for total domination of the drier margins by the large ranchers. Barbed wire, invented in the Middle West by J. F. Glidden in 1873 to protect his wife's garden, altered farming and ranching in every corner of the nation, but reached its greatest importance on the Grasslands. This treeless land finally had its fencing. Strung between trees, posts, or almost any upright object, this fence of twisted, sharp metal barbs attached to extruded wire fencing discouraged grazing animals from rubbing against the wire to eventually push it down. In 1875, 600,000 pounds of barbed wire were manufactured; in 1890, 80 million pounds; and in 1901, nearly 300 million pounds. Barbed wire prices fell from $20 to less than $2 per hundred pounds through the period. Cattle companies could now fence their land, or at least those portions with superior grazing possibility, to create better pasture and improve stock quality by selective breeding.

The windmill was introduced to provide water for steam trains about the time of the Civil War. Recognizing the value of the concept for cattle grazing, smaller, prefabricated models were soon developed that could easily be placed over isolated wells.

Fattened Cattle, 1964, 1992

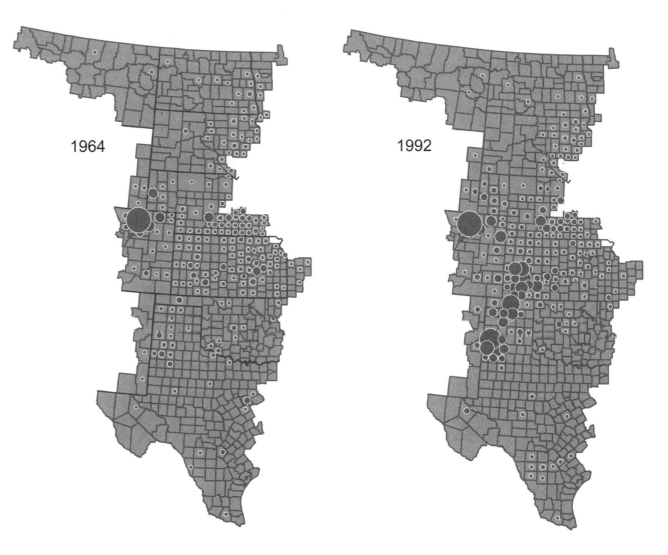

1964

1992

Cattle Sold (Thousands)

786
395
5

Counties with fewer than 5,000 sales not indicated.

Equipped with automatic shut-off valves, these innovations kept watering troughs full and allowed cattlemen to disperse their herds throughout the entire holdings, not just those with surface water. By 1889, there were 77 windmill factories in the United States, eventually manufacturing more than 6 million windmills. Many still stand as sentinels of this early time as they continue to pump water for isolated livestock.

The life span of the Grassland cattle economy was relatively short. Heavy overgrazing quickly overstressed the grassy environment. While efforts were made to improve herd quality, Middle West stockmen, forced to reevaluate their own farming strategies because of the westward moving wheat frontier, began raising high quality grain-fed cattle for the meat packers. Severe competition drove the price of Grassland beef downward in the mid-1880s. Disastrous blizzards in the winters of 1885–86 and 1886–87 killed millions of cattle, and the drought that followed forced these cattlemen to sell many of the survivors at low prices in a poor market. The period of the great cattle barons and smaller ranchers trying to join their numbers had passed. Ranching would survive on the Grassland, but it would no longer enjoy the grand domination it had once held.

The heyday of the Grassland cattle kingdom lasted barely twenty years. Still, that relatively brief history is responsible for defining the nation in the eyes of much of the world. The cowboy, great herds of cattle being driven across the plains, and the rough and exciting life of the cow town—these are our dominant images of the Grasslands. These themes have been presented to Americans endlessly over the past century through dime novels, larger works of fiction, countless movies, "singing cowboys," and most recently a spate of Western television shows. The Grasslands *are* America to much of the world's population.

Farming Enters the Grasslands

Farmers pushed back the eastern margins of the Grasslands, even as the cattlemen were expanding their empire northward. Nebraska's population tripled to 500,000 in the 1870s; Kansas's population passed one million in the early 1880s. The advance of the farm frontier across the more humid eastern plains was complete by the late 1880s. The nonirrigated portion of the central and western plains was first settled by farmers in the 1890s, but that section has sustained a series of agricultural expansions and retreats associated with cyclic shifts in precipitation and market conditions.

The expansion of agriculture onto its drier margins was ultimately made possible by a series of innovations that overcame the dryness. Chief among these were barbed wire, windmills, changed land laws, Turkey wheat, dry farming, large-scale irrigation projects, and railroads. If barbed wire improved the lot of the rancher, it was essential for the Grassland farmer. Grazing animals could be separated from croplands at an affordable price, but more important, ownership sovereignty could be stamped on the earth in a way that no stockman could deny. The windmill minimally assured adequate water for the farm family and their livestock and the development of an irrigated kitchen garden. Land law modifications, most importantly the Homestead Act of 1862 and the Timber Culture Act in the early 1870s, allowed settlers legally to acquire 320 acres at a very low price, thus at least partly accommodating the almost inevitably lower yields associated with plains farming.

Mennonite farmers brought a hard red winter wheat from the Black Sea area of

Weld County, Colorado

The construction of irrigation canals in 1871 set the stage for Weld County to become one of the nation's richest agricultural counties. The resulting initial agricultural growth based on irrigated crops spawned one of the nation's modern feedlots, which today finishes almost a million cattle annually for slaughter. More recently cattle have been joined by hogs, laying chickens (1.7 million), and turkeys (4 million) utilizing the same locational advantages. Almost 300,000 sheep also winter in the county.

Large acreages of irrigated sugar beets, alfalfa, corn, and other field crops support these livestock operations by providing large quantities of locally produced feed, much of it byproducts from processing.

1992 Value of Agricultural Sales: $1,180,067,000

	1992	1982
Avg. farm size (acres)	717	648
Cattle and calf sales	925,210	943,575
Hog and pig sales	210,167	51,087
Sheep and lambs	289,605	188,002
Sugar beets (acres)	21,807	17,269
Potatoes (acres)	4,875	4,323
Alfalfa (acres)	88,430	80,717
Corn (acres)	143,961	142,912
Wheat (acres)	159,269	183,622
Dry beans (acres)	30,270	25,244

Russia in the early 1870s that increased crop reliability in this arid land. Called Turkey wheat for its supposed area of origin, this grain was unusually well adapted to the climatic conditions of the central and southern Grasslands. Most precipitation on the plains falls during spring and early summer. Traditional European winter wheats had struggled in the aridity of the Grassland—the new Turkey wheat thrived. By early in the twentieth century Turkey wheat constituted over 90 percent of the country's winter wheat crop, with most of that crop grown on the plains south of the Platte River.

The success of dry farming in the region was assured with the introduction of the new wheat. Growing a crop of grain on a field every other year and cultivating unused fields to control weeds and maintain a loose soil to conserve moisture, dryland farmers were able to push the agricultural frontier farther and farther into the arid margins. At first the process usually involved "dust mulching," repeated disking, and harrowing to maintain a very fine soil surface that would inhibit the movement of water to the surface where it would be lost by evaporation. The Dust Bowl experience of the drought years 1933–39, when 150,000 square miles of southwestern Kansas, southeastern Colorado, the Oklahoma and Texas panhandles, and northeastern New Mexico went up in giant dust clouds, convinced most farmers that dust mulching was inappropriate to this dry region, though it is still practiced in the Palouse Hills of eastern Washington.

Irrigation was the key component of the early expansion of agriculture onto the dry Grasslands. Horace Greeley, owner of the New York *Tribune*, was an avid farm pro-

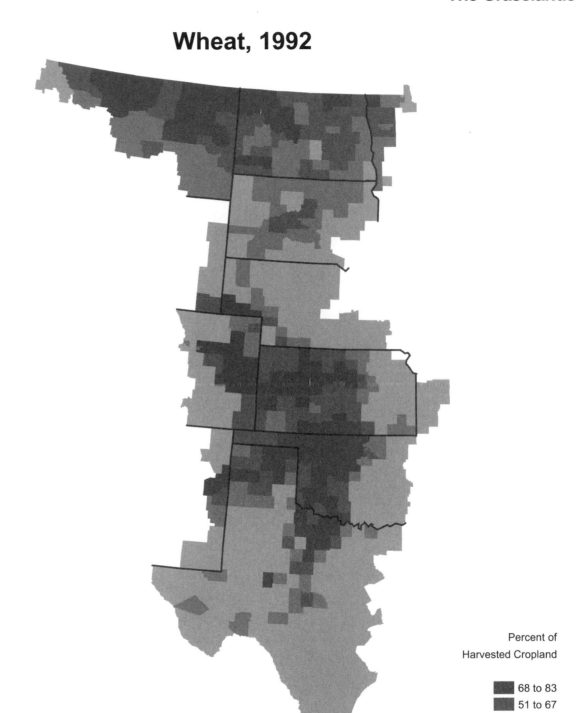

Wheat, 1992

Percent of
Harvested Cropland

68 to 83
51 to 67
34 to 50
0 to 33

moter. Lured to Colorado by the Pike's Peak gold rush of 1859, he became fascinated with the agricultural possibilities of Colorado's high plains. Greeley sent N.C. Meeker west to establish a cooperative irrigated farming venture. Selecting a site of more than 100,000 acres on the South Platte River about fifty miles north of Denver, Meeker established the Union Colony in 1869. A twenty-seven mile irrigation ditch had been dug by 1871 to carry water to irrigate the lands of the colony's 700 shareholders. The Union Colony, second only to the Mormon experiments in Utah, strongly influenced the development of large-scale irrigation in the dry West. The colony's project engineer moved on to develop another large irrigation system for a European syndicate east of Fort Collins. Both projects are in Weld County, today one of America's richest agricultural centers. The irrigation canals remain part of the core of the county's vast irrigation system today.

A web of rail lines was constructed in the last decades of the nineteenth century to create the requisite network needed to allow farmers to haul produce by wagon to central collection points. Hundreds of small communities soon sprang up along the track serving their surrounding areas as transshipment points and storage centers to provide basic services to the surrounding region.

These communities thrived for a half century or more until paved roads began allowing grain to be hauled farther distances and farm families to have access to several for the purchase of their services. Increasing farm size and declining rural populations intensified their decline. Today even the grain elevators are sometimes closed, the victim of enlarged 100-ton grain railcars that could not use the rail sidings of hundreds of these small towns. Some of these service communities are now abandoned. Most have boarded store next to boarded store, surviving more out of habit than any important role they play.

Grasslands Agriculture Today

The American Grasslands today is a region of tremendous agricultural diversity in the midst of seeming continuity. The region's diversity is not so much the result of a mixing of activities as it is the accumulation of a variety of large and internally homogeneous areas.

There are four broad explanations for this regional diversity. One is differences in the length of the growing season. The Winter Garden area of the lower Rio Grande valley in south Texas averages around 300 days of frost-free weather, long enough to support semitropical sugarcane and citrus fruit. By comparison, the "High Line" region of northern Montana, as well as much of northern and central North Dakota, has a growing season of no more than four months. A second factor is the quality and depth of soils. Most of the major surviving areas of large-scale cattle grazing are in areas of either very thin soils (the Flint Hills of Kansas and the Osage Hills of Oklahoma) or relict sand dunes covered by grass (the Sand Hills of north central Nebraska). Variations in precipitation, and more particularly of moisture availability, is a third critical component. While the southern plains may receive more precipitation, higher temperatures and the longer season of high temperatures result in far higher rates of evaporation and transpiration. Finally, irrigation has enabled the farmers of large sections of this dry region to cultivate crops commonly associated with humid areas. Most dramatic is the Llano Estacado of the Texas panhandle region, which leads the nation in the production of upland cotton, with about 15 percent of the nation's total crop grown within about a 100-mile radius of the panhandle city of Lubbock.

Crops

Wheat continues to be the signature crop of the Grasslands. Almost all the Grasslands' wheat is grown using dry farming methods. Regional wheat cultivation has long been divided into two broad areas, a Winter Wheat Belt from southwest Nebraska and northeastern Colorado southward to the Llano Estacado, and a Spring Wheat Belt extending from north central Montana eastward to the Red River Valley along the North Dakota/Minnesota border, with a southern extension into east central South Dakota. Kansas west of the Flint Hills was the regional leader in winter wheat production (364 million bushels) in 1992, followed by Oklahoma (171 million bushels) and Texas (129 million bushels). North Dakota leads in spring wheat and in overall production (464 million bushels), followed by Minnesota (138 million bushels) and South Dakota (86 million bushels). Between the two Belts is an area north of the Platte River in Nebraska and extreme southern South Dakota where little wheat is grown. The Sand Hills of north central Nebraska and neighboring badlands areas in South Dakota do not support any significant cultivation. The lower Platte

River valley is home to a corn economy similar to that of the western Corn Belt.

Farmers have taken advantage of newer, more cold-hardy varieties to push winter wheat cultivation northward into the Spring Wheat Belt in recent decades. Winter wheat yields are higher, and the product (primarily used for bread) has a greater market demand. Montana now produces nearly as much winter wheat as spring wheat.

Several other crops, all also commonly grown using dry farming, compete with wheat. Grain sorghums have surpassed wheat in acreage on the Texas panhandle (where the crop is irrigated), and clearly dominate cultivation in southeastern Nebraska and portions of Kansas. The principal use of sorghums is as stock feed. Improved varieties, notably the milos, now come close to rivaling corn as a quality animal feed, while their drought adaptability makes them highly suitable for this environment. The grain, introduced from semiarid northern Africa, does well during periods of sparse precipitation and is a desirable dry farming alternative to winter wheat in areas with numbers of cattle.

Barley, rye, sunflowers, and sugar beets compete with spring wheat across the northern Wheat Belt. North Dakota, South Dakota, and Nebraska rank 2-3-4 in rye production nationwide. North Dakota's farmers grew 35 percent of the country's $950 million barley crop in 1992, with Montana (concentrated in the Judith Basin in the north central part of the state) and Minnesota (the Red River Valley again) each contributing an additional 12 percent. Eastern North Dakota supplies about half of our sunflowers, followed by South Dakota (19 percent) and Minnesota (12

Commercial sunflower production is almost entirely concentrated on the Grasslands. (JF/RP)

Sunflowers, 1992

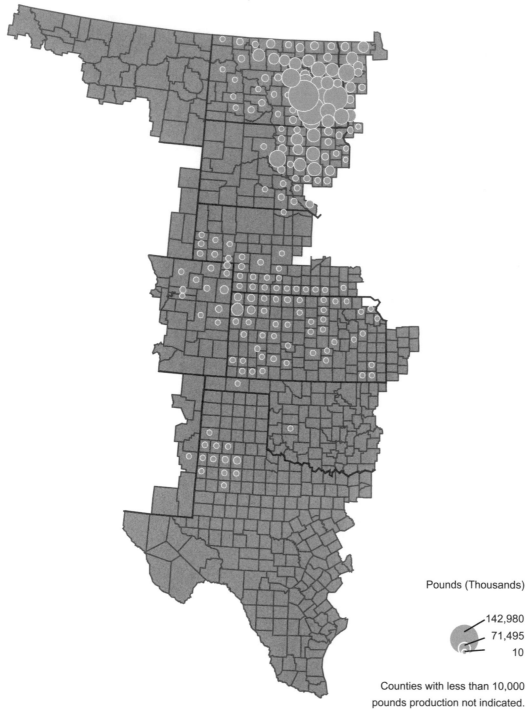

Pounds (Thousands)

142,980
71,495
10

Counties with less than 10,000
pounds production not indicated.

percent). Sugar beets and a small grain (either wheat or barley) are commonly grown in rotation in the Red River Valley.

Irrigation has greatly expanded the variety of crops that thrive on the Grasslands, especially in recent years. The Winter Garden area of the Lower Rio Grande valley in south Texas (locally referred to as "the Valley"), with its ten-month growing season, produces a rich mix of semitropical and off-season crops, ranging from sugarcane to citrus to onions to cultivated dandelions. Citrus cultivation, especially grapefruit, expanded in the valley during the 1930s and 1940s. Texas soon rivaled Florida in grapefruit production. A series of disastrous freezes in the winters of 1947–48, 1950–51, 1961–62, 1983–84, and 1989–90 each destroyed most of the valley's citrus trees, resulting in severe immediate loss and a gradual long-term decline. The grapefruit harvest declined from a high of 24 million boxes in the 1945–46 season to 11 million in 1982–83, and 0 in the postfreeze years of 1984–85 and 1990–91. This reduced harvest is also partly the result of land use competition from the many thousands of "Snowbirds," mostly retired seasonal migrants who now winter in the Valley.

A vast, irregular, 225,000-square-mile swath of the Grasslands from Nebraska's Sand Hills to the Texas panhandle is underlain by the Oglala aquifer. The aquifer is recharged primarily at its northern end, where its porous sedimentary beds are thickest and close to the surface, and are overlain by the poor consolidated soils of the Sand Hills. Farmers began to mine this vast water reservoir in the 1930s. Today over 100,000 wells tap the aquifer, most notably on the Llano Estacado of the Texas panhandle and adjacent areas. This extensive use has resulted in a continued lowering of the level of the aquifer, especially on the Llano, where wells are often triple the depth of fifty years ago. This overuse represents a major long-term problem for all farmers who use the reservoir, and is of immediate concern to farmers in parts of the Llano.

Oglala water was used to allow the introduction of cotton, alfalfa, sorghums, corn, and sugar beets to the panhandle. In the neighboring Pecos Valley of eastern New Mexico cotton and sorghums are joined by alfalfa and peanuts. Alfalfa and corn dominate use farther north. Indeed, alfalfa, the basic hay crop of the West, has the largest irrigated acreage on the Grasslands everywhere north of southern Colorado. Corn is the major irrigated crop of southern Nebraska, where the circular cropping pattern of center pivot irrigation systems drawing from the Oglala is a dominant image for anyone who has flown over the area.

Streams flowing eastward out of Colorado's Rocky Mountains provide irrigation for two grassland areas. The Arkansas River Valley of eastern Colorado produces sugar beets, corn, and a significant crop of cantaloupes. In northeast Colorado the Colorado–Big Thompson Project and the Platte River together provide irrigation opportunity over a wide area of northeastern Colorado. The Big Thompson Project captures water from the Colorado River, which flows southwestward from the western flanks of the Front Range, and sends it eastward through a tunnel under Rocky Mountain National Park, where it joins the South Platte River. Weld County, which benefits most from the

Sugar Beets, 1964, 1992

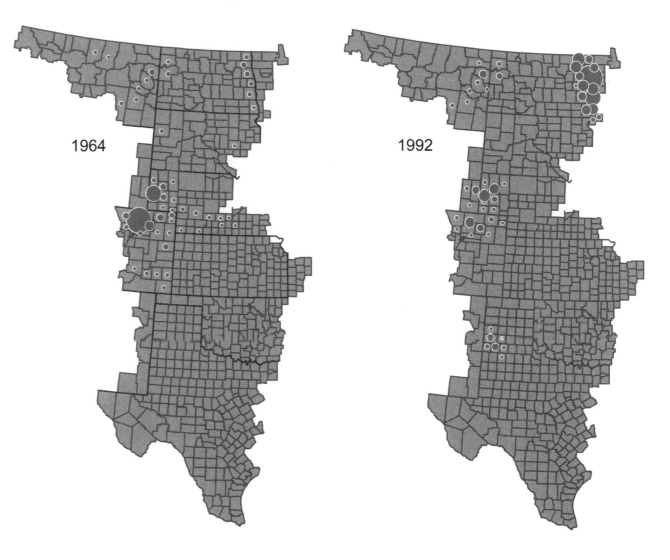

1964

1992

Tons (Thousands)

1,661
643
5

Counties with less than 5,000 tons production not indicated.

project, leads Colorado in the cultivation of irrigated dry beans, alfalfa, and sugar beets, and ranks a close second in corn.

Animals

Animals remain a key component of the agricultural economy of the Grasslands. Cattle are the most important stock animal across the region. The major exception is the hill country of the Edwards Plateau north and east of San Antonio. Sheep, cattle, and angora goats, each consuming a different portion of the vegetation of this mixed grassland/woodland, often share the same pasture. Edwards Plateau goats provide 85 percent of the nation's mohair. The hill country is also home for the nation's largest collection of exotic animal farms and ranches, from ostriches to African big game.

Summer grazing of cattle remains common in Grassland areas that do not support cultivation. The Sand Hills of Nebraska, Flint Hills of Kansas, and Osage Hills of Oklahoma are given over nearly entirely to grazing, as are drier and rougher uplands throughout the region. Supplemental feeding with hay in winter is often necessary because of snow cover. The ranchers usually raise cattle for a year or eighteen months, when they are sold to feedlots for fattening.

The dramatic change in the cattle economy of the Grasslands has brought a boom in the number and size of the region's feedlots. Most range cattle had formerly been shipped out of the region for fattening, generally to the Corn Belt. The capacity of Grassland feedlots surged between the mid-1960s and mid-1980s. Nebraska feedlot operators, for example,

marketed 1.8 million cattle in 1962 and 4.2 million in 1984. More than a million cattle can be accommodated at one time in the large Texas panhandle feedlots centered around Hereford, while those around Weld County, Colorado, can handle half that number. This expansion has paralleled the growth in irrigated cotton, corn, sorghum, and sugar beets. Cottonseed cake, sugar beet pulp, sorghum, and corn silage all provide economical and efficient seed for these giant operations.

An equally dramatic shift of the nation's cattle slaughter industry away from the traditionally large centers to small Grassland communities is associated with this change. Nebraska now leads the nation in the number of cattle slaughtered, followed by Kansas, Texas, and Colorado. These four states now account for 65 percent of the country's annual cattle slaughter of approximately 32 million. Between 1975 and 1988, metropolitan counties in Iowa, Illinois, Minnesota, and Wisconsin lost 14,000 meat packing jobs, more than half their 1975 total. Over 13,000 jobs were added in Nebraska, Kansas, and Texas, which more than doubled their total.

This shift of the red meat packing industry onto the Grassland can be explained by three factors: the regional shift in feedlot operations, industrial cost cutting—land and labor is generally cheaper on the Grasslands—and innovations in the packing industry that have allowed significant savings in transportation costs. The development of boxed beef (in which beef is packaged and shipped in smaller, more easily transportable boxes) and improved refrigeration technologies have made the industry more footloose. Vertical integra-

tion has increasingly joined feedlot and processor. More recently, the industry has seen a gradual evolution of protein companies, such as ConAgra, that pack beef, pork, poultry, and even catfish in massive operations around the country. A secondary move of these companies into processed foodstuffs such as frozen dinners for home and institutional use is now taking shape.

A Mixed Future

The Grasslands, like so much of American agriculture, is undergoing substantial change. The region faces a series of major problems. Rapid withdrawal of aquifer water, especially on the Texas panhandle, has significantly increased the cost of pumping and has led to the possibility of a significant near-future reduction in irrigated activities. Some have suggested that the drier, more rugged portions of the central and northern Grasslands should be returned to the buffalo. Still, the growth of the animal feedlot and slaughter industries, continued expansion of center pivot irrigation on the northern Oglala, and an increasing national demand for sunflower products all help maintain islands of agricultural prosperity.

The Lowland South: Belts and Islands

The Lowland South agricultural region stretches from the Sea Island Atlantic shoreline eastward until the dryness of the Texas interior brings an end to traditional southern ways of agriculture. The South has always been as much myth as reality: identified as an individual place by the colonials for its warm, salubrious climate; defined by its crops and plantation system; delineated by the founding of the Confederate States of America; and today maintained because it is more convenient for outsiders to describe this little-known region as a single place than to understand it. Southerners have always been a proud, independent people, and the South has always been a proud, independent region. There has never been a single South, culturally, economically, or agriculturally, and there certainly isn't one today.

The statistical heterogeneity of this region, however, does not mean that agricultural activities are scattered helter-skelter through it. Agriculture is spread across a series of broad activity belts and smaller islands of production. The Tobacco and Cotton Belts are the oldest of these, while the Peanut Belt is a twentieth-century phenomenon. The islands, in contrast, are pockets of concentrated crop production of relatively recent origin. Catfish, peaches, pecans, and the remainder of the islands are points of specialized production that have often become concentrated at those places as much because of historic accident as real locational advantage. Whether one focuses on either cultivation belts or islands, it should always be kept in mind that

much of the modern rural South is not cultivated at all; rather the chief rural activity over much of the past fifty years has been land abandonment.

Woodland is the dominant rural image throughout almost all the Lowland South. At least 60 percent of every Lowland South state is classified as woodland, and less than 17 percent of each is in crops. Only 7 percent of Alabama is cropped. Vast areas of the Piedmont and Gulf Coast lowland are virtual continuous forest, much planted

to fast-growing slash and loblolly pines that often mature to harvest size in less than twenty years. The Coastal Plain forests, planted as carefully as any annual crop, are distinctive in their single-minded devotion to individual tree species that line the rural roads with exactness of height and distance for mile after mile. Piedmont woodlands tend to be in smaller parcels with wide variations of species and height broken with scattered shacks, abandoned tenant homes, and newer "factory-built" homes

Lowland South

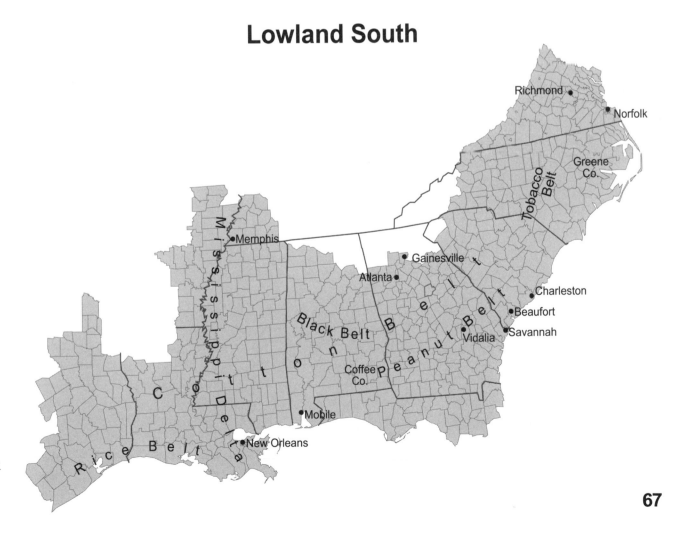

scattered along the wayside. Cornfields also dot these Piedmont woodlands. Woodland is an important part of almost every southern agrarian landscape, even in those areas that are more intensely farmed and known for the production of specific crops. One is almost never outside of view of a stand of pine in the rural Lowland South today.

Plantation agriculture dominated the early economic agricultural landscape of most of the region. Tobacco was the first of these crops, concentrated on the margins of the Chesapeake in the early years, but ultimately spreading in a discontinuous belt westward to southwestern Virginia/Tennessee and Kentucky and southward into the

North Carolina Piedmont and inner Coastal Plain. A smallish relict area of production still survives in northern Florida/southeastern Georgia. Rice plantations dominated agricultural production in the Atlantic marshlands in a narrow belt from southern North Carolina to northern Florida until the Civil War destroyed the fields, traditional labor system, and ultimately the belt itself. Rice cultivation moved westward to the lower Mississippi basin and the prairies of western Louisiana. A thriving sugar plantation culture had been a part of the Cajun culture prior to the Civil War, and it was reborn again after a short period of competition with the

newly arrived rice. Finally, King Cotton held reign over a broad belt of discontinuous cultivation across almost all the Gulf Coastal South from about 1820 until its death about a century later. In each case, however, it should be remembered that significant levels of cultivation of other crops were always present, that few areas were actually numerically dominated by large plantations (though many were economically and areally), and that woodlands always represented a significant land use even on the great plantations.

Historical Evolution

The Lowland South has long been an agrarian landscape dominated by dichotomies. Economically the region was dominated by a system of plantation agriculture characterized by extensive operating units, a high degree of product specialization, and labor-intensive cultivation systems that utilized large numbers of slaves on all but the smallest operating units. The region has also always had large numbers of farmers operating small, general farms that cultivated small acreages of cotton, tobacco, and other crops as well for cash sale. In 1860 the farms of the rice-growing districts of the Atlantic Coastal Plain and the Mississippi Delta averaged more than 800 acres, while those of central North Carolina, much of Alabama outside the Black Belt, and parts of northern Georgia were 300 acres or smaller. This disparity is even greater in terms of the intensity of cultivation of those farms. The Tobacco and Cotton Belts had the highest percentages of improved land, though few areas in the Lowland South reached one half of farmland improved or a third of the county

Tobacco is the oldest commercial crop of the Lowland South. Note the modern bulk curing barns across the back of the field. (RP)

area improved. Piedmont farms typically had much less cultivated land.

While slaves provided the bulk of all farm labor in 1850 on large holdings, 71 percent of all slave holders owned nine or fewer slaves. The planter aristocracy was composed of fewer than 10,000 families. Slaves were concentrated as a result in the areas of labor-intensive plantation cultivation, accounting for more than half the entire population in much of the coastal rice areas of South Carolina and Georgia, the Mississippi Delta, portions of the Alabama Black Belt, and the Louisiana sugar districts.

Southern agriculture has undergone massive restructuring since those days, but the disparities between large and small farmer continue. The average farm of most Lowland South states continues to contain less than 250 acres, yet when only farms with sales of $10,000 or more are considered a very different picture emerges. Great areas of these states have farms averaging 500 or even a thousand acres, especially in the Mississippi Delta, Alabama Black Belt, and portions of the Atlantic Piedmont. Many of these large holdings date from the nineteenth century, but even more have been assembled since 1930 as bankrupt small farms were amalgamated to create ever larger holdings. More land was incorporated into these large-scale, efficient, farming enterprises as even more tenants and small farmers left the region after 1940. Low-intensity farming, most notably trees and livestock, has become dominant, though cotton has been increasing in importance over the past few years.

The southern plantation has undergone two major structural transformations since the Civil War. The traditional plantation focused on a central core of buildings, which included a larger owner's house (the proverbial plantation house), a variety of small sheds and barns for work animals, equipment, and storage, one or more moderate-sized homes for hired managers, and a cluster of houses for the slaves. Operations focused on this central complex as gangs of slaves moved through the annual cotton, tobacco, rice, or sugar cycle. The Civil War brought the first transformation as owners turned to tenant farmers and sharecroppers to provide labor to continue cash crop production. In the cotton and tobacco areas this meant converting slave quarters to tenant housing and then moving the individual houses onto the new, smaller cultivation units. The central core portion of the plantation shrank in importance, and the number of outbuildings dwindled along with the worker housing. This system continued for more than half a century

Change in Tobacco Acreage, 1982-1992

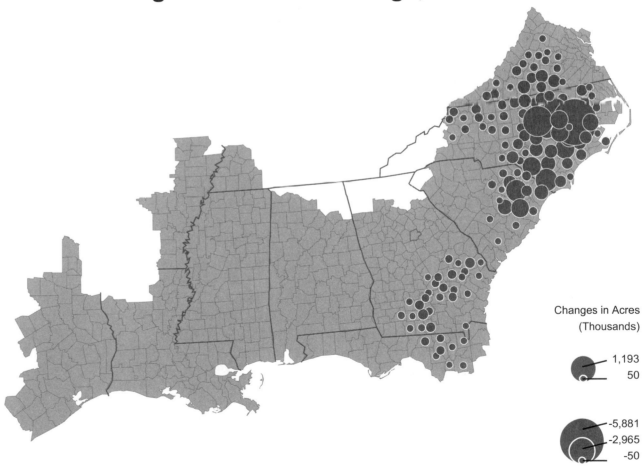

Changes in Acres
(Thousands)

1,193

50

-5,881
-2,965
-50

Counties with fewer than 50 acres change in acreage not indicated.

with the tenant and sharecropper families often moving from plantation to plantation searching for new opportunities every few years as the disadvantageous rental system brought economic ruin to the tenants with every bad year.

The mechanization of the cotton fields in the 1930s, and the attendant departure of thousands of tenant families to the cities to find work, brought the second transformation. One or two hired field hands with tractors and mechanized harvesters could do the work of thirty or more tenant farm-ers. The focus of the farm returned to the headquarters where new equipment sheds were built, and the barns fell into disrepair as the mules and other work animals disappeared. The tenant houses scattered about the plantation fell into disrepair, eventually collapsing to leave little more than a free-standing chimney as a reminder. Soybeans, tree farms, livestock operations, and new cash crops became typical of this now sparsely populated landscape.

The mechanization of cotton cultivation and declining production drove thousands of tenant farmers from the land during the 1930s. (RP)

Principal Crops

Corn, soybeans, and winter wheat are the most widely distributed crops in the modern Lowland South. A third or more of the Atlantic Coastal Plain and often half the Mississippi Delta are taken up with soybean cultivation most years. Corn takes about another third to a half of the harvested cropland on the Atlantic and Gulf coastal plains, but is comparatively unimportant on the Mississippi Delta. Winter wheat is significantly less important than corn or soybeans and often is sown as a winter cover. Woodland and pasture cover much of the remainder of the rural South outside those areas devoted to highly specialized crops such as tobacco, cotton, and peanuts, or livestock.

The Cotton Belt

Cotton production dominated much of the Lowland South landscape until the combined efforts of the federal government, the boll weevil, and low-cost western cotton ultimately drove thousands of farmers out of business or to more profitable crops. Peanuts, livestock, tree farming, catfish farming, broilers, and swine have been the largest growth crops in traditional cotton areas. Higher prices and rising production costs of western farmers have made cotton increasingly profitable in recent years. The early 1990s have brought a remarkable expansion of cotton production in areas that had not produced a cotton harvest in more than twenty years. New gins are being built, and it appears that while King Cotton may be dead, Prince Cotton may be coming of age.

The Tobacco Belt

Tobacco was one of the most reliable crops for southern farmers for more than two centuries. The Virginia/North Carolina flue-cured region was the largest tobacco production area in the nation, and after the invention and popularization of the ready-made blended cigarette, the most profitable. Concerns about tobacco's role in cancer and heart disease have lowered American cigarette consumption by more than 100 billion per year in the last decade alone. All Lowland South production areas have been hit by these declines, despite continuing government subsidization and the increasing importance of foreign markets. Thousands of tobacco farmers in the Carolinas, Georgia, and Florida have all been forced to find other activities.

The Peanut Belt

Peanuts are concentrated in two areas of the United States: the coastal plain from southeastern Alabama to near Augusta, Georgia, and in southeastern Virginia/northeastern North Carolina. Peanuts are often grown in these areas in association with soybeans, corn, winter wheat, and, increasingly, cotton. Georgia is the largest peanut producing state with more than a third of the nation's annual crop.

The Islands

Rice. Arkansas continues to lead the nation in rice production with about 40 percent of the annual crop, while the three-state Mississippi basin rice area contributes 56 percent by value. Almost all rice cultivated in this region is of long and extra-long grain

varieties that are preferred in the American and European markets. California's relative importance as a rice production area continues to provide little competition for the region because of its reliance on short and medium-grain varieties that are almost entirely exported to Asia. The Louisiana/Texas district is the third largest district in the nation with a $303-million crop in 1992. Southern rice growers tend to concentrate heavily on this single crop with high levels of automation and a minimum of dependence on imported labor.

Sugar. The sugarcane district of southern Louisiana continues largely because of federal government subsidies to the cane industry. Concentrated in a narrow belt from Lafayette to Bayou Lafourche west of New Orleans, almost all production takes place on large plantations. The Louisiana industry has long been heavily automated with relative minor use of seasonal labor. The Louisiana crop of five to seven million tons annually typically is only one half to a third the size of the Florida harvest.

Change in Cotton Acreage, 1982-1992

Changes in Acres
(Thousands)

133
67
.5

-67
-.5

Counties with fewer than 500 acres change in acreage not indicated.

Fruits, nuts, and vegetables. Scattered areas of the Lowland South have long been used as centers of fruit, nut, and vegetable production. Areas near Norfolk, Charleston, and Beaufort (South Carolina) on the Atlantic Coast and several counties in Mississippi and Louisiana were the first centers of early season vegetable production in the South and continued in that role until the rise of large-scale, off-season vegetable production in Florida came to dominate the market by the middle of the twentieth century. Only remnants of these production zones are still visible today.

Peaches also prospered in the nineteenth century in Georgia and South Carolina, especially after severe freezes cut production in northern orchards. Georgia to this day calls itself the "Peach State," even though statewide production has fallen below that of some individual counties in California. The fragility of this fruit, how-

Coffee County, Alabama

Most famous for its statue to the boll weevil in the town square of Enterprise, Coffee County agriculture has undergone a massive restructuring over the past decade. Cotton (544 percent), laying chickens (489 percent), and broilers (129 percent) increased significantly, while soybeans and hogs declined. Poultry now accounts for more than two thirds of all sales, followed by peanuts, and a distant cattle.

	1992	1982
Farms (number)	760	944
Average farm (acres)	231	206
Marketings ($ thousands)	94,608	60,529
Poultry sales ($ thousands)	64,546	34,882
Broilers (number)	37,163,342	28,842,548
Layers (number)	350,370	78,055
Peanuts (acres)	27,940	21,261
Cattle sold	9,966	11,991
Cotton (acres)	4,811	978
Hogs sold	10,157	32,105
Soybeans (acres)	426	24,258

Change in Broiler Sales, 1982-1992

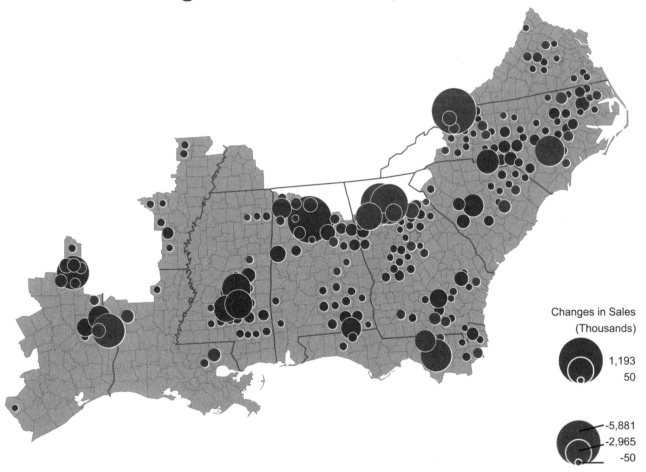

Changes in Sales
(Thousands)

1,193
50

-5,881
-2,965
-50

Counties with fewer than 50 acres change in acreage not indicated.

ever, has given some advantages to these farmers who are able to pick fruit in a riper state and ship shorter distances than western shippers to enable them to provide a more flavorful product, much in the same manner that fruit from the surviving orchards in New York and New Jersey has continued to thrive. The expansion of peach orchards across the South Carolina Piedmont has sent that state's production past Georgia in the last decade, coincidentally making Interstate 85 one of the most scenic spring highway drives in the Lowland South. Pecans are the most important of the nut crops. Georgia is the largest producer, with crops exceeding $80 million in some years.

Poultry. Broiler production has cascaded out of the Uplands into the Lowland South as Gold Kist and smaller processors encouraged production throughout the region. The North Carolina broiler industry is in the process of being transferred to the coastal plain as both Holly Farms (now a Tyson Foods subsidiary) and Gold Kist continue to expand their broiler operations on the state's coastal plain.

North Carolina has become the largest producer of turkeys in the United States with a broad band of production stretching almost all the way across the state's southern piedmont and coastal plain. Like broilers, turkeys are housed in low poultry houses located on secondary or even gravel roads to discourage casual visitors. The birds' susceptibility to disease is legendary, and tens of thousands of fowl can be lost in weeks once an epidemic is unleashed. The continuing decline of tobacco and increasing consumption of turkey nationally suggests that turkey production will continue to expand for the foreseeable future, primarily because of the area's lax environmental constraints, relative nearness to market, and low price for locally produced feeds.

Swine. Pork traditionally has been the most important meat in the southern diet. Though hogs were primarily raised for local consumption, swine husbandry was well suited for expansion in the postwar era. The development of mass production through the adaptation of poultry technology to swine production has allowed the creation of animals with less body fat and faster growth rates. Grow-out operations, similar to the largest poultry operations, are increasingly important on the North Carolina Coastal Plain as the nation's largest pork producers rapidly expand restructure production to take advantage of the low feed, labor, and energy costs of southern producers.

Cattle. As much as a third of southland farms are now devoted to permanent pasture for beef cattle. Almost 5 million calves were bred in the Lowland South in 1992, primarily for shipment to western feedlot operators. Cheap lands, warm winters, and inexpensive supplemental feeds make this region a perfect area for the production of calves, especially with increasing pressure on western cattle operations' use of federal grazing lands.

Change in Hog Sales, 1982-1992

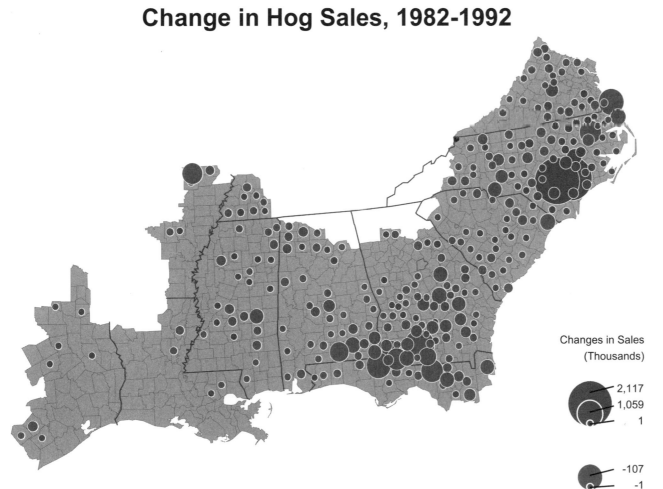

Changes in Sales
(Thousands)

2,117
1,059
1

-107
-1

Counties with less than 1,000
change in sales not indicated.

Greene County, North Carolina

This inner coastal plain farming area reflects both the past and the future in its varied agriculture. Total agricultural marketings increased 63 percent over the past decade as county farmers shifted from their traditional tobacco-corn-soybean cropping pattern to increasingly concentrate on broilers, hogs, turkeys and cotton. Hogs are now the most important source of income ($53 million) while tobacco follows ($31 million), but the relatively new poultry industry has become the third ($19 million) most important source of agricultural income.

	1992	1982
Farms (number)	407	670
Average size (acres)	276	160
Agricultural sales		
($ thousands)	118,128	72,254
Hogs sold	728,661	255,556
Broilers sold	5,003,375	45,000
Tobacco (acres)	25,674	18,027
Corn (acres)	30,177	42,547
Soybeans (acres)	25,674	18,027
Wheat (acres)	4,842	6,317
Cotton (acres)	8,330	0

Catfish

A parked car or two and a cluster of fishermen hanging over a bridge railing has long been a common sight throughout the Gulf Coastal South, especially in the lower Mississippi basin. Freshwater fish have been a dietary mainstay for many rural southerners, probably beginning as a dietary supplement for the plantation slaves. Bream, buffalo, catfish and other species have been stocked in ponds for individual harvest throughout the twentieth century. Large-scale commercial production of pond-raised fish began in the 1960s. The catfish soon became the species of choice for commercial pond operators and the Mississippi Delta of Louisiana, Arkansas, and Mississippi the largest center of pond catfish production. Almost six million pounds of processed catfish were produced in the region by 1970 and 457 million pounds in 1992. Catfish have become an important profit center for former cotton producers with acreages too small to be efficient in today's automated industry. Alabama farmers, seizing on the success of their neighbors, have pushed production upward to make the state the second largest producer today, followed by Arkansas and Louisiana.

Several major poultry producers are also involved in this industry today. ConAgra, already a major saltwater fish packer, became interested in catfish processing in the late 1980s. Gold Kist, a major supplier of aquaculture feeds, built a research facility to breed the perfect pond catfish in Mississippi in 1994. It is anticipated that these and other large-scale food processors will continue to intrude into catfish and other aquaculture production as they perceive these and other flesh products simply as protein to be distributed to the public. Pond-raised catfish fillets are already commonly used in fish fillet sandwiches and as "unidentified" whitefish fillet restaurant entrees. This trend is likely to expand as North Pacific fishing grounds continue to experience declining harvests of white-flesh trash fish.

The Upland South: Coves and Broilers

The blue haze covering the region like a downy blanket softens the mountains, but has never been able to soften life in the Upland South. This cultural and economic backwater is characterized by high poverty, low rates of agricultural return, rural depopulation, and the highest percentages of rural nonfarm population in the nation. Agriculture thus has always been an important activity here, but often not the most important one, even among those still living on the farm. Part-time farming and hobby farms were important elements of agricultural life here before anyone thought to give these concepts name. The land is a pervasive part of these people's self-image; making a living from those images has been difficult. This is a place where horse-drawn plows, hayricks, and home meat slaughter share space with glass-lined silage tanks, high-tech chicken production, and some of the richest thoroughbred horse farms in the world.

The Evolution

The initial development of this region in many ways presaged the entire settlement history of this region. European settlers moved into the Valley of Virginia in the early eighteenth century. Some came with great land grants to develop some of the largest plantations of the period; others came with nothing but their determination. The result was a landscape divided into the haves centering on the best lowland areas and the have-nots living in the backwaters

and broken lands of the region. A dual economy soon evolved with only limited interchange and interaction.

The earliest settlers moved both westward from the margins of the Chesapeake Bay into the Uplands carrying the tobacco and plantation cultures with them and southward out of southeastern Pennsylvania with that general farming, egalitarian heritage. Two conflicting lifestyles evolved—one of extreme wealth with an

emphasis on property and a second focused simply on making a better life. The main "highways" west—Braddock's and Forbes Roads to the north, the Great Valley Road carrying traffic southward into Tennessee, the Jonesboro and Saluda Roads linking the Carolinas to the interior and the Wilderness and Nashville Roads opening the Kentucky Bluegrass and Nashville basins— were important arteries to the development of early commerce, but so too were the

Upland South

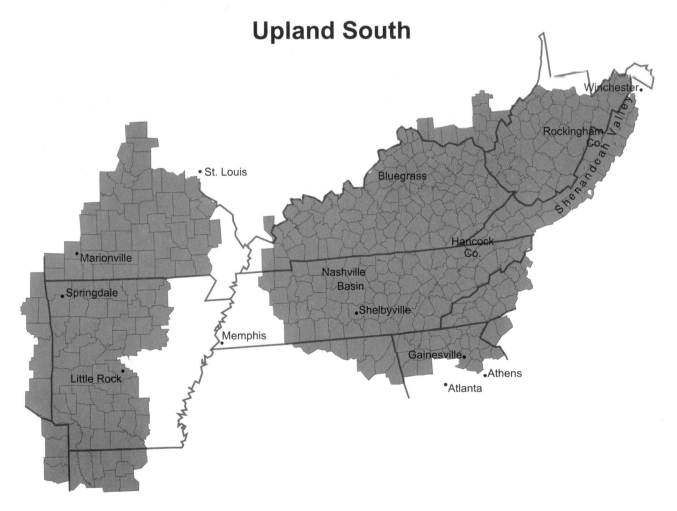

Most cultivation in the Upland South is concentrated in coves and along river valleys. (RP)

more westward trek would bring them financial security.

Life for the typical Upland South farmer has almost always been precarious. Most arrived with little, settled on land that had little to offer, and were often unable to ship their small surplus to market. Often poorly educated, these people left only fragmentary records of their lives, much of which is assembled more from conjecture rather than from reliable statistical data. The poorer farmer on the agricultural margins in 1840 probably possessed a horse and mule, a bull and a few cows, a ram with a handful of ewes, a larger number of pigs, and a rooster and a few hens. The family planted corn as their staple crop, though much of this in turn was fed to the animals. Most hogs foraged for themselves on the mast from the hardwood forest with the only human assistance coming from annual burning of the forest floor to encourage the growth of forage plants. These semiferal hogs' ears were clipped to identify ownership when the excess animals were caught for slaughter in the fall to preserve for the winter table.

Commercial agriculture for these farmers was difficult. Muddy, rutted roads made the transport of large quantities of agricultural goods more expensive than the value of the goods themselves. Corn was the most important crop, even though harvests were often fewer than thirty bushels an acre from these typically unfertilized fields. The problem was shipping surplus to market. The distillation of whiskey to reduce the cost of transportation from the region certainly was important, if a bit overstated in the literature. Southwest Pennsylvania was a major center of dissent during the Whiskey Rebellion, though thousands of

local tracks following less important Amerindian pathways into the backwaters of the Appalachians.

The beginning of the nineteenth century found much of the better land claimed—the Shenandoah Valley; the Kentucky Bluegrass; the Nashville Basin; and the lowlands of the Kentucky, Tennessee, and other rivers where land was rich and economic potential high. Less well-suited lands were only available to later arrivals who filtered through the gaps and up the rivers to find new homes beyond the edge of the settled world. Having little more than their determination and strong

backs, these later settlers filled in the remainder of the lands, even though they were unsuited for settlement.

The western margin of the Upland South, the interior uplands, had a different settlement history. Although individuals settlers were scattered throughout the Ozarks and Ouachitas in the late eighteenth century, continuous settlement did not become a reality until the nineteenth century. Filling in of the land was not completed until midcentury. The settlers themselves came from many areas, but most were from Kentucky and Tennessee—still moving and still believing that one

Upland Tobacco, 1859-1889

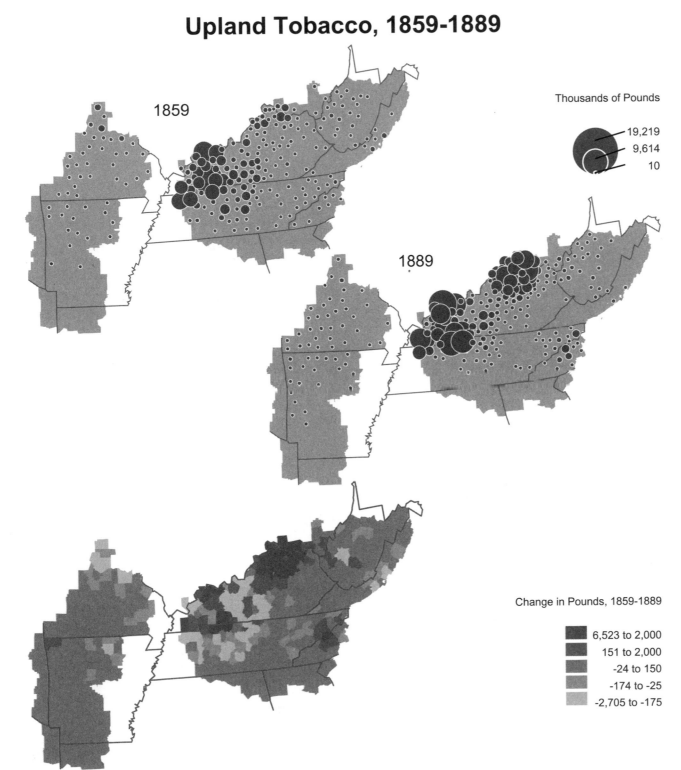

1859

Thousands of Pounds

19,219
9,614
10

1889

Change in Pounds, 1859-1889

6,523 to 2,000
151 to 2,000
-24 to 150
-174 to -25
-2,705 to -175

stills were scattered throughout the entire region. It was reported that more than 10,000 stills were operating in northern Appalachia in 1810, and it is estimated that at least 19 million gallons of whiskey were produced in the eleven Appalachian states. While these numbers were deduced with a degree of creativity, it is clear that this was an important source of income for many.[1]

Shipping surplus meat to market was impossible; walking the live animals to market was not. Reports of cattle, hog, and other livestock drives from isolated Appalachian centers into the larger markets are common for the late eighteenth and early nineteenth centuries. The rise of the canal era largely destroyed this market for northern Upland South farmers in the late nineteenth century. Livestock production for sale to external markets declined after that period, except in those areas where rail or water transport was available.

Many in eastern Appalachia planted tobacco for home use. Tobacco generally was sold in thousand-pound hogsheads until the early twentieth century. These quantities were too heavy and bulky to be transported over the poor roads in the region, and large-scale tobacco farming was found only in areas with access to good transportation. A small trade in plugs and other partly processed items did exist, but again most isolated farmers were not able to participate in this trade.

Specialty Crops

Tobacco

Tobacco production began in the late eighteenth century in the Upland South and was well advanced by the middle of the nineteenth century. The inland tobacco farmers could not economically ship their hogsheads overland to eastern markets, but were able to open a vigorous trade downriver after trade barriers through New Orleans were lifted in 1787. Kentucky had become the second largest tobacco producing state by 1850 and was within 14,000 pounds of equaling Virginia in 1860. Virtually all this tobacco was shipped downriver to New Orleans until the Civil War.

Western Kentucky was the most important early center of tobacco cultivation in the Upland South. Several varieties of burley, especially "Red Burley" and "Little Burley," were grown and heat cured with hickory fires down the centers of their near air-tight barns. This tobacco was especially popular for the plug and twist chewing and other oral tobaccos that were the most important manufactured tobacco products for the first half of the nineteenth century. Cigar and pipe use followed in importance, while cigarettes were relatively rare. Thousands of brands of chewing tobacco, snuff, and cigars were manufactured here to serve local markets because of the high cost of importing products from the East.

The Civil War forever altered the Kentucky tobacco market by blocking trade through New Orleans and forcing the adoption of the newly constructed railroads as the principal method of moving raw and manufactured tobacco to eastern markets. Even more important was the discovery of the mutant white burley strain in 1864. Cultivation of this lighter leaf tended to remain concentrated in eastern and central Kentucky because of the well-developed rail system available to move it inexpensively to the centers of cigarette manufacturing in

Virginia and North Carolina. Smokeless tobacco consumption was severely wounded in 1913 by the introduction of the first modern blended cigarette (Camel), which was manufactured with a 30 percent Kentucky white burley content. With some notable exceptions cigarette manufacturing moved eastward, but the future of Kentucky tobacco cultivation was assured until the safety of the very use of tobacco came into debate in the 1960s.

Thoroughbred Horses

Thoroughbred horse racing in America began in Virginia in the early eighteenth century. The Virginia Piedmont became, and still remains, one of the nation's most important equestrian centers. Concerns about the evils of horse racing in the Virginia legislature ultimately convinced many horse breeders, who also had interests in the tobacco industry, to move westward into the Kentucky Bluegrass, which became the dominant center of thoroughbred horse breeding in the nation. Horses are bred in

The Upland South dominates the fast growing broiler industry. (Karla Harvill, Gold Kist)

almost every state today, yet Kentucky has remained the most important center for training and sales and continues to host the Kentucky Derby, the single most important thoroughbred race in the nation.

The Nashville Basin became a secondary thoroughbred center, though it was never as famous. The Tennessee walking horse developed south of Nashville near Shelbyville. The area continues to be the largest center of walking horses.

Broilers

Chickens have been an integral part of every American barnyard since colonial times. Large-scale commercial production, however, did not become important in the Uplands until after World War II. The factors underlying the localization of broiler production in the Upland region are not clear, though low feed costs, inexpensive labor, and low land costs are generally cited. It is unlikely, however, that the success of these companies was actually based on any physical or economic advantages found in this region. It could be argued that if John Tyson and the other early entrepreneurs that developed much of the technology that drives this industry had been living on the Gulf or Atlantic Coastal Plains or the southern Corn Belt, that an entirely different geography of the industry would have developed.

Four areas stand out as centers of broiler production in the Upland: (1) a discontinuous band along the western border of Arkansas extending from southwestern Missouri into northeastern Texas; (2) a band extending across northern Alabama into Georgia, including small

Upland Broiler Cores, 1954-1992

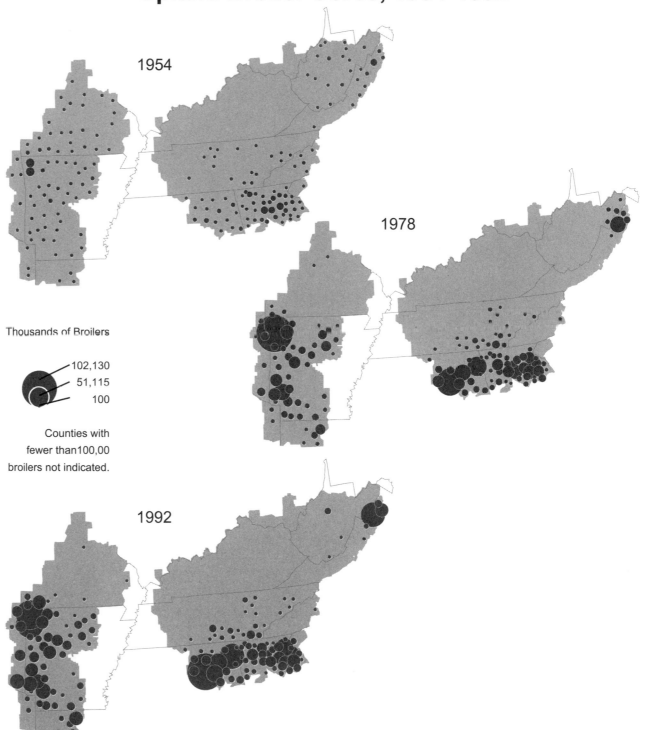

Thousands of Broilers

102,130
51,115
100

Counties with
fewer than100,00
broilers not indicated.

1954

1978

1992

parts of Tennessee; (3) the uplands of North Carolina; (4) and a smallish relict area in the northern Valley of Virginia, which is again seeing some growth as overspill from production in the eastern shore of Maryland. Broilers represent one third of all Arkansas farm marketings, and poultry (including eggs and turkeys) account for 43 percent. The growth of the Arkansas area has paralleled the expansion of Tyson Foods, based in Springdale, and ConAgra, based in Omaha, Nebraska. Both companies have concentrated on acquiring

Hancock County, Tennessee

Tucked in the mountains of east Tennessee, Hancock County farmers crop only about 11 percent of their farms. Like much of the Upland South, farms are small, farm incomes low, and activity concentrated on at least one specialty crop. Changes in Hancock County agriculture, like most Upland South counties, parallel those changes taking place on the urban fringe. Cattle, hay, and other activities requiring low labor and capital inputs are increasingly important, while traditionally more intensive crops are declining. The exception, tobacco, continues because of its high returns from relatively small acreages, but ultimately it, too, will be affected by this trend.

	1992	1982
Farms (number)	736	868
Average size (acres)	109	93
Farmland cropped (%)	11	10
Agricultural sales ($ thousands)	8,376	5,936
Tobacco (thousands of lbs.)	3,017	2,623
Cattle (inventory)	14,656	12,489
Hay (acres)	6,802	5,708

their competitors, which has fueled both their growth and the overall concentration of broiler production in the region. The northern Georgia/Alabama area was the largest broiler production area in 1970, largely encouraged by Gold Kist, a regional cooperative, and the Pillsbury Company, which invested heavily in facilities based in Gainesville, Georgia. The area slipped in relative importance in the 1980s, although Culman County, Alabama, remains the highest producing county in the nation with more than 100 million broilers per year. This area is bifurcated into two centers, with the eastern core slipping southeasterly onto the piedmont near Athens, Georgia. Although Gold Kist is the largest processor in this area, Tyson, ConAgra, Seaboard, and a host of smaller processors still account for a significant part of the area's annual output. Holly Farms (now owned by Tyson Foods) is the dominant broiler processor in North Carolina. Their decision in the early 1990s to concentrate future broiler production in eastern North Carolina will have a serious impact on operations in the western part of the state.

Fruit

Two of the largest fruit production areas in the United States were located in the Upland South at the turn of the twentieth century. The upper Shenandoah Valley around Winchester, Virginia, is one of the oldest important apple areas in the United States. Much of the surplus production was processed into applesauce, juice, and cider. Changing tastes, competition from Washington state growers, and rising land values as a result of suburban growth from Washington, DC, have lowered the relative

position of this area's fresh fruit in the national market. The lower value of the fruit, slightly more than one half that of Washington fruit, has not encouraged expansion.

The southwestern Missouri and northwest Arkansas area was one of the most important fruit centers in the nation in the late nineteenth century, but has virtually disappeared today. Railroad promoters encouraged large apple and peach plantings along their lines for easy shipment to St. Louis and beyond. Marionville, Missouri, farmers, for example, had about 2,500 acres of orchards capable of producing 100,000 barrels of apples annually in 1923. Labor shortages, disease, bad weather, and market competition all hurt the industry, that had largely disappeared by 1950.

The Welch Company established about 2,500 acres of Concord grapes during the 1920s to be processed in a juice plant in Springdale, Arkansas. Labor shortages, weather problems, and a declining market all eventually brought an end to this part of the fruit industry. Several small wineries predate the Concord grape era, and a handful of minor operations still produces a variety of wines.

Strawberries were also important in the same region of southwestern Missouri and adjacent Arkansas and Oklahoma beginning in the 1890s. Production remained important through the 1920s to decline during the Depression. It never recovered, primarily because of labor shortages and cheaper berries from the West Coast, where plentiful labor from Mexico was available.

The decline of fruit production in the western Uplands is poorly understood, but seems to stem as much from the aggressive marketing skills of producers and distributors from other areas as to problems endemic to this region. Upland producers seemingly were neither able to keep up with the changing industry or to access capital for expansion during the crucial periods of rapid growth. The problems generally cited for the decline of the region were also found in the more successful areas as well.

Specialty Areas

Shenandoah Valley

The Shenandoah Valley developed an important small grain tradition during the early nineteenth century. By 1840 it was the largest single wheat production area in the Old South. Lesser amounts of rye were also cultivated. Competition from the western Corn Belt and later the Great Plains forced farmers to seek other crops in the late nineteenth century. Large quantities of livestock, dairying, and poultry began appearing in the 1890s and have remained the dominant agricultural activities. Apple production areas also developed around Winchester, which continues to be one of the nation's larger centers of processed fruit. Rockingham County, once the nation's largest center of turkey production, has largely been supplanted by other lower-cost production areas. Broilers, on the other hand, have been of increasing importance as Delmarva producers search for expansion locations. Creeping suburbanization from the Washington, DC, metropolitan area brought large numbers of estate farms, mostly raising thoroughbred horses or specialty pure-breed cattle after World War II. More recent industrialization in the eastern panhandle of West Virginia and surrounding areas has brought rising land prices and the other changes characteristic of the near urban fringe.

Kentucky Bluegrass

The rolling hills, pastures, and rock fences of the Kentucky Bluegrass are one of the most distinctive agricultural landscapes in America. The area was the second most important wheat center of the Old South in the early nineteenth century. Hemp was introduced about midcentury and became a dominant activity until sisal began replacing burlap as the preferred covering for cotton bailing in the Mississippi basin. White burley tobacco was discovered just north of the Bluegrass, grew in importance throughout the last half of the nineteenth century, and continued into the twentieth century as it became an important ingredient in the standard "American" cigarette.

Thoroughbred horse breeders began migrating into the Bluegrass during the late eighteenth century and the region is famous for its breeding stock. Many of the nation's most famous thoroughbred racing stables and breeding farms are still located in the Bluegrass. Today's inner Bluegrass landscape remains dominated by thoroughbred farms and associated activities with increasing numbers of livestock and tobacco farms farther away from Lexington.

Nashville Basin

The Nashville Basin largely parallels the Kentucky Bluegrass, though much of its development came a few years later. On the northern edge of the Cotton Belt, Nashville was an important regional center. Small grains, corn, and even a bit of cotton were grown in the nineteenth century. Horses were also important, though never to the degree as in Kentucky. Urban expansion and the growth of the country music industry in Nashville have supported the development of an extensive estate landscape, though many make little attempt to make a profit at farming.

Notes

1. While ascertaining meaningful numbers for stills and output is impossible for the early nineteenth-century Southern Upland, it is clear that the nation's tastes were changing as the relatively expensive rum was being replaced by a variety of locally distilled grain whiskeys, primarily corn and rye.

Migrant labor continues to play an important role in Florida agriculture. (RP)

South Florida Industrial

The South Florida Industrial Agricultural region extends northward to about the 50 degree Fahrenheit January average minimum temperature. The three most important factors controlling recent agricultural development in the region are short periods of severe weather, water, and a rapidly expanding population. Numerous killing freezes have struck the state, the worst occurring in 1835, 1894–95, and 1984–85. Each was responsible for killing millions of citrus trees. Hurricanes are the second most destructive weather events with storms such as Andrew in 1992 not only destroying thousands of homes and businesses, but flooding agricultural lands and destroying crops as well.

High annual rainfalls have meant that excess water has plagued much of south Florida throughout its history. Artesian wells were common, while shallow wells easily provided water in other areas. It was widely believed in the nineteenth century that regional drainage programs were necessary to foster economic growth in the region. The Florida legislature appealed to the federal government as early as 1845 to study the feasibility of draining the Everglades to expand development. It was also believed that cutting through the limestone ridge along the east coast would drain millions of acres of prime agricultural land near Miami. Work began in 1906, but it was only after hundreds of miles of canals were dug that significant results were obtained.

The Kissimmee River and Lake Okeechobee were believed by many developers to be the key to solving the Everglades "problem." In the 1970s the Army Corps of Engineers was given the tasks of straightening the Kissimmee River to speed water flow and constructing a 10-meter-high levee around Lake Okeechobee to slow the southward flow of water into the Everglades. Much of the past twenty years has been spent trying to undo the damage of these projects. The shortened Kissimmee River does not naturally cleanse its waters of the high levels of nutrients and agricultural chemicals. The entry of these highly organic materials into Lake Okeechobee speeds the eutrophication process (reduction of oxygen levels below necessary levels to maintain life) in the lake and ultimately the Everglades. Algae blooms starved oxygen from both water bodies while the highly enriched waters entering the Everglades have promoted the rapid spread of water hyacinth, creating a variety of drainage problems, further eutrophication, and a multitude of related problems. The battle over the maintenance of proper lake levels has become gargantuan as farmers, recreationalists, hydrologists, land developers, and environmentalists all have differing views on the proper approach to effective drainage control, the maintenance of the Everglades habitat, how most efficiently to recharge the local aquifers that play a crucial role in supplying most of south Florida with water, increase recreational opportunities, and expand farming. The rapidly increasing population in both Miami and along the west coast has placed further demands on the environment and the conflicting needs of the developers.

The population of Florida doubled between 1970 and 1992. The Miami/Fort Lauderdale metropolitan area population increased by almost three-quarters. The impact of this growth has brought massive increases in water demand, value of agricultural lands, and problems in environmental management. Water withdrawals in the state as a whole increased from 17 billion to 17.9 billion gallons per day between 1985 and 1990. Public water supply withdrawals increased from 1.9 to 2.2 billion gallons per day. Despite widespread increases in the use of drip irrigation and other water-efficient irrigation techniques, agricultural use increased from 2.9 to 3.7 billion gallons per day. Florida is the second

South Florida Industrial

most important agricultural state in the nation. Agriculture by value is dominated (1991) by five crop complexes: citrus ($4.7 billion), cattle ($365 million), sugar ($485 million), off-season vegetables ($748 million), and nursery/greenhouse crops ($590 million).

Citrus

Feral citrus groves from early explorations were found throughout much of the state when permanent settlement was initiated in 1565. Jesse Fish, the state's first orange baron, began shipping juice and fruit to London from the St. Augustine area in 1776. Douglas Dummett is generally given credit for establishing the Indian River citrus area in 1830. Dummett pioneered

the grafting of sweet oranges to sour orange root stock that created the hearty, sweet product still identified with the Indian River. The second center of early orange production was the "ridge," an elevated upland running about half the length of the state from southeast of Gainesville to near Lake Placid.

The frozen orange juice concentrate process was developed by the Florida Citrus Commission in 1942 and is the heart of the state's industry today. Frozen concentrate is created by dehydrating fresh orange juice and blending the concentrate of several varieties to create a consistent product. The finished product generally also has some added fresh juice, for color and flavor, and some orange pulp solids. More than two thirds of the 1994 orange crop was pro-

cessed into frozen concentrate.

Imports from Brazil today account for more than half of all orange juice concentrate consumed in the United States. Brazilian concentrate was sent to Florida ports to be blended with American juices for many years, but declining market demand for frozen concentrate and increasing sales of single strength "fresh from concentrate" juice have made it more efficient to ship it directly to northeastern and west coast ports for repackaging.

Florida produced 175 million boxes of oranges, 42 million boxes of grapefruit, 5.5 million boxes of tangerines and tangelos, and 1.6 million boxes of limes in the 1993–94 season. The center of production has been steadily moving southward for more than a century. Periodic killing freezes since 1835 have repeatedly destroyed groves and forced growers southward to seek safer production areas. New grove development today is concentrated from Lake Okeechobee to Fort Meyers, although environmental restrictions and water controls have made this last migration slow and expensive. The typical new orange grove is characterized by ten-feet-deep, water-filled ditches on the perimeter, containment ponds for runoff, and drip irrigation to minimize water and chemical applications.

Grapefruit, 1954-1992

1954

1992

Thousands of Field Boxes

15,437
7,788
138

Counties with sales less 100,000 field boxes not indicated.

Cattle

Commercial cattle ranching began in Florida in the nineteenth century. These early ranches raised a smallish, rangy mixed-breed animal that could thrive on the poor scrub forage of the thickets and was immune to the heat and ubiquitous cattle tick. European breeds could not survive the

Oranges, 1954-1992

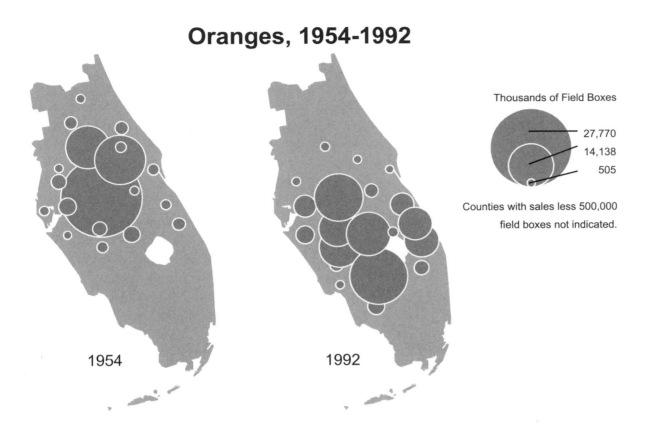

1954   1992

Thousands of Field Boxes

27,770
14,138
505

Counties with sales less 500,000
field boxes not indicated.

ward. More than 450,000 acres of cane produce 1.6 million tons of refined sugar from seven mills. The three largest growers control more than half of all production, and the consolidation rate of plantations is increasing. Two mills and one of the largest plantations in the state were recently acquired by a refugee Cuban sugar plantation family.

The Florida sugar harvest has traditionally utilized large numbers of laborers who hand cut the cane. These men have primarily been brought to the area from Jamaica because of the growers' inability to attract American workers because of the low wages offered and the dangers of working with heavy machetes cutting the heavy cane.

Cattle and Calf Sales, 1992

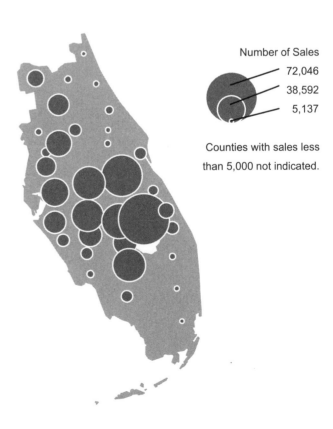

Number of Sales

72,046
38,592
5,137

Counties with sales less
than 5,000 not indicated.

heat, the tick, or the poor forage. A state-sponsored tick eradication program and cross-breeding of European breeds with tick- and heat-resistant Brahmans have allowed the redevelopment of an extensive cattle industry in recent years, although some breeds still do not fare well when grasped on rough scrub forage. The typical animal today is about one-quarter Brahman and three-quarters Angus, Hereford, or Shorthorn.

Almost a million cattle and calves are marketed annually from Florida ranches. The state has become the second largest producer of feeder calves in the United States, shipping most to the southern Great Plains, though almost 100,000 are shipped annually to California. The value of the calves is comparatively low, and the state's

$365-million cattle industry is only twenty-seventh in receipts. The state has only a single packing house.

Sugar

Commercial sugarcane was introduced into South Florida in the late nineteenth century, but did not become important until the 1920s. The Celotex Corporation bought the Southern Sugar Corporation in 1925 as a source of raw cellulose for the manufacture of wallboard and immediately began expanding operations. Today the renamed United States Sugar Corporation is the state's largest cane producer.

More than half the nation's sugar now comes from Florida's cane fields along the east side of Lake Okeechobee and south-

Vegetables, 1992

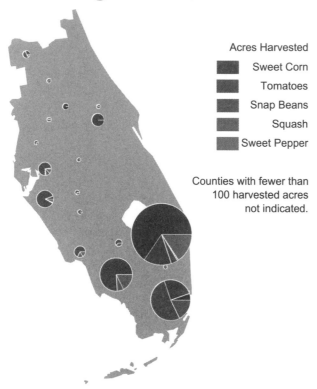

Acres Harvested

- Sweet Corn
- Tomatoes
- Snap Beans
- Squash
- Sweet Pepper

Counties with fewer than 100 harvested acres not indicated.

Early attempts to introduce mechanized cutting on a large scale failed because the soft muckland soils could not support the heavy mechanized cutters. Recent labor issues, however, have brought new approaches to mechanized cutting and virtually all is cut by machine today.

Cane growers are currently facing a multitude of problems. Florida environmentalists frequently criticize their water management practices, which include dumping large quantities of drainage water into Lake Okeechobee part of the year and withdrawing large quantities during the dry season. Smoke from burning cane fields is a major source of air pollution throughout Central Florida in late winter. Finally, increasing competition from fructose raises questions about the future profitability of this highly subsidized industry, which may no longer be essential to the national welfare.

Vegetables

Florida is the nation's second largest producer of fresh vegetables with 13 percent of acreage and 19 percent of value (1993). The high value of Florida vegetables stems from its concentration on off-season crops, and the state leads the nation in fresh tomatoes ($593.8 million), bell peppers ($201.9 million), winter season strawberries ($102.7 million), snap beans ($92.8 million), cucumbers ($76.8 million), sweet corn ($69 million), watermelon ($66.6 million), eggplant ($15.6 million), and escarole/endive ($8.5 million). It is the second largest producer of celery ($58.7 million) and the third largest producer of fresh cabbage ($41.7 million), head lettuce ($19.4 million), and carrots ($18.5 million). Less than 1 percent of the state's vegetable crop is processed.

Vegetable production has declined slightly over the past few years in the face of increasing imports. Dole and other multinational produce distributors have encouraged the development of large off-season vegetable farms in Costa Rica, Peru, Chile, and elsewhere with increasing impact on American producers. Imported green peppers, melons, onions, and other crops are often offered to consumers at lower prices than American produce because of differences in production costs. The recent passage of the North American Free Trade Agreement (NAFTA) will further this process, especially for tomatoes.

The distribution of vegetable production has shifted significantly as pressures from development have lured vegetable farmers into selling their farms at tremendous profits. Florida agriculture receives property tax relief with rates reflecting usage, but the lure of prices passing $30,000 per acre for lands that were purchased for as little as $100 per acre has been very persuasive. The largest concen-

Polk County Florida

The expansion of Florida's population has been felt in every corner of the state, especially near Orlando. Polk County, southwest of Orlando, is no exception as farmland declined by more than 100 square miles during the 1980s. Citrus fruit continues to dominate revenues, but more than 20,000 acres went out of production in the 1980s due to the combination of freeze damage, competition for land with developers increasing land prices, and increasing orange juice imports. Nursery/greenhouse crops, especially outdoor plants and sod, and dairy significantly expanded over the past decade in response to the growing local market for these products, but total agricultural sales declined.

	1992	1982
Farms (number)	2,294	2,357
Average size (acres)	266	289
Agricultural sales ($ thousands)	203,350	231,195
Fruit ($ thousands)	137,996	172,601
Oranges (acres)	89,694	107,393
Grapefruit (acres)	13,789	26,052
Cattle sold	44,496	75,604
Dairy sales ($ thousands)	8,745	5,894
Nursery/greenhouse ($ thousands)	29,613	15,105
Nursery ($ thousands)	17,144	12,166
Sod ($ thousands)	5,941	n/a

Nursery/Greenhouse Sales, 1992

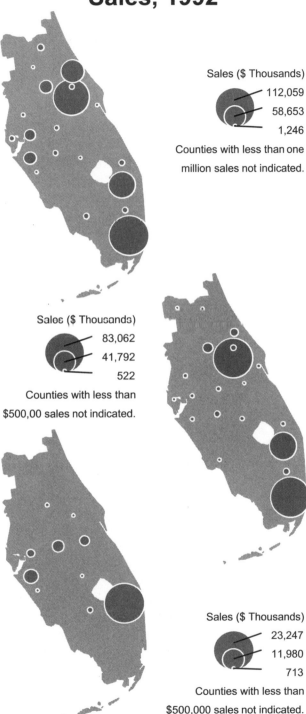

Sales ($ Thousands)

112,059
58,653
1,246

Counties with less than one million sales not indicated.

Sales ($ Thousands)

83,062
41,792
522

Counties with less than $500,00 sales not indicated.

Sales ($ Thousands)

23,247
11,980
713

Counties with less than $500,000 sales not indicated.

trations today are found south of Miami, south of Lake Okeechobee, along the limestone ridge paralleling the southeast coast, and near Fort Meyers and Naples on the southwest coast.

Nursery/Greenhouse Crops

More than a billion dollars of nursery and greenhouse products were harvested from Florida farms in 1992. More than one-quarter of that amount ($297 million) stems from the sale of foliage plants, especially palms, dracaena, ficus, and spathiphyllum. Nursery plant growers focus on evergreens, deciduous shade trees, and azaleas to account for an additional $309 million in sales. Potted flowers account for an additional $79 million, bedding plants, $92 million. The cut flower harvest in Florida has increased to $103 million annually. Florida is the nation's largest producer of unfinished plant materials, primarily cuttings used by growers to produce their finished products, and is second in the production of sod ($64 million).

Most nursery/greenhouse operations occupy relatively little acreage and are comparatively footloose. The first concentration was on the ridge west and north of Miami, but expanding urbanization has lured many elsewhere. The location of operations most often relates to individual growers' success and expansion, rather than normal locational considerations.

The hot humid climate and flat lands surrounding Lake Okeechobee provide an ideal environment for nursery crops, turf farming, and winter vegetables. This crew is planting sod on the nation's largest turf farm near Belle Glade, Florida. (RP)

Alfalfa hay is the most important crop in the irrigated sections of the Ranching and Oasis region. (RP)

Ranching and Oasis

The Ranching and Oasis agricultural region is the largest and most physically diverse of all the nation's agricultural areas. This is a landscape dominated by desert scrub lowlands and coniferous uplands, interspersed with bright green splotches of irrigated agriculture. Government agencies control millions of acres of land crucial to the survival of the region's ranching, while crop farming is almost entirely at the mercy of governmental irrigation projects and their policies. Long regarded as an inalienable right, access to water and land is increasingly becoming a battleground pitting ranchers, environmentalists, and developers in a battle for survival.

Governmental Land-Use Policies

Land policy problems began almost as soon as the first Americans arrived in this region. Much of the Rio Grande and California had been distributed by Spanish and Mexican land grants prior to the United State's acquisition of the Far West. While supposedly protected in the original treaties, many claims were not settled to the satisfaction of the claimants, especially in the Hispano communities of New Mexico.

The disposition of the remainder of the region's government-owned acreage has only been slightly less contentious. The original 160-acre grants of the Homestead Act of 1862 were clearly inadequate in a region where it often took 80 to 100 acres to support a single cow. John Wesley Powell, western explorer and head of the U.S. Geologic Survey, recommended to

Federal Grazing Permits, 1992

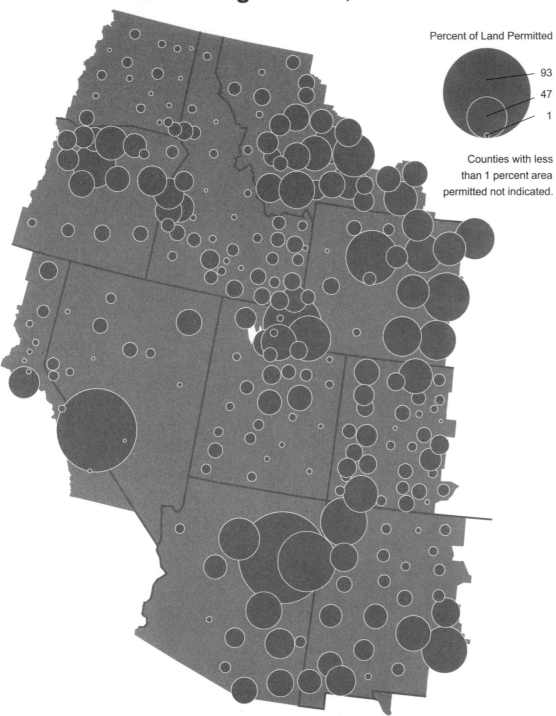

Percent of Land Permitted

93
47
1

Counties with less
than 1 percent area
permitted not indicated.

Congress in 1879 that grants be increased to at least 2,560 acres. Congress waited until 1909 to enlarge units to 320 acres of nonirrigable land and to 1919 to raise them to 640 acres.

The late nineteenth century brought the first beginnings of environmental conservation in the United States. The establishment of a system of national parks awakened interest in maintaining large western acreages in governmental hands in perpetuity. Almost 355 million acres (47 percent) of the eleven western states still remain in National Forests, Bureau of Land Management Lands, National Parks, and military reserves and bases. Of these lands, the management of grazing lands by the National Forest and Bureau of Land Management lands has consistently been the most controversial.

Congress passed the Creative Act in 1891, which authorized the designation of forest reserves. National Forests began to be designated within the year and more than 190 million acres are currently held, with 140 million of those acres being located in the eleven western states. The concept of grazing permits was introduced in 1910 to control overgrazing of federal lands with fees assessed to pay for revegetation and rodent control.

Control, permitting, and fee systems were extended to other federal lands in 1934 with the passage of the Taylor Grazing Act. Grazing control of these General Land Office lands was initially administered by the Grazing Service, and after 1946 by the Bureau of Land Management. Private grazing on public lands has become an increasingly volatile issue as the public debates the irreparable ecological damage of overgrazing, the desirability of returning these lands to their original state, the fairness of the issuance of permits, the original disposition of preexisting land grants, the smallness of the fees paid by grazers, the attitude of many permittees that they own those lands in perpetuity, and ultimately the growing demand for increased recreational space. The scope of the problem is massive. More than a million sheep and 1.3 million cattle are grazed on National Forest lands annually, while simultaneously the public logs 263 million recreational visitor days. The West is becoming crowded and many see only mutually

White Pine County, Nevada

Much of the Ranching and Oasis region is dominated by alternating north/south trending basins and mountain ranges situated in the heart of a severe desert environment. This is mining and ranching county with animals trucked many miles between seasonal grazing areas. White Pine agriculture is dominated by great sheep and cattle ranches with virtually all crop agriculture focused on irrigated alfalfa to feed the livestock during the winter months.

	1992	1982
Farms	120	127
Average size (acres)	1,931	1,841
Marketings ($ thousands)	8,687	6,908
Cattle and calves sold	14,080	14,663
Sheep	17,381	15,258
Hay (acres)	17,426	17,168
Irr. alfalfa (acres)	12,261	12,943
Wild hay (acres)	2,591	2,925

exclusive solutions to these thorny public-use questions.

Water

Virtually all of the Ranching and Oasis region is desert, or at least water deficient. Large rivers traverse thousands of miles of arid lands, but the cost of developing their irrigation potential is beyond the reach of even the largest user. The modern irrigation era began on July 23, 1847, when a group of Mormon settlers diverted a small stream near the site of the Mormon Temple in Salt Lake City to plant five acres of potatoes. The Mormons rapidly expanded their irrigated fields to 5,000 acres in the spring of 1848, 16,000 acres in 1850, and more than a quarter million acres by 1890. Mormon success in irrigated agriculture stemmed not only from their determination, but from church encouragement of group settlement and cooperative solutions to community problems.

The legal basis of today's reclamation policy began with the passage of the Act of July 26, 1866. While most provisions of this act eventually were reversed, its de facto acceptance of Spanish water law has remained the basis of most western water policy. The Desert Land Act (1877) carried policy a step further by providing that grants of 640 acres would be made to those who brought water to desert lands, and the Carey Act (1894) granted states 1 million acres of federal lands if they caused irrigated lands to be created and taken up within the following twenty years.

Direct federal participation in reclamation of desert lands began with an appropriation of $30,000 in 1891 for the construction, purchase, and use of irrigating

machinery on Indian reservations in three western states. The Geological Survey studied 147 possible reservoir sites for general use between 1888 and 1900, but was unable to act until the passage of Teddy Roosevelt's Reclamation Act of 1902. The Reclamation Service was immediately founded as a part of the U.S. Geological Survey. Four projects were authorized during the first year, seven in 1904, and a total of twenty-four in the first five years of the bill's existence.

The Army Corps of Engineers did not become involved in western reclamation until the inclusion of flood control projects on the Sacramento and upper Mississippi Rivers in the Harbors and Rivers Act of 1917. More projects quickly followed. Although many Corps projects provide water for irrigation, the management and dispersal of irrigation water is done by other agencies. Federal multipurpose reclamation projects were not initiated until the passage of the Boulder Canyon Project in 1928. Virtually all federal reclamation projects conceived and constructed thereafter were designed to provide some combination of flood control, navigation, municipal and industrial water, salinity, sedimentation control, sanitation management, recreation, fish and wildlife preservation and propagation, and agricultural irrigation.

Multipurpose projects have been plagued with conflicts of intent from the outset. Best management of recreation, navigation, and sanitation practices calls for full reservoirs as much of the time as possible; best flood management practices call for the reservoirs to be as near empty as often as possible. Water storage practices operate best when dams are at capacity at

Ranching and Oasis cattle and sheep are often transported hundreds of miles to distant upland meadows for summer grazing. (RP)

ing animals in the lowlands in the winter and into the mountains during the summer became a part of the annual cycle, and later, as the intervening lands were appropriated by others, a continuing source of irritation.

Cattle are more important than sheep in most areas. Initially introduced by Spanish colonists, cattle ranching was widespread by the nineteenth century. Colonial ranches in California often concentrated on cattle for their hides and tallow. Meat production in California was not possible until the mining boom created a local market. The discovery of gold, silver, and other minerals in the Great Basin and Rockies spread both production and market demand to those areas as well.

Though somewhat less traveled than sheep, cattle typically are moved in late spring from winter pastures both in the region and in California, Oregon, and Texas to mountain pastures. Generally, the calves are born and weaned during the summer. The significantly enlarged herd may be moved several times during the summer, but ultimately is returned to its winter pastures in the fall when most of the young animals are sold.

Modern ranching is far more complex than in the past. Not only must grazing contracts for summer pastures be negotiated, trucks obtained, herds transported hundreds of miles, and sales of heifers made, but much of the summer is a mad dash to produce sufficient winter hay for the breeding herd. Alfalfa is the hay of choice in most of the West, although some timothy is also harvested. The Humboldt, the Snake, the Deschutes, and almost every other riverine or lake environment in the region turns green with alfalfa soon after

the end of spring and almost empty at the beginning of winter. Resolution of the conflicts over how to best meet these mutually exclusive needs is ongoing.

Grazing

Grazing is the single most widely practiced agrarian activity in the region. Cattle are the most visible of all of the grazed animals, although sheep, horses, goats, and llamas are also found in large numbers. Virtually all grazing in the region calls for

the movement of herds over long distances in search of seasonal forage in several variations of transhumance grazing. Llamas and goats are the least likely to be moved to seasonal pastures.

Sheep were probably the first grazing animals herded in the region. Local Indians quickly learned that sheep were an asset that could quickly improve their quality of life. Indian flocks never reached large numbers because of the pervasive poverty of the people and the land, but sheep soon became an integral part of their life. Herd-

summer begins. Roads become covered with flatbed trucks hauling mountains of baled hay to wherever herds winter. Ranchers in some areas, however, are increasingly seeing the advantages of growing higher value crops and purchasing hay from others. The paradoxical vision of hay trucks passing in opposite directions is becoming an increasingly common sight today.

Large numbers of Angora goats are found in Texas, New Mexico, and Arizona. Virtually all the Angora goats in New Mexico and Arizona belong either to Indian or Hispano herders. The Navajo tribal government has invested heavily in upgrading its citizens' sheep and goat herds. Declining markets in recent years, however, have dramatically increased the percentage of cattle in Navajo herds.

Subregions

Eastern Basin and Range

The agrarian landscape of the eastern Basin and Range was made distinctive by the pervasive role of the Mormon church, and its practices eschewed isolated farmsteads whenever possible. Hundreds of towns were laid out in accordance with the guidelines of the *Book of Mormon*. Expansive grid streets, great in-town building lots sufficient for house, barn, home orchard, truck garden, and other accoutrements of farm life, and a distinctive architecture soon gave these communities a visual character that still survives. Plagues of locusts, droughts, and other problems during the early years convinced the church to call on the membership to keep large larders of food for future emergencies, as well as adopt a mixed farm crop strategy. The result was

Sheep and Cattle, 1992

Inventory (Thousands)

211
106
1

Counties with fewer than 1,000 sheep are not indicated.

Sheep

Cattle

Potatoes, 1954, 1992

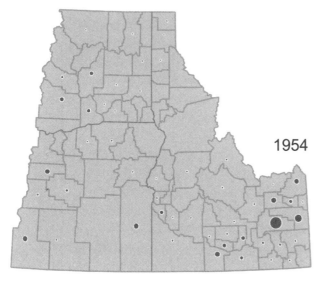

1954

Hundredweight Harvested

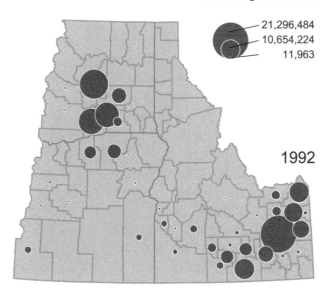

21,296,484
10,654,224
11,963

1992

the development and continuation of one of the most diverse farming landscapes of the Far West. Though the sum of the crops is much the same as elsewhere in the Intermontaine, individual farmers tend to engage in a wider variety of activities than is typical elsewhere in the Far West.

Livestock grazing is the most important agrarian activity with $283 million in sales by Utah farmers in 1991. Though locally important, dairy products trail with $149 million in sales, and sheep accounted for $16 million that year. Isolated from fluid milk markets, most dairy products are processed into cheese, milk, and cottage cheese.

Alfalfa hay is the highest value field crop in Utah ($122 million in 1991), followed by corn for grain and silage ($28.7 million), wheat ($17.8 million), barley ($17 million), and other hay ($15.5 million). Utah farmers have engaged in commercial fruit production for more than a century. Pears are the single most important in volume, while apples are the most valuable ($10.4 million). Utah is second to Michigan in tart cherry production. Most commercial orchards are concentrated on the terraces of Lake Bonneville north and south of Salt Lake City. Summer storage onions are the only other important crop in the region.

Snake River Plain

The Snake River Plain has been transformed over the past thirty years from a sleepy backwater to a modern industrial agricultural scene shipping crops throughout the world. Idaho farm marketing receipts totaled $2.8 billion in 1992. Although noted primarily for its potatoes,

Idaho is a top producer of ten major crops and livestock, including trout (1), mint (3), processed sweet cherries (6), summer storage onions (3), and processed sweet corn (6).

Potato cultivation was introduced in 1882 by Mormon farmers from Utah. Idaho potatoes, however, did not take on their current mythic character until the 1960s when changing American tastes largely replaced boiled and mashed potatoes with baked potatoes and French fries, especially in restaurant dining. Food critics soon proclaimed the Idaho potato as the perfect baker. Idaho potato processor J. R. Simplot's creation of the frozen French fry product almost guaranteed Idaho's dominance of that market.

Bingham County, Idaho

Lying in the heart of Idaho's potato country, Bingham County agriculture reflects the complex symbiotic relationships of the potato, sugar beet, small grain (wheat and barley), and alfalfa crop rotation and utilizes the byproducts to support dairy and beef cattle feedlots.

	1992	1982
Farms	1,282	1,447
Average farm size (acres)	1,070	696
Agricultural sales ($ thousands)	215,446	173,901
Potatoes and sugar beets		
($ thousands)	100,614	60,357
Potatoes (acres)	67,007	54,320
Sugar beets (acres)	9,975	7,709
Wheat (acres)	145,119	140,371
Barley (acres)	24,528	41,659
Alfalfa (acres)	43,125	42,489
Cattle and calves sold	63,668	91,693
Dairy cattle	15,830	13,177

The Snake River Plain is not an endless sea of potato fields. Potatoes are grown on three- to five-year rotations. Wheat, barley, dry beans, alfalfa, and sugar beets are grown in the alternate years, producing a constantly changing mosaic of crops in the irrigated sections of the valley. Idaho is the nation's second largest producer of sugar beets, with receipts increasing by more than a third between 1986 and 1992. Potato and sugar beet byproducts in turn make excellent cattle fodder. The Snake Valley has become a feedlot center, first for beef cattle and more recently for dairy herds. More than 75 percent of Idaho cattle marketings came from 125 feedlots.

The expansion of the dairy industry has been hampered by the Snake River's isolation from the nation's main markets. Large dairy operations on the Pacific Coast have effectively made expansion of the fluid milk sales outside the state difficult. State dairy operators thus concentrate on solid products, especially American cheese (157 million pounds in 1992). Most skim milk is dehydrated into dry milk products.

A variety of lesser known products adds elaboration to the agricultural landscape of the Snake River Plain. Hops, mint, and sweet corn (for processing) are the most common of these crops, while mink and trout are the most important "other" animal raising activities. Idaho's trout industry provided 27 million pounds of trout ($37 million) to the nation's restaurants in 1992. The state produces 75 percent of the nation's food size (over twelve inches) trout from forty-two commercial trout farms along the Snake River mostly near Twin Falls.

Great Columbia Plain

The Great Columbia Plain lies primarily on a plateau of extrusive lava thousands of feet thick and covered with varying depths of loess, a wind-blown glacial sediment. Situated in the rain shadow of the Cascade Mountains, the region is a near-desert environment where unirrigated agriculture is possible only because loess soils retain soil moisture from year to year if left in clean fallow. The result is a peculiar landscape of strips of grain interspersed with strips of cultivated barren fallow fields, which are more intensely tended than those being farmed. Typically, the top several inches of each of the fallow fields are carefully cultivated four times a season to keep all water-sucking weeds from the precious soil moisture.

The Great Columbia Plain has been one of the world's great wheat production regions for more than a century. Smaller acreages of barley, oats, and rape (source of canola oil) are also grown. The development of irrigation in the twentieth century transformed large sections of agricultural landscape from golden to green. Walla Walla sweet onions are the best-known crop, but alfalfa is the most important in acreage and value. Recent years have brought an explosion of cool weather vegetables for freezing, especially peas, green beans, sweet corn, asparagus, and carrots. Vegetable production initially focused on Walla Walla/Richland, but expanded both northward to Quincy/Ephrata and southward into Oregon.

The Yakima Valley region, including the Wenatchee and Okanogan Valleys, is the most important soft fruit district in the world. Apples have been at the core of the region's production and marketing cam

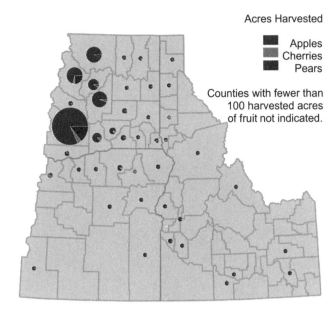

Fruit, 1992

Acres Harvested

- Apples
- Cherries
- Pears

Counties with fewer than 100 harvested acres of fruit not indicated.

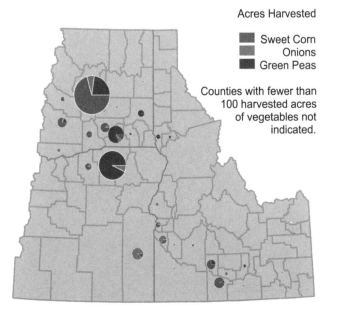

Vegetables, 1992

Acres Harvested

- Sweet Corn
- Onions
- Green Peas

Counties with fewer than 100 harvested acres of vegetables not indicated.

paign for more than seventy-five years. The 1990 crop produced more than 4 billion pounds of fresh apples and 1 billion pounds of processed apples valued at $784 million. The smaller (but more valuable) frost-inhibited crop of 1991 was valued at almost $1 billion. The region's growers have advertised heavily to make Washington synonymous with apples throughout the world. They currently are using their marketing clout to popularize several new "foreign" apple varieties, primarily Fuji, Braeburn, and Gala, less commonly grown by growers in the eastern United States. Pears, sweet cherries, prunes, and apricots account for an additional market value of $180 million. Grapes have long been grown in the area, and Washington leads the nation in grape juice production today. Boutique wineries began in the 1970s with increasing national recognition, especially for the area's white Riesling, merlot, and white Chardonnay wines.

Hops are a perennial vine with cone-shaped fruit that is the flavor agent for beer and other malt beverages. They were introduced into Washington in the 1860s and into the Yakima Valley in 1875. More than three-quarters of the nation's hops are now produced in the Yakima Valley. Washington farmers have concentrated on bitter hops, the source of beer's distinctive flavor, though aromatic varieties are of increasing importance. Most are grown on long-term contracts (up to seven years).

A last interesting element of the human agricultural landscape is the large number of permanent Mexican–American residents throughout the Great Columbia Plain and the Snake River regions. Most were originally recruited as migratory laborers by sugar beet, hop, and fruit farmers as tradi-tional local help began moving to the Puget Sound area for factory work during World War II. Farmers were so concerned about finding reliable labor that many created permanent jobs to keep the migrant workers in the area year around. While many of the original Mexican–Americans no longer work in agriculture, their presence is felt in even the smallest towns of the northern Intermontaine West through numerous mercados and Mexican (not Tex-Mex) restaurants.

Other Agricultural Specialty Areas

Several other less important agricultural areas still thrive in the Ranching and Oasis region, including the Klamath Lake, Humboldt River, Truckee River, Uncompahgre, and Rio Grande valleys. The Lower Colorado, Imperial/Coachella Valley areas, and Gila/Salt River Valley areas are also physically located within the Intermontaine physiographic province, but function economically as a part of the Pacific Industrial region and are discussed therein.

The Klamath Lake (Oregon/California) district has long been a wintering ground for cattle grazed in the highlands during the summer. Intense hay production covers most of the arable land in summer. Potatoes were locally important for many years, but have declined. Nevada's irrigated agricultural areas similarly focus on traditional irrigated desert crops such as alfalfa. Potatoes have been increasing in the Humboldt River Valley, as have onions. More than a thousand acres of garlic are grown north of Pyramid Lake.

The Uncompahgre Valley of the Upper Colorado basin in western Colorado developed irrigated agriculture around the turn of the century. Approximately 70,000 irrigated acres were cropped in alfalfa, fruit, pasture, oats, wheat, and barley until the 1960s. Fresh fruit production was hurt in the 1970s as national distribution consolidated within the most efficient production areas, but growth is again taking place in recent years based on supplying local markets with apples, pears, and cherries. Direct sales to summer campers and winter skiers play an increasingly important role in this area. Nurseries, specializing in bedding plants and cut flowers, have also now moved into the valley.

Traditional general farming in the Rio Grande Valley has begun to change in recent years. The Las Cruces area today concentrates primarily on the production of irrigated cotton, onions, dairy, alfalfa, chiles, fall and spring lettuce, pecans, and apples. Apples have been increasing in recent years, especially in the northern valley. Chile production also has dramatically increased as the Mexican cuisine revolution swept the nation, though typically the descendants of the original Hispano farmers who introduced and nurtured this crop largely have not participated in its success. The San Luis Valley of the upper Rio Grande in southern Colorado remains one of the least changed areas in the region with alfalfa, other hay, livestock wintering, and some potatoes dominating agrarian life. Some cool season vegetable production has also begun to appear.

Pacific Industrial

The Pacific Industrial agricultural region is formed from two seemingly different subregions, the Pacific Northwest and California/Central Arizona, held together as much by history as by contemporary agrarian reality. There are no boundaries, rather the region forms a continuous transition from the huge, corporate farms of the San Joaquin Valley to the highly industrialized smaller farms of the Sacramento and San Joaquin River deltas to the predominantly family-farmed Sacramento Valley to the smaller, but highly specialized operations of the Willamette. The California/Central Arizona and Pacific Northwest subregions will be discussed separately primarily because of the special role of irrigation in the South.

California/Central Arizona

The transformation of California agriculture from livestock ranching and grain production into one of the world's most industrialized agrarian landscapes began in the 1890s with the introduction of widespread mechanization of production, elaborated into "factories in the fields" by the 1950s, and crystallized as the archetype global farm region in the 1990s. Spanish missionaries brought agriculture to California in the late eighteenth century, and the state's rich soils soon yielded a cornucopia of crops. Commercial exploitation followed more slowly because of the region's isolation. Cattle ranching dominated the scene until the discovery of gold created a market demand that sent thousands of would-be miners to the agricultural fields, rather than

to the gold fields. Wheat and barley were the first notable export crops as the Central Valley became a veritable sea of grain in the 1880s. Operations were so big that one farmer, a Dr. Hugh J. Glenn from what is now Glenn County, had the goal of growing and harvesting a million bushels of wheat from his ranch in a single year.

Faster, more reliable rail service, refrigerated railcars, and new methods of packing fruit brought large-scale production of fruit and vegetables to the region. California was only the fifteenth largest agricultural state in the 1909 agricultural census, yet had already become the nation's largest producer of fruits and nuts.

Characteristic Elements

Agriculture is made distinctive by five characteristic elements: the strong role of specialty crops, bimodal farm structure, intense irrigation, the importance of hired labor, and innovation.

Specialization. While specialization and concentration are the hallmarks of industrial agriculture, California farmers have carried this trend to new extremes. California leads the nation in the production of sixty-five crops and commodities ranging from the prosaic (dairy products) to the exotic (oriental vegetables and rabbits). The resultant landscape is one of the most diverse crop milieux on earth, created almost entirely from some of the most specialized niche crops cultivated anywhere.

Pacific Industrial

Bimodal Farm Structure. While large holdings have long been the keynote of California farming, agriculture in this region has always supported large numbers of small, intense commercial enterprises concentrating on high-return crops. The great agricultural operations, however, lie at

97

the core of the state's mystique. The largest are so great that the Southern Pacific Railroad, apparently the largest nongovernment landowner in California with holdings of more than 1 million acres, barely makes the top ten of agricultural landholders. The Kern County Land Company, controlling more than 200,000 acres of the richest agricultural land in the nation, seems to be the largest. Other major owners include those of Standard Oil, the Chandler family, J.G. Boswell ranches, and the Gallo Wineries. Conflicts with attempts to organize agricultural labor in the 1960s and more recently efforts to circumvent water regulations limiting irrigated acreages by individual land owners has created a situation where many of the state's largest land owners have purposely obscured the size and location of the entirety of their holdings. Land ownership, however, is just a single element of large farming in this region of industrial agricultural. Production and distribution of many crops are controlled by only a handful of packers and distributors. Virtually all tomatoes, peaches, prunes, citrus, and numerous other products are grown under production contracts, which, in a sense, transfer the traditional power of the farmer (producer) to the distributor. Many of these are international food conglomerates with global production and distribution networks. As a result, the size of individual producer holdings is increasingly overshadowed by the nature and power of the distribution

Left: California's Central Valley is one of the nation's most intensely farmed areas. Note the irrigation canals cutting through this scene of the Sacramento delta. (RP)

The nation's finest wine grapes are grown in the coastal valleys north of San Francisco where high summer and cold winter temperatures are moderated by cool sea air. (Pat Pillsbury)

Butte County, California

Located near the northern end of the Central Valley, Butte County reflects the dichotomies of California agriculture. Grain fields, wheat, barley, and rice dominated much of the county's agriculture throughout the twentieth century with minor production of fruits and nuts centered on Chico, the region's processing center. Recent increases in market demand have brought an expansion of specialty orchard crop production, most notably prunes, walnuts, almonds, peaches, along with a coincidental decline in grain and field crops. The county has had a tradition of agricultural experimentation in new specialty crops. A major producer of kiwifruit in the 1980s, local farmers are currently expanding into guava, Asian pear, and similar fruits.

	1992	1982
Number of farms	1,944	1,785
Average farm size (acres)	233	262
Agricultural sales ($ thousands)	182,470	151,745
Orchard crops ($ thousands)	114,766	76,751
Almonds (acres)	36,057	41,550
Walnuts (acres)	15,846	16,888
Prunes (acres)	11,219	9,447
Peaches (acres)	2,742	3,205
Olives (acres)	2,358	2,529
Kiwifruit (acres)	1,746	974
Rice (acres)	70,476	105,679
($ thousands)	47,898	51,690
Wheat (acres)	7,624	12,006

Vegetables, 1954, 1992

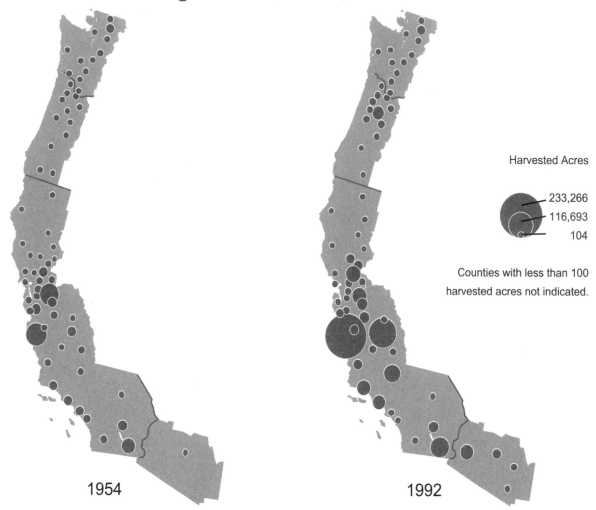

Harvested Acres

233,266
116,693
104

Counties with less than 100 harvested acres not indicated.

1954

1992

system, which can enhance the fortunes or crush even the largest producer within a few years.

The average size of the California farm was 367 acres in 1992, below the national average and well below the state's own peak of 474 acres in 1974. Part of this statistic is illusionary as many of the state's largest agriculturalists farm their holdings as corporately independent, smaller ranches to reduce their visibility and circumvent a variety of regulations. Land leasing to

create larger effective holdings is also common. Even considering these points, however, the number of viable small units is rising. The state is a land of very large and very small farms. Floriculture/horticulture, Asian pears, kiwifruit, and a multitude of other specialty crops bring large returns from small acreages. While the behemoths dominate media coverage, total farm income, and the public mind, thousands of farmers prosper while cultivating small acreages of very high-value crops.

Intense Irrigation. Most of the California/Central Arizona agricultural region is arid by any standard. The three principal sources of water for irrigation include surface streams, groundwater, and artificial impoundments with canal distribution systems. Most surface streams in California are perennial, but the long dry period from May to September brings severe shrinkage in flow through the crucial growing period. Farmers began tapping groundwater supplies in the nineteenth century, but it was not until the 1920s that surface and submersible pumps became widely used. Groundwater withdrawal currently exceeds recharge by 2 million acre feet or more annually in California alone.

Irrigation from artificial impoundments by means of canal systems is the dominant source of agricultural water in most areas. The construction of a 230-mile aqueduct from the Owens Valley to Los Angeles approved in 1905 was the first modern large-scale project in the region. Seeing the need for additional water, the Los Angeles Metropolitan Water District claimed an additional 1 million acre feet of Colorado River water in 1924. The size of this entitlement was challenged by Arizona, and in 1963 the Supreme Court cut it by more than half. The All-American Canal carrying Colorado River water to the Imperial Valley was also built about this time. The Central Valley Project was initiated in the 1930s to collect excess water from the upper Sacramento River and to transport it to the upper San Joaquin Valley.

Modern irrigation in Arizona's Salt River Valley began with the construction of Swilling's Ditch in 1868. The development of the Salt River followed as the first project under the National Reclamation Act of 1902. The two key changes in Salt River irrigation were the phenomenal growth of Phoenix, which placed additional pressure on the limited water supplies, and the 1963 Supreme Court decision expanding Arizona's share of the Colorado River impoundments. The construction of the Central Arizona Project that followed allowed continued expansion of the system.

The California Water Project was approved by voters after the disastrous 1955 flooding of the Feather River. The original $1.75 billion projected cost of the system was exceeded by many times before it was completed. The California Water Project captures water flowing to the sea in the most northerly section of the state and transports it through a system of canals and pumping stations more than 700 miles down the Central Valley, over Tejon Pass (4,183 feet), and into the preexisting Metropolitan Water District distribution system in southern California. This task is roughly the equivalent of shipping water from Cleveland, Ohio, to New York City across a veritable desert and over the

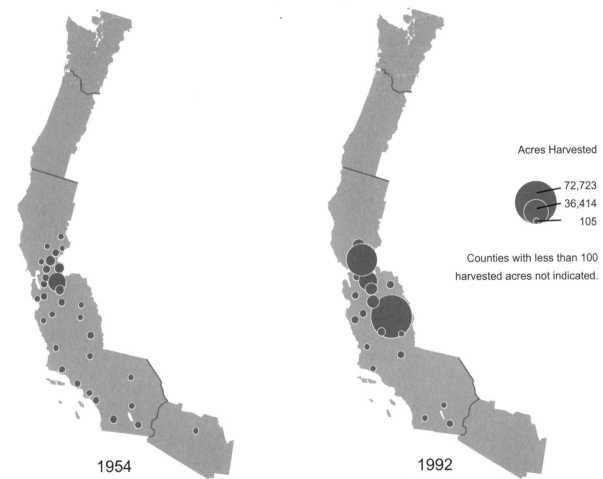

Tomatoes, 1954, 1992

Acres Harvested

72,723
36,414
105

Counties with less than 100 harvested acres not indicated.

1954

1992

Smoky Mountains. Whether one views this project simply as an example of the high costs of providing basic amenities to large parts of the Far West or as a massive raid on the public coffers by rapacious corporate farmers, the California Water Project must stand as an astounding engineering marvel.

Shortages, increasing water costs, and salinization are the greatest problems facing western irrigation today. Many argue that irrigation water is subsidized by an unrealistic cost accounting system, and their arguments are slowly forcing the revaluation of water pricing and allocation. Just as pressing is the problem of soluble salts in most of the irrigation water. More than 3,000 pounds of salt are typically dissolved in a single acre foot of water in the Imperial Valley, with lesser amounts in other areas. Soil salt contamination is an inevitable outcome of large-scale irrigation in all arid areas, and salt levels are increasing throughout the arid, irrigated West. Agriculturalists are turning to drip irrigation systems to slow contamination, but even this expensive investment can only slow its inevitable arrival.

Dairy Cows, 1954, 1992

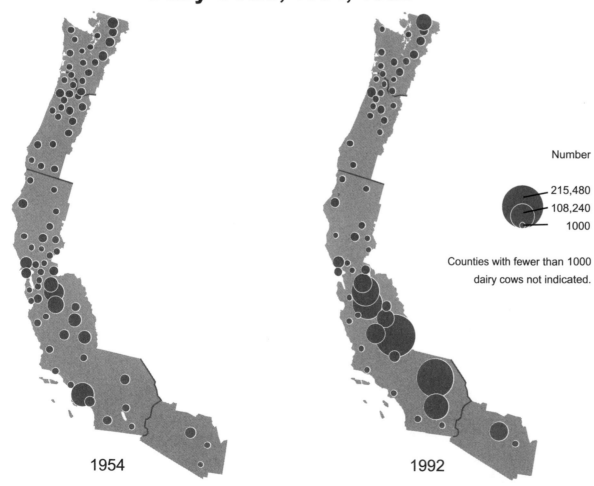

Number

215,480
108,240
1000

Counties with fewer than 1000 dairy cows not indicated.

1954

1992

Importance of Hired Labor. The region's orientation toward labor-intensive specialty crops has meant that farmers have been dependent on large numbers of seasonal farm labor. More than 10,000 unemployed Chinese laborers idled by the completion of the transcontinental railroads were the initial source of cheap labor. Their growing economic independence culminated in the passage of the Exclusion Act of 1882. Japanese laborers replaced them in the 1890s. By 1920 Japanese farmers had bought 74,769 acres of California farmland and leased an additional 383,287 acres. Their success, made visible by their domination of the production of a handful of crops, most notably berries, onions, asparagus, green vegetables, and celery, helped foster latent anti-Japanese sentiment that brought passage of California's Alien Land Law (1913) and the legal cessation of all Japanese in-migration in 1924. Filipino, Italian, Portuguese, and Mexican laborers replaced them during the 1920s and were in turn replaced by the stream of Dust Bowl refugees in the 1930s. These "Okies" worked the fields, gained a certain degree of fame through *Grapes of Wrath* and other stories of their hardships, and changed the cultural texture of the Central Valley forever. Many left agriculture at the beginning of World War II to work in urban factories or join the military, but their cultural presence still lingers, especially in the San Joaquin Valley.

Mexican laborers returned to California's fields after Mexico's entry into World War II in June of 1942. These workers were temporary, and it was not until the 1960s that the rush of laborers from Mexico to California as permanent residents became overwhelming. Most went to the cities, but enough went to the fields to continue the supply of cheap labor for agriculture. Workers of Mexican origin still dominate agricultural work in the region, but increasing numbers of recent emigrant workers from elsewhere in Latin America and southeast Asia are beginning to appear in larger numbers. The cycle of labor exploitation is beginning to rotate one more time.

Innovation. There is a restless energy among California agriculturalists, who seemingly feel little of the need for continuity so characteristic of farmers elsewhere. Well educated, often well financed, and always well supported by the agricultural establishment in the state, these farmers seem to search endlessly for new, more profitable crops, regardless of their previous experiences. Significant mechanical innovation began in the late nineteenth century as Central Valley farmers modified eastern grain combines to work the Central Valley's adobe clay soils and then adapted them further to harvest rice and other crops as well. Intense research on the development of refrigerated railcars, the creation of new marketing strategies, and new irrigation methodologies followed. The crawling tractor (Caterpillar) was created to work in the heavy clay Delta soils. The pace of mechanization increased even further with the appearance of new automated harvest equipment in the early 1960s in response to the farm labor movement led by Cesar Chavez.

More recently, rising water and labor prices, coupled with increasing problems with salinization, have turned attention to more efficient water use. Traditional flood irrigation of fields and orchards started becoming too expensive in the 1960s and

Floriculture, 1954, 1992

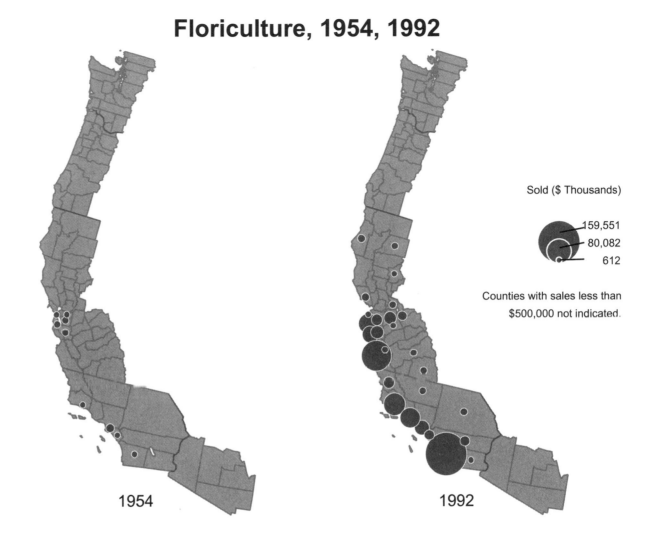

Sold ($ Thousands)

159,551
80,082
612

Counties with sales less than
$500,000 not indicated.

1954

1992

was replaced by movable sprinkler systems. The increasing costs of labor to constantly reset the sprinklers has more recently led to their replacement by permanent buried sprinkler systems or drip irrigation systems. Modern drip systems are tied to automatic computer-driven dispensers of fertilizers, insecticides, and herbicides, which have further reduced both costs and the environmental impact of agriculture on the state.

Like elsewhere in the nation, computerization is generally sweeping the industry in

every way from obtaining on-line data on the latest crop and weather forecasts to predictive modeling of agricultural production and prices. Computers have become an integral part of daily life for many of the larger ranches just to track equipment, livestock, livestock feed prices, and the latest agricultural publications.

Grapes, 1954, 1992

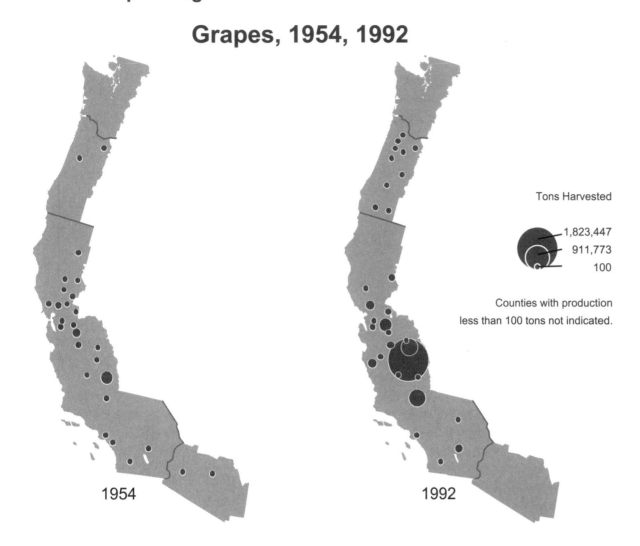

Tons Harvested

1,823,447

911,773

100

Counties with production
less than 100 tons not indicated.

1954

1992

Agricultural Regions

California/Central Arizona may be broken into four major agricultural regions and numerous subregions. While all share aspects of the defining elements of the region, each has a distinctive agricultural landscape.

Central Valley. California's Central Valley, a 450- by 50-mile trough lying between the Sierra Nevada and Coast ranges, is composed of three distinct sections: the San Joaquin (River) Valley in the south, the Sacramento (River) Valley in the north, and the "Delta" where the two rivers coalesce and ultimately empty into San Francisco Bay in the center. The largest, most intense, corporate, irrigated, and richest farms in the state tend to be concentrated in the San Joaquin section where four of the nation's ten richest agricultural counties are found. The traditional subregional crop patterns of the Central Valley have largely disappeared in recent years, making generalizations

difficult. The southern San Joaquin was long noted for its oranges, cotton, and grapes (for raisins, fresh consumption, and inexpensive wine). The continued suburbanization of the Los Angeles Basin forced many of its dairy operators to move to the Imperial and San Joaquin Valleys in the 1970s and 1980s. The rapidly expanding dairy feedlots around Visalia, utilizing primarily agricultural byproducts as feed, have made California the nation's largest dairy producer and Tulare County the largest single county in total herd size, average herd size, and value of output. Production of almonds, pistachios, peaches, plums, cantaloupes, and tomatoes (for processing) has dramatically increased in the past few years, while cotton has tended to decline with increasing water costs.

The central Delta section traditionally has been oriented toward smaller farms concentrating on relatively high-value crops such as vegetables, garlic, and poultry. The suburbanization of the San Francisco Bay area has also brought urban fringe agriculture into the region in recent years, as well as increasing population levels. Field crops continue to be much more evident than farther south, especially for the production of fresh and processed vegetables, garlic, and dry onions. Literally millions of chickens, eggs, and turkeys are also concentrated here, as well as a variety of orchard crops, especially cherries, pears, and figs.

The Sacramento Valley has been less intensively cultivated than either of the regions to the south, although that has begun to change in recent years. Increasing numbers of farmers capitalizing on the suburbanization of the San Francisco Bay and even the Delta have purchased these less expensive lands and are importing their

Strawberries, 1954, 1992

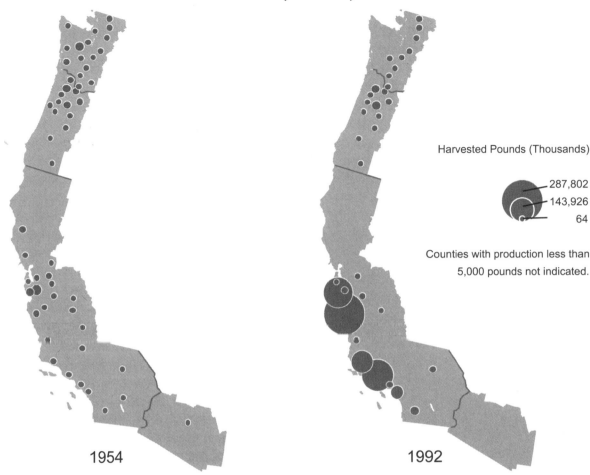

Harvested Pounds (Thousands)

287,802
143,926
64

Counties with production less than
5,000 pounds not indicated.

1954

1992

Marion County, Oregon

Situated at the northern end of the Willamette Valley, Marion County farmers
are increasingly concentrating on specialty
crops. Nursery/greenhouse crops have
become the leading activity over the past
five years, followed by vegetables, hay,
and dairy. Cool weather vegetables, especially sweet corn, have now climbed to
second in importance. The county is the
nation's second largest producer of filberts
(hazelnuts). A small wine industry has
been developing in recent years with some
success. The Willamette Valley has long
been a center of grass seed development
and the region provides lawn and hay
seeds throughout the nation.

	1992	1982
Farms	2,494	2,825
Average farm size (acres)	121	110
Agricultural sales ($ thousands)	313,155	190,789
Nursery/greenhouse ($ thousands)	96,484	38,958
Nursery plants ($ thousands)	48,621	9,789
Sod ($ thousands)	3,480	n/a
Veg./flower seeds ($ thousands)	3,817	2,120
Vegetables ($ thousands)	44,715	33,876
Dairy sales ($ thousands)	32,989	24,332
Fruits, nuts, berries ($ thousands)	32,029	14227
Vegetables (acres)	39,882	40,949
Sweet corn (acres)	16,620	18,461
Green beans (acres)	11,850	13,824
Green beans (acres)	2,298	2,915
Orchards (acres)	9,479	8,060
Filberts (acres)	6,393	4,774
Grass and field seeds (acres)	75,892	55,228
Hay (acres)	24,658	19,744

more intensive agricultural style into this
somewhat bucolic section of the Central
Valley. The valley's traditional crops were
peaches, nuts, rice, wheat, barley, and
tomatoes. All these crops have continued to
be important, but some changes have taken
place. Grain production has generally
declined, except for rice, which stretches
from Sacramento to Chico and has actually
expanded because their clay soils are of less
value in the cultivation of other crops.

Prunes (mostly relocated from the now
suburbanized Santa Clara Valley), almonds,
walnuts, and more exotic crops such as
pistachios, kiwifruit, and recently guavas
and Asian pears have all seen significant
expansion. Indications of the comparatively
lower agricultural intensity of the area are
quite evident in the continued large-scale
cultivation of large areas of lower value
crops.

Coastal Valleys. The coastal valleys contain

the nation's largest producer of ryegrass, fescue, Kentucky bluegrass, and orchard grass seed. An important Christmas tree industry supplying Douglas fir and other western conifers nationally has also developed.

Livestock is the leading source of agricultural income in Oregon, but relatively unimportant in the Willamette Valley. Broilers are locally important south of Portland. About three-quarters of the state's dairy cattle live in Willamette Valley counties, but most are found along the foothill margins. About 150,000 sheep and 300,000 cattle are inventoried in the valley at the beginning of each year, but most summer elsewhere.

Puget Sound Lowland

Cool summers, heavy year-around rainfalls, and soggy soils have made agricultural development in this region difficult. Puget Sound agriculture originally developed to supply local markets and the Alaskan gold fields, but has been hurt by global marketing. Dairy farming, beef cattle, and poultry are the three largest activities remaining. Vegetables and nursery crops have long been important on the Bellingham plain, though urban growth and competition have challenged production. Christmas trees are important in several areas.

Rogue and Umpqua Valleys

The Rogue and Umpqua Valleys are comparatively small agricultural valleys in southern Oregon near the California border. Comparatively insignificant in the national scene, these areas have become famous partially because of the Fruit of the Month

Club and the Jackson and Perkins Nurseries based there. While both companies now go well outside their home regions for product and distribute their wares internationally, the area remains famous for its high-quality fruit, nursery crops, and cheeses.

Alaska

The extent and importance of Alaskan agriculture is severely limited by the state's northerly location and harsh environment. Nearly all of the state's farm economy focuses on serving the small, local market. The declining relative costs of importing foodstuffs has brought widespread deterioration of agriculture in the region.

Over one half of Alaska's population and most of its agriculture are concentrated in and around Anchorage on the state's south central coast. The Matanuska Valley, at the head of Cook Inlet, has received widespread national attention for decades for its oversized produce because of the region's long days during the growing season. Despite this fame, however, these fertile and well-drained soils today support fewer than fifty farms.

Hay is the state's most important crop (by area) and occupies about three-quarters (22,700 acres) of its harvested area. Nursery and greenhouse crops are the state's most important crop group by value with $6,639,000 in sales in 1992. Floriculture specialties account for 81 percent of these crops and greenhouse vegetables the majority of the remainder. Vegetables, the state's most famous crop, are declining with only 262 acres harvested, primarily lettuce, cabbage, carrots, and peas in declining acreages. Potatoes, oats, and barley constitute the remainder of the state's important crop.

Livestock, once the state's most important agricultural activity, has been rapidly declining because of low-cost imports. The state's surviving thirty-four dairy farms house only 715 animals. The laying chicken population has declined from 33,000 to under 2,000 since 1978. There are only a few more broilers and fewer than a thousand turkeys raised annually.

Alaska

Agricultural Sales, 1992

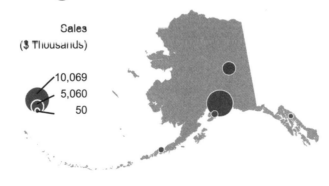

Sales
($ Thousands)

10,069
5,060
50

Field Crops, 1992

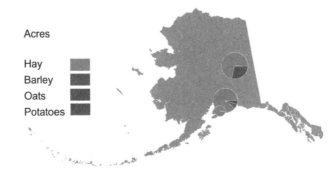

Acres

Hay
Barley
Oats
Potatoes

Livestock, 1992

Acres

Milk Cows
Beef Cattle
Hogs

109

Increasing Hawaiian acreages are devoted to producing food for the local market as is seen in these taro fields on Kauai. (Robert Pillsbury).

Hawaiian Islands

Hawaii's semitropical climate, isolated mid-Pacific location, and continuing association with the United States encouraged the development of a plantation agricultural economy based on export of tropical products to the mainland as early as the middle of the nineteenth century. Although only a tenth of the state's land is arable, the generally adequate soils, sufficient water supplies, and lack of frost encouraged the cultivation of long-season crops that could not be grown elsewhere in the United States. Tobacco was the first important crop of the islands, but soon was replaced by the tropical crops that became the hallmark of this island chain's highly specialized agriculture.

Commercial sugarcane began in the 1830s, although the crop had long been grown there after its introduction by Polynesian migrants prior to the arrival of the Europeans. Beginning in the 1850s, contract laborers, first from China and later Japan and the Philippines, were imported to provide the backbreaking labor of harvesting the crop. By the 1860s sugar had become the backbone of the Hawaiian economy. Commercial cultivation of pineapples, also introduced by pre-European Polynesians, did not begin until late in the nineteenth century. Production exploded after Sanford Dole's development of improved varieties after 1903. The development of automated peeling, coring, and slicing machinery a decade later secured the future of the crop, which today is immutably tied to both the place and the crop's early innovator.

Sugarcane and pineapples dominated the Hawaiian economy through the first half of the twentieth century. More than 50,000 laborers tilled the cane fields and another 5,000 worked in the pineapple plantations during the 1930s, while more than 5,000 were employed in the processing plants—40 percent of the state's total labor force. The islands supplied 80 percent of the world's pineapple harvest in the 1940s and their cultivation dominated the economies of Lanai and Molokai and was very important in the lowlands of Oahu and Maui. Sugarcane dominated the agricultural economies of Oahu and Maui, however, as well as that of Kauai and Hawaii during this period.

Pineapple production peaked after World War II and has declined from just over a million tons to 557,000 tons in 1992. Hawaii now harvests about 25 percent of the world's crop. Pineapples are now absent from Molokai and Lanai because of the higher labor costs found there. Ninety percent of Hawaii's pineapples are currently raised by just three conglomerates that in recent years have increasingly shifted processed production of fruit to the Philippines, Costa Rica, and other areas with lower labor costs. The market for fresh pineapples, however, has continued to expand rapidly to support continued production in Hawaii.

Sugarcane production expanded until the late 1960s to peak at nearly 11 million tons. Congressional refusal to renew the Sugar Act in 1974 brought declining production, which was less than 5.8 million tons in 1992. Most sugarcane is grown on large holdings that refine and market their output through the California and Hawaiian Sugar Corporation, a producer's cooperative.

The decline of sugarcane and pineapples has encouraged increased production of a variety of higher value, but less labor-intensive, specialty crops. Nursery and greenhouse crops are the third most important crop group in the state with $66 million in sales. Cut flowers, primarily orchids, anthuriums, roses, and heliconias, are the most important of these with $36 million in sales (1992). Foliage plants, nursery crops, and potted flowers (often the same species as the cut flowers) constitute the remainder of the important nursery and greenhouse crops.

Macadamia nuts are now the third most important crop in the Hawaiian Islands. About 90 percent of the state's 22,634

The Hawaiian Islands

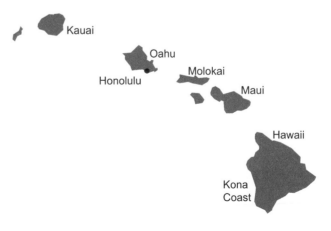

acres of macadamia orchards are concentrated on the island of Hawaii. The "Big Island" is also home to about three-quarters of the state's 87,000 beef cattle. The island chain supported a total of 200,000 cattle of various kinds in 1991. Cattle ranching dominates agriculture land use on the big island where most livestock is grazed on pasturage with little use of supplementary feeds. Apocryphal stories of marijuana as

Nursery and Greenhouse Crops, 1992

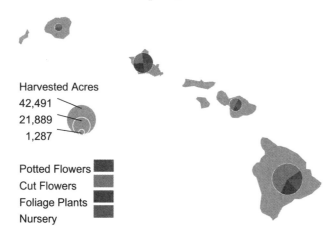

Harvested Acres
42,491
21,889
1,287

Potted Flowers
Cut Flowers
Foliage Plants
Nursery

Fruit and Nuts, 1992

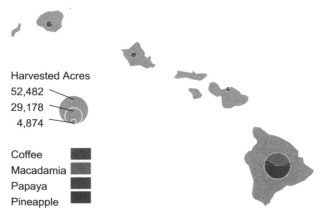

Harvested Acres
52,482
29,178
4,874

Coffee
Macadamia
Papaya
Pineapple

the state's most important agricultural crop abound in the general media, but little documentation exists either to confirm or refute these allegations.

A variety of other less well-known specialty crops have also become locally important in recent years. While coffee has long been grown in Hawaii, the recent explosion of boutique coffeehouses and producers has rapidly increased consumption of exotic coffees in general, and Hawaii's Kona coffee in particular. Coffee acreages have tripled since 1987 to 7,783 acres due to the high producer prices ($2.05/pound). Papaya production, in contrast, has been relatively stable, despite growing consumption on the mainland.

Vegetable crops have also diversified significantly in recent years. Taro acreage (495), the mainstay of the traditional Polynesian diet, has recently been matched by lettuce (494), emphasizing the changes that have been taking place in the ethnicity of the islands in recent years. Similarly, the state's other important vegetables are primarily oriented to the contemporary mainstream American diet, including sweet corn (371 acres), tomatoes (281 acres), and sweet peppers (253 acres). The state's Asian population and visitors are represented in the cultivation of Chinese cabbage (399 acres) and daikon (233 acres).

Agriculture in Hawaii has generally declined over the past forty years as tourism and other industries have become increasingly important. Less than 1 percent of the labor force is currently employed in agriculture, and the situation is not likely to change. Most land in Hawaii is owned by a small group of early families who have traditionally placed their property in long-term leases, rather than selling it. This has

Livestock, 1992

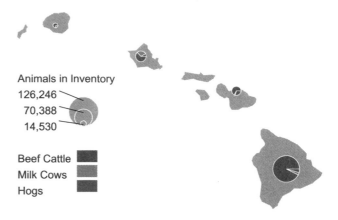

Animals in Inventory
126,246
70,388
14,530

Beef Cattle
Milk Cows
Hogs

Vegetables, 1992

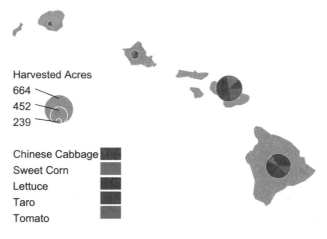

Harvested Acres
664
452
239

Chinese Cabbage
Sweet Corn
Lettuce
Taro
Tomato

given this small group the power to effect relatively rapid change in land use in short periods of time. For example, the relatively low cost of Hawaii as a vacation destination and its large number of golf courses and other facilities have attracted large numbers of Japanese tourists in recent years. This demand has encouraged these large landowners to take increasing amounts of land out of agricultural production and convert it into golf courses and other recreational facilities.

III
Cornucopia's Abundance

America's massive agricultural potential has promoted the development of automated production of even relatively minor crops. This equipment harvests peas in Washington, Oregon, and Idaho during a typical summer season. (RP)

America's Agricultural Diversity

American agriculture supplies a rich diversity of crops and animals to the national and global dinner table. This country provides the most diverse national agricultural output in history. The amazing variety of American agricultural output stems from many influences including the nation's environmental and demographic diversity, a diverse and changing set of eating habits and food preferences, and the globalization of agricultural production and marketing. The wide range of climates found in the United States stems from its great size (well over 3 million square miles), north–south extent in the midlatitudes (over 1,500 miles excluding Alaska), and long coastline exposed to the modifying impact of great oceans, which, in turn, is balanced by large interior areas with characteristic climates that are colder in winter, hotter in summer, and frequently drier than those nearer the sea. Much of the eastern half of the country is relatively moist, while the western half is generally quite dry, except for a narrow belt along the Pacific Coast. The growing season (the period between the average dates of the last spring frost and the first fall frost) ranges from essentially year round in Florida south of Lake Okeechobee and eleven months in the Imperial Valley to no more than five months in northern North Dakota and northeastern Montana. While southern Florida is a producer of such subtropical crops as oranges and sugarcane, the Red River Valley (along the border between North Dakota and Minnesota) concentrates on such short-season crops as barley, sunflowers,

and sugar beets. Farmers of the coastal valleys of California use their peculiar climatic situation to dominate the American market for such fresh vegetables as cauliflower and broccoli, while farmers in the equally dry, but cooler, valleys extending

eastward from the Cascades of western Washington use their equally specific environment to supply the world with blemish-free Red Delicious apples. States such as Wisconsin and Vermont are famous for their dairy products, largely because

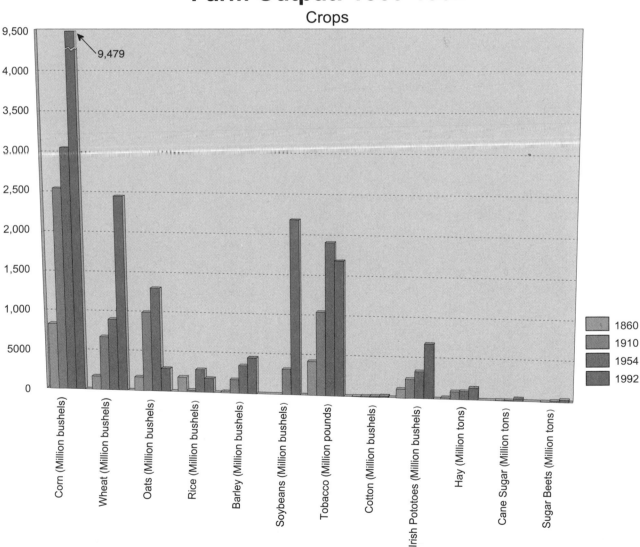

Farm Output: 1860-1992

Crops

115

most common breeds of dairy cattle thrive in cooler environments and because their short, cool, fairly moist growing season is ideal for most hay crops. Corn thrives in the long days and hot, humid weather of summertime in the upper Middle West.

America's ethnic diversity is another important contributor to our agricultural cornucopia. Foodways are fairly conservative—many of us tend to continue the dietary habits of our ancestors. It is not surprising that Milwaukee and St. Louis, with large early German populations, remain centers for the manufacture of pilsner beers and sausages of all kinds. African imports such as yams and okra have long been southern favorites. The recognition of ethnic cuisines exploded after the 1870s, when new great European migrations to the United States brought immigrants from previously untapped home areas from central, eastern, and southern Europe. A second and more dramatic expansion began in the 1960s, when the immigration focus shifted from Europe to Latin America, Africa, and Asia. As pizza, a classic introduction from earlier migrations, has become a dietary staple for virtually all Americans, it is now becoming "normal" for a small-town shopkeeper in North Dakota to stock and personally consume a frozen burrito dinner and to find a Chinese restaurant in some of the smallest, most isolated communities of the nation. These dietary additions often created significant new opportunities—
from prickly pear cactus to bok choy to truffles—for the nation's farmers.

In a somewhat contrary trend, the diet of many Americans has simultaneously been in a period of transition in opposing directions as well. We eat out far more

frequently, often at fast food restaurants. This has brought major new markets for the foods that form the core of their business: ground beef, frozen potato products, and head lettuce in the mainstream; Chinese vegetables, jalapeño peppers, and tofu in the new areas of culinary growth. In the process, Americans' concern about their health and the food that they consume has become important. These patterns have brought less demand for high-fat-content foods such as ice cream, while they have promoted an unparalleled growth in the consumption of frozen yogurt, turkey, and chicken. While some of these changes are driven by economics (the price of chicken

is significantly less than that of beef) much of it is health related. It will be interesting to see the impact on the nation's diet as the meat packers become protein processors as they move into the vertical integration of massive hog factories and fish farms.

Dramatic changes in storage and transportation technology over the last century are another significant contributor to our changing diet. The development of economical canning late in the nineteenth century followed by the development of viable frozen foods for home use in the middle of the twentieth century has meant that we can consume most vegetables and fruits throughout the year, not just during

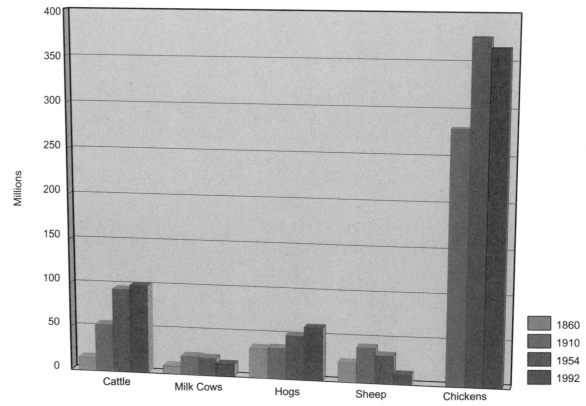

Farm Output: 1860-1992

Livestock

Table III.1 **Leading Agricultural Commodities, 1992**

	Market Value (in $ million)	Leading States (by value)
Cattle and calves	37,862	NE, KS, CO
Milk products	20,105	WI, CA, NY
Corn (for grain)	19,378	IL, IA, IL
Soybeans (for beans)	11,843	IL, IA, IN
Hay (all)	10,506	CA, PA, IA
Hogs	10,088	IA, IL, NE
Broilers	9,156	AR, GA, AL
Wheat	7,979	ND, KS, OK
Greenhouse/nursery	7,635	CA, FL, PA
Cotton (lint)	4,184	CA, TX, MS
Eggs	3,389	CA, GA, AR
Tobacco	2,996	NC, KY, TN
Turkeys	2,387	NC, MN, AR
Potatoes	2,336	ID, WA, ND
Tomatoes	1,819	CA, OH, MI
Sorghum	1,693	TX, KS, NE
Oranges	1,544	FL, CA, AZ
Apples	1,421	WA, CA, NY
Grapes	1,422	CA, WA, NY
Peanuts (nuts)	1,285	CA, TX, AL
Rice	1,099	AR, CA, LA
Sugar beets	1,086	MN, CA, ND
Barley	951	ND, ID, MT
Sugarcane	876	FL, LA, HI
Lettuce	822	CA, AZ, FL
Strawberries	685	CA, OR, NY
Almonds	670	CA
Mushrooms	669	PA
Horses	647	KY, FL
Onions	614	CA, OR, CO
Cottonseed	607	TX, CA, MS
Sheep (meat and wool)	516	TX, CA, CO
Dry edible beans	467	MI, ND, CA
Corn (sweet)	462	WI, MN, WA
Cut flowers	418	CA, FL
Peaches	373	CA, GA, NJ
Grapefruit	294	FL, CA, TX
Oats	391	SD, MN, ND
Carrots	338	CA, MI, FL
Broccoli	280	CA, AZ, OR
Pears	273	WA, CA, OR
Walnuts	268	CA
Lemons	261	CA, AZ

harvest season. Refrigerated railcars from the nineteenth century, refrigerated trucks from the midtwentieth, and air freight from the late twentieth have allowed the importation of fresh produce from distant growing regions and reasonable prices. All these developments have tended both to increase overall consumption of exotic foods and to change the geography of production.

Finally, America is easily the largest international supplier of farm products. While our national consumption of tobacco has declined substantially over the past twenty-five years, a growing international demand (now threatened by improving quality and lower prices for tobacco grown elsewhere) has buoyed American tobacco production. The variety of American exports is amazing. While American farmers are most famous for exporting billions of dollars of soybeans, corn, and wheat each year, the nation also exports large quantities of less frequently recognized products such as cattle hides (more than $1 billion) and horse meat ($100 million).

The early permanent European settlers along the eastern shoreline of the United States were almost immediately dependent on the foodstuffs they could produce for survival. Distances were too great and the cost of shipment too high for them to depend on foreign food sources. The new occupants arrived in search of valuable export crops, but were soon forced to concentrate on survival as well. The good fortune of those who settled New Spain, where gold and silver provided riches, was denied to their counterparts in Brazil, most of the Caribbean, and North America. There were other paths to riches, however, and settlers in these areas soon turned to

agriculture as a means of becoming wealthy. While sugar was the usual export of the subtropics, North Americans sorted through a variety of farm products and technologies to identify the foodstuffs they could profitably export. Their decisions were based on a mix of three influences—the agricultural traditions of their homelands in northern Europe, the adaptation of the newly introduced plantation economic system, and local Amerindian crops and technologies.

The diverse heritage of the settlers spurred the emergence of a rich array of crops. Amerindian crops, notably "Indian" corn and various squashes, were quickly adopted as a source of food for the poorest and for those on the smallest farms. Farmers with larger holdings and more resources often clung to the thousand-year-old European three-crop rotation of food, feed, and fallow. Wheat was the desired food crop and either barley or oats the normal feed crop. The surplus wheat was often exported to the Caribbean sugar plantations. By the American Revolution, especially in the rich limestone lands of southeastern Pennsylvania, the inefficient European medieval three-field system had been replaced by a four-year rotation of corn, wheat, oats, and clover. Wheat was still the important cash crop, but animal fodder had become the core of the new business of farming. Cattle sales, much improved through careful breeding, replaced wheat as the principal source of farm income. Cattle were driven to southeastern Pennsylvania from as far away as Ohio, New York, and Virginia to be fattened for sale in the eastern urban markets. This mixed farming system with continued modifications was ultimately carried westward to form the core of the farm economy in the agricultural interior.

Farther south, the longer growing season and hotter summers meant that some plantation crops might be profitably cultivated. This strategy had formed the cornerstone of the rich colonial economies of many Caribbean islands and it was believed that it could on the North American mainland as well. A series of plantation crops was introduced, each with its own timing and geographic focus. The first tobacco export left Jamestown in 1613, and soon Maryland and North Carolina were vying with Virginia to fill and often glut the international market for the leaf. Rice cultivation was brought from Africa to the coastal lowlands of South Carolina late in the seventeenth century. The sugarcane industry started in Louisiana in the last decade of the eighteenth century. Indigo was an important South Carolina crop in the mideighteenth century, but overproduction elsewhere brought an end to this colorful industry by 1800. It was cotton that eventually became king of the region's agriculture. Coastal, or sea island, cotton was grown in Jamestown in the first years of the colony, although it was not until after 1793 and the invention of the cotton gin that the crop became important to the southern economy. Cotton production reached 50,000 bales in 1800, then doubled every decade until raw cotton became the dominant national export by the start of the Civil War. Cotton remained the most important southern cash crop through the 1930s, although a variety of food crops, notably corn, actually occupied more acreage than these plantation exports.

U.S. Census data about the pattern of production of the agricultural economy did not become reliable until the middle of the nineteenth century. Most of what has been written about American agriculture prior to that time is based on fragmentary evidence. After that time, with some exceptions, information on agricultural output and activity is well documented. The census years 1860, 1910, 1954, and 1992 fall in key periods of change in American agriculture and have been used as often as possible to illustrate the state of agriculture and agricultural change to provide continuity and comparability between crops. The three dominant themes that connect the early periods include (1) increasing overall growth of the agricultural industry through expansion; (2) surging productivity brought on by increasing yields; and (3) the almost constant westward shift of the nation's agricultural core. Most growth prior to 1910 stemmed from expansion of acreages. Since 1910, improved genetics, more intense planting patterns, and increased use of agricultural chemicals have played the most important role in explaining expanding production.

The westward movement of production has characterized agriculture throughout the nation's history in the largest sense. Massachusetts was the wheat capital of the colonies. Through time, the center of production moved westward to Pennsylvania and New York to Ohio, Illinois, Iowa, and eventually Kansas and North Dakota. Cotton pressed westward into the dry country of western Texas and California's Central Valley. While California has surged ahead of Wisconsin as the nation's leading dairy state, Wisconsin is moving to pass Massachusetts as the country's principal supplier of cranberries. California has replaced the East as the major source of a wide variety of fruits and vegetables. This westward movement can be broadly interpreted as a search for those regions where the climate and soil especially encouraged particular activities, supported by transportation and storage improvements that eased the problems of shipping to market. Competition and the cost of land and labor were also important driving forces. Wheat yields are higher on good farmland in Illinois than in central Kansas, but the Illinois farmland yields even greater returns from corn and soybeans. Many fruits and vegetables grow well in the Corn Belt, but the paucity of large numbers of inexpensive agricultural laborers has forced production out of many areas. Finally, despite the diversity of production, the farm economy has been dominated by an ongoing mix of key products. Corn, wheat, cotton, tobacco, cattle, hogs, dairy products, and poultry were important in 1860; all remain among the ten leading

Table III.2
Leading Agricultural Exports, 1992

Product	Value ($Millions)
Oilseeds and products (principally soybeans)	7,156
Corn	4,593
Wheat	4,318
Meats (including prepared)	3,236
Vegetables (including processed)	2,790
Fruits (including processed)	2,786
Raw cotton	2,183
Tobacco	1,568
Poultry and products	1,193
Nuts	1,155
Cattle hides	1,107
Sorghum	839
Rice	757
Dairy products	638

Source: *Agricultural Statistics*, 1993, Table 671.

agricultural products today. Some dramatic shifts are nevertheless important. Soybeans did not become a significant crop until early in this century; nursery and greenhouse crops have boomed, as have the fruit and vegetable industries; oats, mostly used as food for horses, declined precipitously as the horse disappeared as an important farm animal by midcentury.

The geography of American agriculture is a composite of the geographies of hundreds of crops and animals. Almost twenty "billion-dollar products" have been identified as the core of the following discussion, supplemented by discussions on some livestock and crops of lesser value believed to be of either historical or future interest. This group represents a significant share of the total value of the country's farm output, and in their diversity (the group ranges from corn to fruit to aquaculture) provides a good introduction to the major influences on our agricultural geography. Where are they found today? Why? How has their pattern of output varied historically? What roles have changing market demand and distribution systems played on their fortunes? These are some of the questions addressed.

Idaho trout farms along the Snake River produce almost three quarters of the nation's trout for restaurants. (RP)

Aquaculture

Aquaculture, recently achieving the status of a billion-dollar crop, is often called the fastest growing segment of American agriculture. While almost all of us have eaten farm-raised fish, few of us are aware of either the breadth or the extent of fish farming in the United States and worldwide. More than one-quarter of the world supply of shrimp is now raised on fish farms, while almost a half billion pounds of catfish are farm raised each year in the United States.

Chinese farmers began raising fish in ponds as early as 2,000 B.C., establishing a tradition that has continued as an important source of protein to this day. The Romans began fish farming about 50 A.D., while fish ponds were common in monasteries during the Middle Ages. Although Thomas Jefferson and others experimented with fish farming during the eighteenth century, government programs to restock streams for sport fishermen date from the nineteenth century. Commercial aquaculture for food did not begin until the 1950s in the United States. The more recent explosive growth of aquaculture is related to the nation's increasing interest in controlling dietary fat intake and the efficiency of cold-blooded fish in converting feed into protein. Cultured catfish, for example, gain approximately .75 grams for every gram of food that they are fed. This is about 50 percent more efficient food conversion than for broiler chickens and six times more efficient than for beef cattle.

The distribution of commercial aquaculture largely reflects an industry driven by individuals committed to developing a new food source. The Mississippi Delta is the largest single center of production, followed by the Pacific Northwest and New England. The role of the individual in shaping the geography of this industry is best illustrated by the development of the Atlantic Littleneck Clam Farm near Charleston, South Carolina. South Carolina has long been an important center for the harvest of shrimp and some oysters, but clams had never been important here. A marine biologist became fascinated with the possibility of the controlled cultivation of clams in the South Carolina marshes during the late 1980s and began developing a system of breeding, sowing, and harvesting clams in large cylindric growing environments placed in the marsh. Today his company is the world's largest producer and distributor of clams. Though not joined by competitors as yet, it is only a matter of time before this area becomes a major center of controlled mollusk production, much in the same manner as catfish, trout, and Atlantic salmon areas evolved in the past.

Catfish

Catfish are the most important farmed fish in the United States with 459 million pounds produced in 1993. Commercial fish farming began in the Mississippi Delta in the late 1950s as an extension of earlier buffalo fish operations. The first commercial sale of catfish took place in 1960 in Arkansas. An Alabama grocer also began to

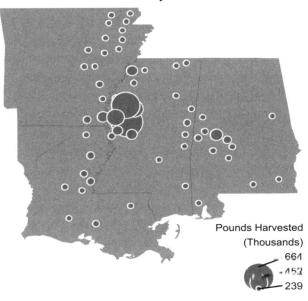

Catfish, 1992

Pounds Harvested
(Thousands)
664
452
239

Table III.3
Leading Catfish Producing Counties, 1992

County	Pounds (Thousands)	Percent of Total
Humphreys, MS	75,809	18.7
Sunflower, MS	67,550	16.7
Leflore, MS	36,758	9.1
Washington, MS	35,496	8.8
Hale, AL	17,879	4.4
Chicot, AR	15,806	3.9
Yazoo, MS	14,610	3.6
Tunica, MS	14,520	3.6
Sharkey, MS	12,026	3.0
Bolivar, MS	8,759	2.2
Ten county total	299,213	73.8
U.S. production	405,371	

sell processed fish from a small pond about the same time. Experiments with reducing production costs, developing more efficient cultivation and processing technologies, and better feeding technology doubled poundage yields per acre by 1976. A shakeout of inefficient and small producers took place in the early 1980s as feed costs increased because of shortages of fish meal. The industry reacted with growers' cooperatives to process and sell their product. The 1980s and early 1990s have brought a stabilization of demand. Total production over the past few years has increased only slightly, less than 2 million pounds between 1992 and 1993.

Commercial catfish production is currently practiced in fifteen states. Mississippi is the center of catfish production with 262 farms at the beginning of 1994 and 91,000 acres of production and the largest average pond acreages per farm (347 acres). Alabama has the second largest number of farms, the largest during 1992 and 1993, but the average pond acreage per

Catfish, 1994

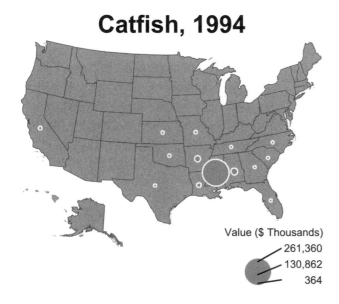

Value ($ Thousands)

261,360
130,862
364

farm is much smaller. Alabama water surface area has dipped to 17,100 (1994) as the smallest operators have gone out of production, but total sales have increased dramatically. Arkansas, slipping to third largest producer, continues to consolidate its operations and increase production levels. Louisiana is the only other significant commercial production area with $18,843,000 of sales for 1994.

Trout

Trout are commercially produced in 461 operations in fifteen states. Idaho has risen in the last few years to dominate the industry with 1993 sales in excess of $29 million and 42 percent of sales. The Crystal Springs Trout Hatcheries in Buhl, Idaho, is the largest commercial trout aquaculture operation in the world, although its total production is unpublished. The ascendancy of Idaho was based initially on the large volumes of cold water from the Snake River so necessary for successful production and the early decision to concentrate on processed fish fillets for restaurants. California and North Carolina are the second and third largest producers, respectively. Most trout producers are comparatively small operations with limited markets and sales.

Salmon

Atlantic salmon farming began in western Norway about 1960 when two brothers began experimenting with raising sea trout in pens attached to floats. The process was patented in 1968. The production of Atlantic salmon was also explored and eventually this species came to dominate the market.

Trout, 1993

Value ($ Thousands)

29,180
14,890
600

About 250,000 tons of salmon were farm raised worldwide in 1991. Norway continues as the center of Atlantic salmon, although it is believed that rapidly expanding operations in Chile are probably the least-cost producers at the present time. The impact of this steady supply of competitively priced fish has brought Atlantic salmon to virtually every fine fish restaurant in America over the past few years.

It is estimated that 26 million pounds of Atlantic salmon were raised in the United States in 1992. Maine produces about 60 percent of the annual harvest and Washington virtually all the remainder. Several attempts have been made to produce Pacific salmon in the United States, but none are currently raised for commercial food fish production. West Coast producers have attempted to expand operations, but have met resistance from commercial fishermen and sportsmen concerned that such farms would reduce their opportunities. Steelhead trout are raised in Maine and North Carolina.

Crawfish

Crawfish have long been a delicacy among Louisiana's Cajun population. The meteoric rise of Paul Prudhomme and his Cajun cookbooks during the 1980s, followed by the popular public television program hosted by Justin Wilson, introduced the region's cuisine to the remainder of the nation. Cajun cooking quickly swept the country, especially as a supplemental menu item in many midpriced restaurant chains focusing on the young urban professional market. This instant popularity put a strain on the supplies of basic regional food supplies and was largely responsible for a rapid expansion of crawfish farming. Wild harvesting had been practiced in southern Louisiana for years, especially in the rice fields in the southwest corner.

Demand for Cajun cuisine declined in the late 1980s with both crawfish pond acreage (130,000) and harvest (65 million pounds) peaking in 1989. Both have been declining since that time, although wide variations occur depending on market price and the volume of the wild harvest, which reached 69 million pounds in 1993. Part of the U.S. market decline is being partially offset by increased exports, primarily to Sweden. The Swedish market, however, demands only large crawfish and lasts only a few months. More than $16 million of crawfish were exported in 1993, an increase of 15 percent over 1992.

Mollusks

Annual farmed mollusk production is estimated at about $150 million annually, though no comprehensive survey for the entire nation has ever been completed.

American mollusk farms largely concentrate on clams and oysters on the West Coast, oysters, northern quahogs, mussels, and scallops in the Northeast, and oysters and clams in the Southeast. The cultivation of other mollusks is also being explored, especially abalone in California and a variety of specialty oysters in Washington. Pollution is a continuing problem for mollusk producers who are forced to operate their farms close to shore where environmental degradation tends to be the greatest.

Mollusk exports continue as an important part of this relatively small industry. Clams are the most important mollusk export, primarily to Japan, Canada, the United Kingdom, and Hong Kong in declining importance. Canada is the largest export market, although exports to Japan increased by 370 percent between 1992 and 1993. Oyster imports far exceed exports. Mussel exports are much smaller than clam and oyster sales, while imports, primarily from Canada (especially New Brunswick and Prince Edward Island) and New Zealand are about $9 million.

Other Fish

A variety of other fish are cultivated in the United States, including tilapia, shrimp, and ornamental fish. The tilapia is the least known, yet is also the fastest-growing product in the cultivated fish market. The tilapia is a mild, white-fleshed fish, increasingly served in restaurants in replacement of more expensive wild harvest species in entrees where the fish species is not designated. This warm water fish is often cultivated indoors in the United States. American production reached 12.5 million pounds in 1993, an increase of 40 percent over the previous year. Imports, however, totaled 25 million pounds. The largest sources of imported tilapia are Taiwan, Costa Rica, Thailand, Columbia, and Indonesia.

More than 633,000 tons of shrimp were farmed worldwide in 1990. China was the world's largest producer of farm-raised shrimp in 1992 and may produce as much as a quarter of the world's crop. Other important producers include Indonesia and Ecuador. The United States produces an estimated 300 million pounds of shrimp per year, but all except 3,000 tons are from wild harvest. Reluctance of sportsmen and environmentalists to allow the penning of estuaries makes the expansion of production difficult. Experiments utilizing raceways and high volumes of underground saline water are currently under way and likely will expand in the future.

The market for ornamental fish has been steadily expanding in recent years, although there are no reliable estimates of total production. Exports over the past five years have risen to $17.3 million (1993), 5 percent greater than in the previous year and double the 1989 production levels. Imports have been rising even more quickly than exports. Exports for 1993 reached $45.2 million.

Yields from these irrigated barley fields near Flathead Lake, Montana, may be double those received by farmers on unirrigated fields on the Grasslands. (JF)

Barley

Barley is the fifth most important (by weight) grain crop in the United States after corn, wheat, grain sorghum, and rice. Barley typically yields between two and three times more tonnage per acre than wheat, but sells at a lower price. The wheat/barley decision generally revolves on what other crops are in the rotation and on soil quality. About 45 percent of the annual crop is used as animal feed, 20 percent is exported, and the remainder is about equally divided among malting for alcoholic beverages, baking, and in the preparation of breakfast cereals.

Barley evolved in the near East and was the least expensive and most abundant grain in many of the ancient city-states of the region. It began to be replaced in bread by wheat about 1000 B.C. in the Mediterranean, although it continued to be widely grown. It was one of the plants tested on Hispaniola for cultivation in the New World on Columbus's second voyage, but failed in that humid climate. It was successfully introduced into eastern North

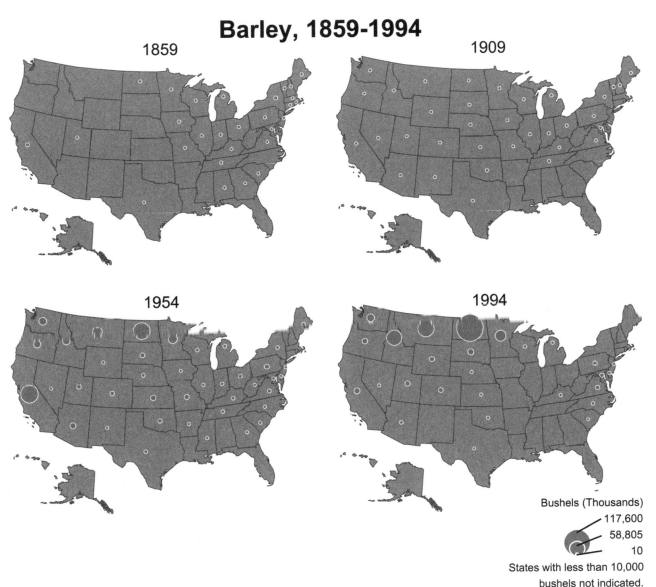

Barley, 1859-1994

1859

1909

1954

1994

Bushels (Thousands)
117,600
58,805
10
States with less than 10,000 bushels not indicated.

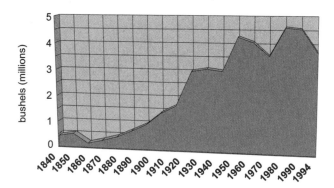

Barley: 1840-1994

America during the seventeenth century and into the Southwest by Spanish missionaries about 1700. Barley was California's first important export crop with large tonnages shipped to British brewers beginning in the 1850s. It was not an important crop in the humid East until new, more humidity-tolerant varieties were introduced.

The geographic distribution of this crop has changed largely in degrees and concentration with few significant shifts during

125

Table III.4
Leading Barley Producing Counties, 1992

County	Bushels	Percent of Total
Cavalier, ND	11,425,836	2.9
Polk, MN	8,628,533	2.2
Barnes, ND	8,467,054	2.1
Whitman, WA	8,292,543	2.1
Grand Forks, ND	8,167,996	2.1
Cass, ND	8,150,557	2.1
Trail, ND	6,223,758	1.6
Bottineau, ND	6,024,854	1.5
Ramsey, ND	5,876,826	1.5
Norman, MN	5,444,847	1.4
Ten county total	76,702,804	19.3
U.S. production	397,245,453	

the twentieth century. The largest concentration of production at the turn of the century was in the northern Midwest. Minnesota farmers harvested 1.5 million acres, followed by North Dakota (1.2 million acres), South Dakota (1.1 million acres), Wisconsin (.8 million acres), and Iowa (.6 million acres). California, especially the Central Valley, was the third largest producing state nationally with 1,195,158 acres, primarily for malting. Lesser amounts were grown in Washington, Idaho, and the high plains of Nebraska/ Kansas/Colorado. A small westward shift of production into the northern Grasslands has made North Dakota the leading state for barley production today, primarily concentrated in and on the edges of the Red River Valley adjacent to Minnesota. Montana, with about one half the output of North Dakota, is now the second largest producing state in a great swath of production westward from its border with North Dakota. Idaho, the most important new

center of production, has witnessed a growth to third with about three quarters of a million acres cultivated each year. Barley is grown there primarily in conjunction with the potato rotation cycle. Minnesota, though much less important than earlier, continues as the fourth largest producing state, primarily in the Red River Valley. California production, the second dramatic change, has slipped to about 180,000 harvested acres, primarily being replaced by

the introduction of irrigated field crops in previously dry farmed sections of the Central Valley. Small amounts of barley continue to be grown along the eastern seaboard, primarily as a rotation crop.

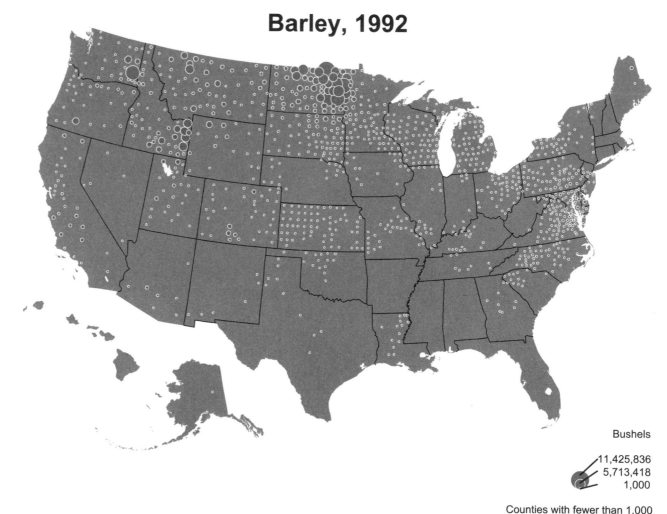

Barley, 1992

Bushels

11,425,836
5,713,418
1,000

Counties with fewer than 1,000 bushels harvested not indicated.

U.S. Production 397,245, 453 bushels

Beef Cattle

Cattle production may be the nation's most ubiquitous agricultural activity. One thousand or more cattle and calves were sold in 92 percent of American counties in 1992. Virtually all nonurban counties had some cattle sales in that year. Simultaneously, this enormous industry is highly concentrated in areas of special advantage. Almost two thirds of the entire cattle slaughter of Nebraska, Kansas, Texas, and Colorado were fattened in local "dry" feedlots. The largest of these sends almost a million cattle per year to slaughter from a single location. In that respect, beef cattle production may be the best illustration of the interrelationship among market demand, access to raw material, and transportation costs that characterize the strategic planning that increasingly dominates contemporary American industrial agriculture today.

The most volatile of these issues is the changing market. Per capita beef consumption peaked in 1976 at eighty-nine pounds (boneless, trimmed equivalent weight), dropped to between seventy and seventy-five pounds during the late 1970s, remained flat through the early 1980s, and then decreased to only sixty-two pounds in 1993. Considering increasing concerns about cholesterol and saturated fats, it is likely that this decline will continue for the foreseeable future.

The nature of purchasing has also changed. Increasing numbers of working couples, single heads of households, and single-parent families have altered the eating habits of Americans. Although the restaurant market has become increasingly segmented in recent years, the largest purveyors of restaurant fare continue to be the fast food chains. The hamburger is the single most important fast food product and hamburger chains are the largest food providers. The fast food hamburger business is dominated by four companies (McDonalds, Burger King, Hardees, and Wendy's). To control food quality these companies and their competitors prefer to use sole distributors whenever possible. These distributors in turn prefer to place their orders with only a few suppliers. The predictability of this demand has allowed the preferred suppliers to invest in ever

Beef Cattle, 1992

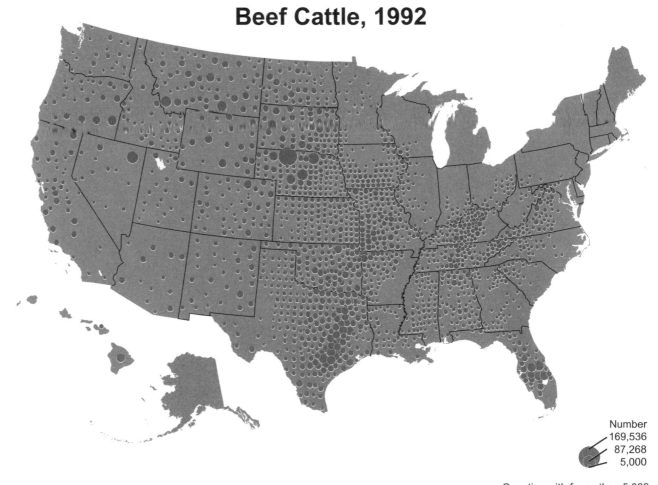

Number
169,536
87,268
5,000

Counties with fewer than 5,000 beef cattle not indicated.

U.S. Inventory: 33,775,000

127

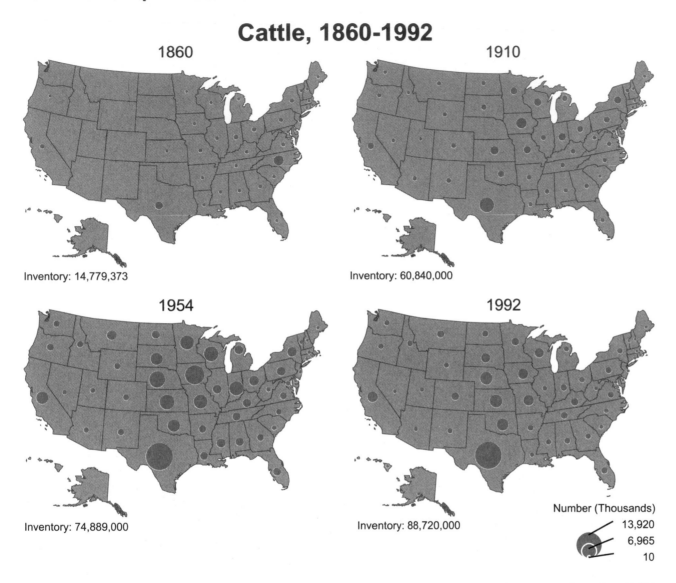

Cattle, 1860-1992

1860
Inventory: 14,779,373

1910
Inventory: 60,840,000

1954
Inventory: 74,889,000

1992
Inventory: 88,720,000

Number (Thousands)
13,920
6,965
10

Counties with fewer than 10,000
cattle not indicated. Milk cows excluded.

national producers, rather than from localized stockyards and meat processors. The combination of these forces, in association with the factors discussed below, inevitably led to an almost complete reorganization of the geography of cattle production and processing over the past thirty years. Ranchers producing the traditionally favored well-marbled beef in the past have been forced to concentrate on raising leaner breeds, while the small-town local stockyard has virtually disappeared. American cattle breeds, ranching, and meat processing have all changed over the past fifty years because of these trends.

Beef Cattle Breeds

The domestication of cattle began before 4000 B.C. Modern cattle breeds are descendants of two species, *Bos taurus*, the wild cattle of early Europe, and *Bos indicus*, the humpbacked zebu cattle of South Asia. Geographic isolation contributed to the early development of substantial genetic diversity globally. Cattle introduced to North America during the colonial and early national period were European descendants of *Bos taurus*. Their progeny have

Cattle and Calves: 1850-1992

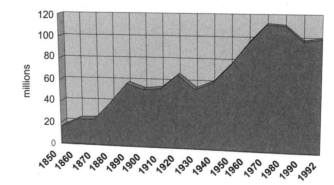

larger production facilities with the knowledge of a predictable demand.

The nature of home purchases has also changed with the changing American lifestyle. Purchases of roasts and other large cuts of meat have declined, while the consumption of the easily and quickly prepared hamburger has increased from 26 percent of all beef sales in 1970 to 35

percent in 1985 and finally to 45 percent in 1993. Further, the increasingly suburbanized consumer has tended to concentrate purchases among a decreasing number of grocery chains. The eventual domination of this market by a handful of retail behemoths in turn has focused demand on a limited number of supply channels who in turn tend to purchase from a handful of

and continue to dominate the cattle herds of the temperate portions of the United States.

The earliest of the European breeds brought to America was the Devon, the "old red cow" from Devonshire in southwest England, which provided early settlers with beef, milk, leather, and ox teams. Robert Bakewell, a farmer in Leiscestershire, England, is credited with developing modern livestock breeding, although less systematic efforts had long existed. The first of these improved breeds brought to America was the shorthorn (then called the Durham), imported as early as 1783. The breed did not become established, however, until large numbers of imports were brought to Ohio and Kentucky in the second quarter of the nineteenth century. Early shorthorn cattle were used for both beef and milk.

A few of the many beef breeds in the United States deserve special mention because of their distinctive quality and importance. Herefords were developed on the grasslands of county Hereford in England. They were first used as draught animals, but eighteenth-century breeders began to develop their meat qualities. Henry Clay is credited with bringing the first Herefords to this country in 1817, although large-scale importation did not begin until after 1850. The popularity of the Hereford stems from its ability to withstand a relatively wide temperature range, to require less care than many large breeds, and to put on weight rapidly.

The Aberdeen-Angus breed, usually simply called Angus, evolved in southern Scotland and was introduced to the United States in 1873. Angus meat is well marbled (the fat is distributed in striations through-out the meat), traditionally a desired meat trait in the American market. Angus have led all other American breeds in total annual registrations since 1963. The Charolais, a French breed, was introduced via Mexico in 1936. This fast-growing animal is increasing in popularity because its leaner meat is more attractive to today's increasingly weight- and cholesterol-conscious society.

Northwest European breeds thrive in the cooler climates of the northern United States, but are poorly adapted to the hot, humid environments of the Lowland South and south Texas. Columbus introduced the first long-horned Criollo cattle from Andalusia via the Canary Islands on his second voyage to the Americas in 1493. Criollo cattle were carried to Mexico in the 1520s and eventually reached south Texas. They thrived in the region's hot, humid climate to evolve through several centuries into the famous Texas longhorn, the Florida cracker, and piney woods cattle breeds. Expansion of the open range cattle economy briefly expanded the range of the longhorn northward across the Grasslands after the Civil War, but were soon replaced by the introduction of more efficient European breeds. The few remaining herds of longhorns are maintained for sentimental and tourism reasons, rather than as meat producers. The cracker and piney woods breeds have also been largely replaced by zebu crossbreeds that provide a better quality finished product.

Bos indica, or zebu, cattle were imported for crossbreeding as early as 1849 in the search for heat-tolerant cattle for the South. The Brahman, the most important of these crossbreeds, developed in south Texas and has since been crossed with a variety of northern European breeds. The beefmaster is a cross between the Brahman, the Hereford, and the shorthorn. The Brangus is a cross between Brahman (3/8) and Angus (5/8), joining the hardiness of the Brahman with the meat qualities of the Angus. Perhaps the most famous of the North American crossbreeds is the Santa Gertrudis, a 3/8 Brahman, 5/8 shorthorn mix developed on the Santa Gertrudis division of the King Ranch in south Texas.

Early History

Cattle arrived with the earliest colonists at Jamestown and Massachusetts Bay. Local cattle were raised to meet local demand with little long-distance transfer of either animals or their products. Change began in the mid-1700s when settlers on the southern frontier focused on an open range hog and cattle economy, with excess animals periodically driven eastward to coastal markets.

Farmers in southeastern Pennsylvania began to shift from the crops characteristic of the medieval three-field rotation of food (usually wheat), fodder (usually oats), and fallow after the American Revolution. The fallow year began to be eliminated because the cultivation of corn as a fodder crop enabled them to control weeds, while the cultivation of clover during the fallow year enriched rather than depleted the soil. The new system greatly increased the amount of winter feed available for animals, which enabled farmers to increase the numbers of animals that they could keep over the winter. The larger herds also increased the amount of manure available for fertilizer. Some farmers began purchasing lean cattle that had been driven east from the frontier, fattening them with their large winter feed supplies, and then selling them in the Philadelphia market. The beginnings of a geographically focused American beef industry had been established.

Middle Western Feedlots

The system of purchasing lean cattle for fattening and sale dramatically expanded westward into the Middle West as Middle Atlantic migrants began settling in southern Ohio in the early nineteenth century. They established a rotation of corn, a small grain (wheat or oats), and hay (clover). These crops soon defined Corn Belt agriculture and continued to dominate its landscape for more than a century. Unable economically to ship these crops to market, these farmers soon turned to raising livestock, which could be herded to market as their primary "cash" business. Animals might be allowed to graze for part of the year, but were penned for fattening on corn before they were sold for slaughter.

This so-called feedlot system reached its fullest development in Iowa in the decades following the Civil War. Iowa farmers became the middlemen in the movement of western cattle to eastern markets. The grass-fed western beef, especially the earliest trail-driven animals, were purchased by Iowa and other middle western grain farmers for fattening prior to their shipment to urban markets. Farmers commonly raised hogs in the same feedlot with cattle, with the cattle eating hay and grain and the hogs eating the poorly digested waste that had passed through the cattle.

The passing of the cattle drive did not bring an end to either the production of grass-fed cattle in the Grasslands or the production of finished cattle ready for slaughter in the Corn Belt. Corn Belt farmers were unable economically to supply the majority of their feeder cattle and continued to rely on Grassland and western breeders for their stock. The most important change was the introduction of new breeds. The Texas longhorn was ill-suited to take full advantage of a corn diet and most passed through undigested. The development of new cattle strains of longhorns crossed with northwestern European breeds created an animal more adapted to the feedlot system. Several thousand Hereford cattle were imported by Grassland ranchers in the 1860s and 1870s to provide breeding stock, while Angus were first introduced in the 1880s. The distinctive longhorn was destined to become a memory.

The Rise of the Grasslands Feedlot System

Beef production underwent another geographic restructuring in the decades after the end of World War II. The increasing

Beef cattle production today is largely concentrated in feedlots processing tens of thousands of cattle each per year. (RP)

concentration of beef demand favored the transformation of the traditional Corn Belt farm feedlot into massive systems housing tens of thousands of animals that were better suited to the Grasslands. The greatest concentration of these new mega-feedlots is in a band along the western margins of the Grasslands northward from the Llano Estacado of the Texas panhandle through western Kansas, through eastern Colorado, and into western Nebraska. Weld County, Colorado, now leads the nation in beef cattle sold with almost a million each year from its largest operation.

The rise of these new mega-drylots came as a response to the development of new irrigation technology and a desire by the larger meat packers to move their facilities into new, more efficient facilities away from the labor and urban problems that had plagued their older locations in the Corn Belt. The development of deep-well technology that could tap the Oglala aquifer allowed the development of millions of acres of newly irrigated fields using center pivot irrigation systems. The Big Thompson Project also transferred water from west of the Rockies to irrigate the northern Colorado Piedmont. An airline flight across the region today reveals a dramatic explosion of great green circles of irrigated crops surrounded by unirrigated browning field edges. The expansion of sugar beets and alfalfa throughout the region, and cotton in the Llano Estacado of Texas/New Mexico sector, created millions of tons of agricultural byproducts that could be used as inexpensive feed for these feedlot operations. Most of the labor in these facilities today is provided by Hispanics who have migrated into the region for that purpose,

much as they have migrated into the southern Appalachians to work in the poultry processing plants found there.

Large-scale feedlots have also appeared farther west, especially in areas where large quantities of consumable agricultural byproducts are present, including the Imperial Valley of California's southern desert, the southern Central Valley, Idaho's Snake River Valley, and increasingly in eastern Washington. Imperial Valley farmers concentrate on winter season vegetables, but cultivate cotton, sugar beets, and alfalfa during the hot season. Both dairy and beef feedlots have been developed to utilize these large quantities of byproducts, as has an experimental methane plant to attempt to economically produce natural gas from the feedlot byproducts. Dairy and beef feedlots in the southern Central Valley also primarily utilize cottonseed cakes, alfalfa pellets, and sugar beet pulp, although more exotic items such as almond hulls are becoming common as well. Snake River plain potato farmers typically rotate sugar

Table III.5
Leading Cattle and Calf Counties, 1992

County	Number Sold	Percent of Total
Weld, CO	925,210	1.31
Deaf Smith, TX	782,255	1.11
Texas, OK	587,543	0.83
Castro, TX	540,002	0.77
Parmer, TX	509,239	0.72
Imperial, CA	473,041	0.67
Cuming, NE	409,723	0.58
Fresno, CA	391,324	0.55
Wichita, KS	373,893	0.71
Total ten counties	5,438,719	7.71
Total U.S. sales	70,562,908	

beets and small grains with their more famous potato crops. All these crops generate significant amounts of byproducts for feedlot operators. Though beef feedlots are most important in Idaho, the state has recently risen to seventh in dairy output, utilizing feedlot technology.

Feedlots typically keep their animals from 120 to 240 days. Feedlot operators purchase animals from cattle ranchers who specialize in the production of calves and heifers for this purpose. Traditionally, these ranchers were concentrated in the Ranching and Oasis and western Grasslands, but the restructuring of the industry and increased demand have brought about the development of significant calf production in the Bluegrass and Nashville basins, Alabama's Black Belt, and the Great Valley extending from Virginia to Tennessee. The development of heat-tolerant crossbreeds in recent years has also enabled the development of large herds in south Texas and Florida. While cattle have long been raised in Florida, recent environmental constraints on the expansion of citrus in some areas have made cattle an increasingly attractive activity. More than a million cattle are now raised annually for sale in Florida for shipment to Grasslands and western feedlots. Farmers across the Lowland South are beginning to see the advantage of providing calves and heifers to western feedlots.

Changing Structure of the Slaughter Industry

Beef cattle slaughter was originally a market-based activity with widely dispersed, small slaughterhouses located near almost every urban center in the nation. The increasing concentration of livestock production in the Corn Belt and the development of railroads and refrigerated railcars in the midnineteenth century brought together the necessary ingredients to allow the development of a cartel of packing companies with facilities concentrated in the primary rail hubs of the Corn Belt—Cincinnati, Louisville, Milwaukee, and St. Louis. Though not in the center of raw material production, the superior locational advantages of Chicago allowed it to evolve into the single most important center of meat packing in the nation for a century or more. Chicago slaughterhouses peaked during the 1950s at more than a quarter billion sides of beef in a single year.

The "Big Five" Chicago meat packers (Armour, Swift, Wilson, Cudahy, and Morris) set the standard for the industry. They controlled the city's Union Stock Yard, the central receiving station for animals entering the city. Each operated massive packing houses surrounding the yards. Byproduct processors, making gelatin, glue, fertilizer, and other products, clustered around the packing houses and were critical to the profitability of the entire operation. In the end the city's promoters claimed that the packers and their associates used "everything except the squeal." In 1920 the Supreme Court mandated that the meat packers must sell their stockyard interests.

Similar symbiotic facilities were constructed west of Chicago nearer the sources of supply, first in Kansas City, East St. Louis, South Omaha, and St. Paul, and later in the more distant Wichita, Fort Worth, Sioux City, St. Joseph, and Oklahoma City. All followed the Chicago model of a central stockyard controlled by the packers with one or more of the Big Five dominating each new city.

Smaller meat packers and many livestock feeders were never pleased with the Chicago system and its inherent concentration of power in the hands of a few firms in a few carefully controlled locations. Tentative changes began as early as 1891 when George Hormel began his meat packing operation in Austin, Minnesota. The development of truck transportation in the 1920s intensified the practicality of restructuring the industry away from the older rail-based stockyards. By the mid-1950s truckers accounted for nearly all animal arrivals at the Union Stock Yard. The share of cattle purchased in large stockyards declined from 91 percent to 46 percent from 1925 to 1960. Newer, and usually smaller, packing plants began to spring up closer to animals, most notably in northern Iowa. The heyday of Chicago and the other

Fattened Cattle, 1992

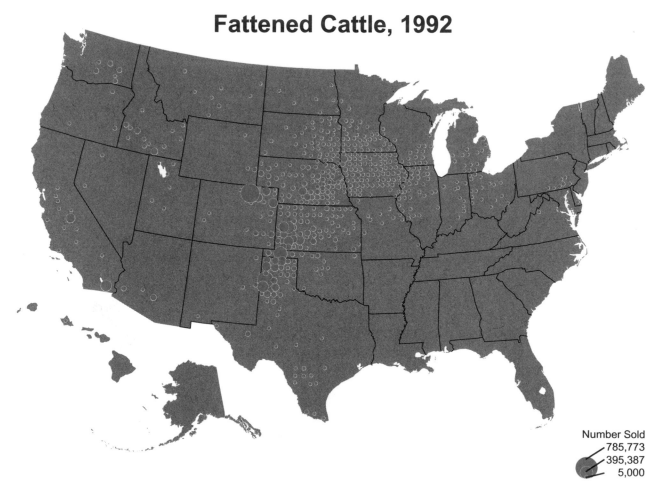

Number Sold
785,773
395,387
5,000

Counties with fewer than 5,000 sales not indicated.

U.S. Inventory: 26,405,739
Value of Sales: $21,121,027,000

giant stockyard concentrations everywhere was over. Swift and Company closed its Chicago plant in 1959, was acquired by a diversified conglomerate in 1972, and ultimately was acquired by the processed food giant ConAgra. Armour also eventually became a subsidiary of ConAgra. Wilson became a subsidiary of Ling, Temco Vought in 1971 and was relocated to Oklahoma City. These old giants thus have been largely replaced by newer packers, most notably Iowa Beef Processors (IBP), with smaller, modern plants located in small cities amid the source areas of the raw materials. Meat is no longer shipped to market as sides of beef or pork, but is butchered into smaller chunks, boxed, and shipped to individual supermarkets where it is cut to local demand. While first concentrated in the western Corn Belt, the industry today is centered in dozens of nonmetropolitan, nonunion, Grassland towns. Nebraska has become the largest center of beef processing with 20 percent of the national total. Nebraska, Kansas, and Texas account for 54 percent of all beef processed in the nation.

Current Trends

Inevitably changing dietary tastes, the increasing domination of the industry by processed food companies, and declining profit margins will continue to bring even greater change to the geography of the beef cattle industry. Meat packing companies have been almost totally replaced by self-styled protein processors who are reevaluating their activities to trim costs further. Conglomerates like ConAgra are not only major beef processors, but process and distribute chickens, turkeys, hogs, catfish, and even saltwater fish, to be able to increase their market share of the nation's food dollar. Leaving nostalgia and conventional locational wisdom behind, these companies constantly search for new, low-cost situations that they can exploit to keep costs down and profits high. Currently, the availability of agricultural byproducts, inexpensive labor, and lax local controls have tended to concentrate their activities to one set of locational advantages. But other places with different combinations of locational advantages and constraints are constantly under consideration. Iowa Beef Processors, for example, is moving a significant part of its hog operations to coastal North Carolina where it believes that it can be more profitable. It is not inconceivable that it may also ultimately move beef operations to that area as well. Simultaneously, areas such as Weld County, Colorado, are expanding from beef production to include hogs and poultry as well. Though seemingly just restructured, it is likely that yet another, grander restructuring is in the offing with the Lowland and Upland South playing a far greater role in beef production than ever before.

These Corn Belt farmers attending a seed company field seminar will select next year's seed from hundreds of alternatives, each with slightly different cob sizes, stalk heights, and maturation rates. (JF/RP)

Corn

Corn, also sometimes called *maize* (*Zea mays*), is easily the most important agricultural crop in the United States. It is grown in every state in the Union and occupies more space (70 million acres), generates more overall income (almost $20 billion), and involves more labor than any other American crop. Its dollar value to the nation's economy approaches the combined worth of wheat, oats, barley, rice, rye, and sorghum. Corn is also the most important cereal in the Western Hemisphere and is exceeded in value globally only by wheat and rice.

Evolution of Corn Cultivation

Modern domestic corn is the progeny of wild species of a genus found in southern Mexico and northern Central America. It is now generally recognized that corn was domesticated from *Zea mexicana*, or teosinte, with the earliest archaeological remains found in Puebla, Mexico, dating from *ca.* 5000 B.C. The domesticate gradu-

ally diffused northward from its southern Mexican core to New Mexico by 1200 B.C. and the eastern woodlands by 300 B.C. These early cultivated corns were probably small-eared popcorns used as supplements to wild foods. Corn did not become an important food in the United States until

after larger-eared flour corns were introduced to the Hohokam in Arizona about 500 A.D. These and other new varieties spread north and eastward to become well established in eastern North America before the arrival of the Europeans in the sixteenth century.

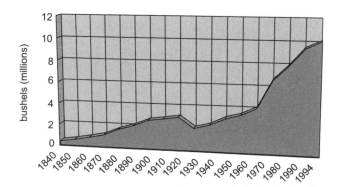

Corn: 1840-1994

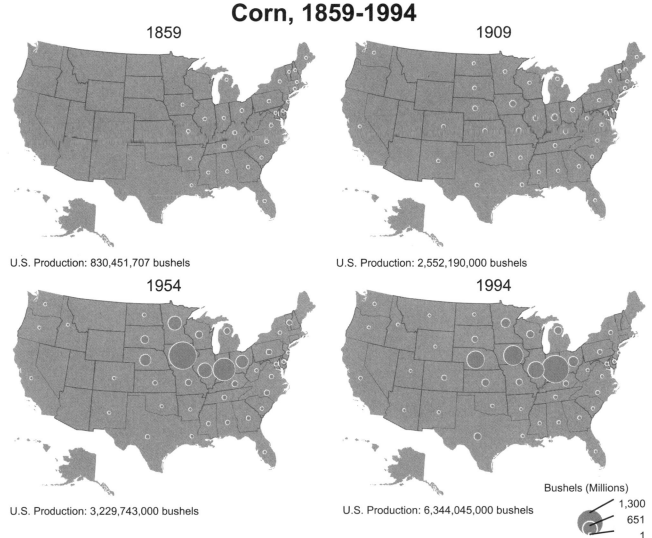

Corn, 1859-1994

1859

U.S. Production: 830,451,707 bushels

1909

U.S. Production: 2,552,190,000 bushels

1954

U.S. Production: 3,229,743,000 bushels

1994

U.S. Production: 6,344,045,000 bushels

Bushels (Millions)
1,300
651
1

Corn, 1992

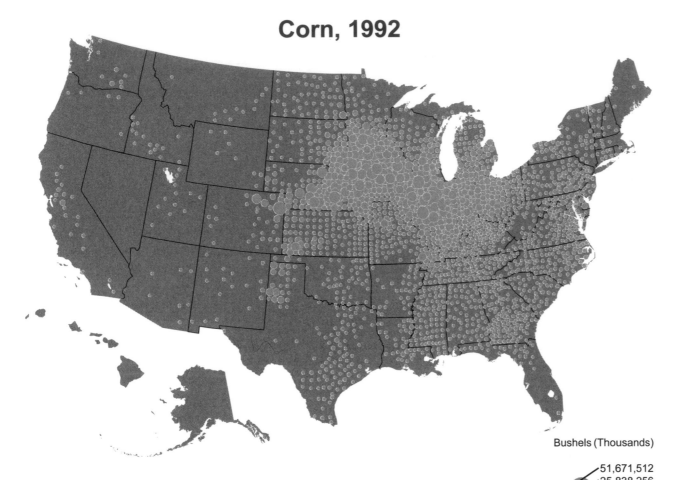

Bushels (Thousands)

51,671,512
25,838,256
5,000

Counties with fewer than 5,000
bushels harvested not indicated.

U.S. Production: 8,697,362,804 bushels

types with hundreds of individual races within each type. Each major type has specific physical characteristics that determine its use and distribution. The most common use of this crop is as a food source for both humans and domesticated animals. While deficient in amino acids, the crop is especially rich in starch (the grain is nearly 75 percent carbohydrates). The high starch content also makes an ideal raw material for the production of alcohol, especially for beer, less commonly for whiskeys, and is the single most common base for industrial alcohol. Corn-based alcohol is of increasing importance as a gasoline additive to lower pollution levels in this country and is used as a pure fuel in many parts of petroleum-deficient Brazil.

The high starch content is also highly adaptable for processing into high fructose liquid sweetener. The corn sweetener companies argue that if the American government removed its subsidies of traditional

Table III.6
Leading Corn Producing Counties, 1992

Counties	Bushels	Percent of Total
McLean, IL	51,671,510	0.6
Champaign, IL	45,486,676	0.5
Kossuth, IA	45,409,171	0.5
Iroquois, IL	44,526,657	0.5
Livingston, IL	43,905,431	0.5
La Salle, IL	42,403,608	0.5
Pottawattarnie, IA	36,053,220	0.4
Bureau, IL	35,818,997	0.4
Vermilion, IL	34,909,100	0.4
Yuma, CO	32,911,969	0.4
Total ten counties	413,096,339	4.8
U.S. production	8,697,362,804	

Columbus carried corn to Spain in 1493 where it received wide interest and ultimately was adopted for at least limited use. Its cultivation soon spread throughout much of the midlatitude and subtropical world in the following century, reaching Africa south of the Sahara during the sixteenth century, south Asia and China by the end of that century, and had become a standard foodstuff among poor southern Europeans during the seventeenth and eighteenth centuries. Early European colonists along the east coast of North America found "Indian corn" nearly everywhere they settled. Corn helped the Pilgrims through their first difficult New England winters and fed Jamestown and Manhattan Island settlers in their early years.

Corn Types and Races

Corn today is categorized into seven major

sugar beet and sugarcane production that the much less expensive fructose would quickly become the dominant form of sweetener in most processed foods. The decision by soft drink companies to utilize high fructose syrups in their products in the 1980s caused a phenomenal growth in production and consumption over the past decade.

About 70 percent of the corn crop, however, continues to be used as animal food in the United States, declining from over 80 percent twenty-five years ago. The rise of great cattle-fattening operations has promoted even more extensive use of corn as feed for feedlots, while cornmeal is almost always the single most important ingredient in dry domestic pet foods. Other uses include the ever popular popcorn, which actually consumes a minuscule portion of the entire American production, and as a summer vegetable for direct consumption from the cob.

Today's widespread distribution of cultivation largely stems from the ease with which new races adapted to local conditions can be created. The Americas at the time of Columbus contained a wide array of different corns. Some had evolved in response to different environments, but most were the consequence of human manipulation to create differences in such things as rate of maturation and grain quality. A single Indian farmer often cultivated a dozen or more races within a single field, each with individual maturation rates and grain quality, to protect himself from drought or other unique environmental conditions. As a result, the crop is grown around the world in such climatic extremes as interior Canada and Siberia to the African and South American tropics.

Corn production has become increasingly sophisticated in recent years as geneticists have developed seemingly endless races and subraces designed for specific conditions. One seed company, for example, offered thirty different corns for 1995 plantings for farmers in southeastern Minnesota alone. The differences were a continuum of fifteen variables relating to yield, growing period, ear configuration, height, and so forth. Maturation periods, for example, ranged from 79 to 110 days. Plant heights ranged from four to eight feet. When multiplied by the more than dozen seed companies offering comparable products, today's corn farmer is presented with a seemingly endless array of choices in which he is able to control the rate of harvest while maximizing returns.

Five of the world's seven major corn types are grown in this country: (1) *dent corn*, the most common form; (2) *flint corn*, a hard, resistant form most commonly seen as the variably colored "Indian" corn near the climatic limits of corn production; (3) *flour corn*, a soft, easily ground type, which is still used by some Amerindian groups in the Southwest; (4) *popcorn*, now almost totally confined to the United States; and (5) *sweet corn*, representing only about one percent of the American crop and largely confined to the United States.

Regional Patterns

While northern commercial grain farmers concentrated on the production of wheat and other small grains during the early nineteenth century, the southern grain farmer focused on corn. Tennessee, Kentucky, and Virginia led the region to produce nearly 50 percent of the nation's crop in 1839. Corn-based foods were so important to the traditional southern diet that they remain as important indicators of southern cuisine to this day. The settlement of the Old Northwest Territory, with its deep prairie soils, shifted production north and westward in the second half of the century. While Tennessee, Kentucky, and Virginia continued to be the leading states in the 1860 census, it should be noted that Ohio production almost equaled that of Virginia. Indiana and Illinois followed Ohio in importance. Middle West production surged during the 1860s, with Illinois more than doubling the corn output of Tennessee that year. The great American "Corn Belt," with its humid, warm summertime climate and organically rich soils, was in formation. Corn could be planted on the newly broken prairie sod where wheat often could not. Production costs were much lower than on the exhausted eastern soils, where lime and imported guano were necessary to increase production.

The years since the Civil War have witnessed a continuing intensification of corn production in the upper Middle West. By 1909 the four Corn Belt states of Illinois, Iowa, Indiana, and Ohio accounted for 43 percent of all corn harvested. By 1993 Nebraska production had surged into becoming the third largest producing state, while corn production in Ohio had declined to less than one half that of Nebraska. The four states of Illinois, Iowa, Nebraska, and Iowa accounted for 58 percent of the nation's 6,344,046,000 bushels produced. The expansion of Nebraska production largely stems from increased irrigated farming along the Platte River in west central Nebraska, which dramatically raised yields. Statewide acreage actually has declined a bit since that time as marginal areas have switched to other crop patterns.

Corn is also an important crop on the Inner Coastal Plain of North Carolina and in southeastern Pennsylvania. Much of the North Carolina crop today is used to fuel the rapidly increasing hog, broilers, and turkey industries of the region. Similarly, the southeastern Pennsylvania output fuels the growth of the local dairy and egg industries. Both are examples of the continuing importance of minimizing transportation costs in the production of smaller corn-fed livestock.

Increasing cotton prices have spurred a resurgence of southern production utilizing the most labor-saving equipment available over the past ten years. (Karla Harvill/Gold Kist)

Cotton

Fifty wild species of cotton, belonging to six groups so different that hybridization between five of them creates sterile progeny, are still found in the tropics and subtropics around the world (Sauer, p. 98).[1] The cotton strain most often used in commercial production, however, is a hybrid between a New World species and an Old World species. It is not known how this cross-fertilization could have occurred between a specie indigenous to Ethiopia and one indigenous to the west coast of South America, although the seeds of both may be submerged in saltwater for long periods without losing their fertility. The earliest remains of this hybrid were unearthed in a fishing settlement located in the northern Chilean desert dating from about 3600 B.C. Additional early sites are found along the South American coast,

primarily in fishing villages where the fiber was used in fishing nets.

Gossypium hirsutum (Mexican and related cottons) was widely spread through the Caribbean and Middle America as a common dooryard and domestic planting when the Spanish arrived in the New World. These New World cottons were taken to Europe with little impact on the cultivation occurring there. Cotton generally was not a practical fiber for large-scale textile use until after the development of a system for economically removing the plant seeds from the bolls. As a result, only about 5 percent of the world's textiles came from cotton prior to Eli Whitney's invention of the cotton gin in 1793. In the United States small amounts of cotton had been

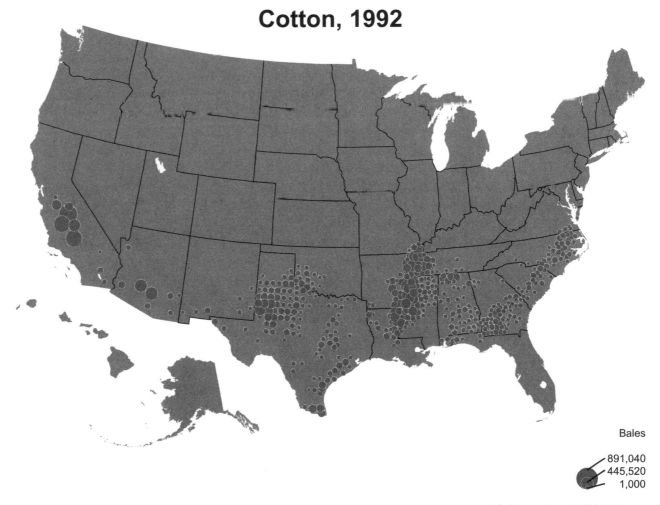

Cotton, 1992

Bales

891,040
445,520
1,000

U.S. Production: 19,728,000 bales

Table III.7

Leading Cotton Producing Counties, 1992

Counties	Production (bales)	Percent of Total
Fresno, CA	891,040	5.5
Kern, CA	655,369	4.0
Kings, CA	627,189	3.9
Pinal, AZ	325,896	2.0
Mississippi, AR	313,224	1.9
Maricopa, AZ	295,992	1.8
Tulare, CA	279,195	1.7
Gaines, TX	252,688	1.6
Dunklin, MO	198,962	1.2
Merced, CA	182,558	1.1
Ten county total	4,022,113	24.8
U.S. production	16,218,500	

planted in the Sea Islands along the southeastern coast prior to that time, but this crop was much less important than the dominant rice culture of the area. The invention of the spinning jenny and other automated textile equipment in the late eighteenth century had dramatically lowered the cost of producing textiles, but no economical raw material for warm weather fabrics had been discovered. The gin not only made the expansion of the cotton textile industry in England and New England feasible, but set the foundations for the rapid expansion of the cotton plantation culture to spread across the American Southeast as well.

The Rise and Fall of King Cotton in the South

Cotton production expanded rapidly from a few thousand bales prior to the invention of the cotton gin to 100 million pounds by 1815. The American harvest reached a billion pounds by 1848, and 2.2 billion pounds by the beginning of the Civil War. Increasing production was tied to the rapid expansion of the western frontier, as a significant amount of southern farmland has always been kept in woodlands, and continued plantings of other crops, especially corn.

Production was concentrated along the Atlantic Piedmont southward from the North Carolina border to Macon, Georgia, as late as 1820. Smaller centers were also found at that time in central Alabama, the Nashville Basin, and in the Tennessee Valley near the Alabama/Tennessee border. About 10,000 bales were grown in the Mississippi Delta, largely in the greater Natchez area. The large holdings and flat fields of the

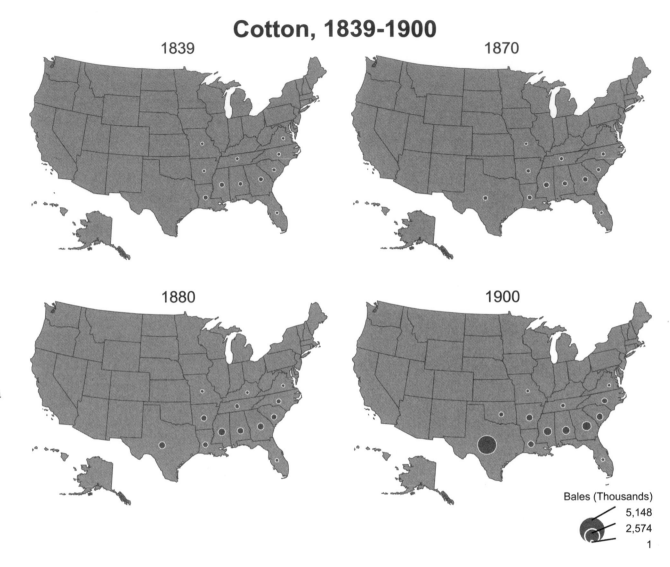

Cotton, 1839-1900

1839   1870

1880   1900

Bales (Thousands)
5,148
2,574
1

Delta served as a magnet for the crop and by 1830 it had become the largest center of production in the nation, followed by the Georgia–South Carolina Piedmont. The crop continued its westward march in the 1840s with the development of the Alabama Black Belt, rising dominance of the Mississippi Delta, and increasing presence in Texas. The Mississippi Delta and Alabama Black Belts were the two largest centers of cultivation at the outset of the

Civil War. The Georgia–South Carolina Piedmont continued to be important as well, while notable expansion occurred on the Inner Coastal Plain of North Carolina, in the Tennessee River Valley of south-central Tennessee, and westward across Louisiana into Texas.

The end of the Civil War brought a massive reorganization of cotton cultivation. Cultivation was labor intensive and had long been based on the heavy use of

Cotton, 1930-1994

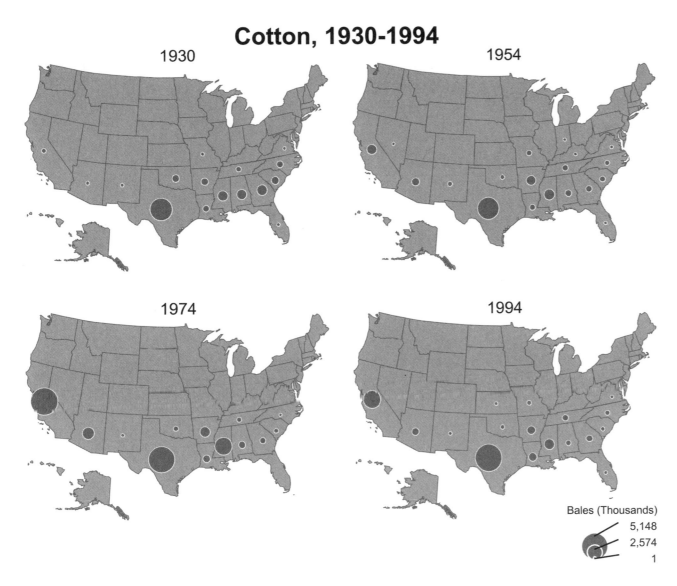

1930

1954

1974

1994

Bales (Thousands)
5,148
2,574
1

presence of a previously unknown weevil that destroyed the unpicked cotton bolls. By 1908 the boll weevil had ravished crops as far northward as Arkansas and crossed the Mississippi River into the Delta. By 1913 the Mississippi crop had been cut by a third. The weevil was found in Georgia in 1916. This new pest was controllable through diligent work by the farmers in spraying and catching the eggs in the larval stage. While the boll weevil did not preclude the continuation of cotton production in the South, and large quantities of cotton continued to be produced in the region for several more decades, it did severely alter cultivation in the region for several decades.

Agricultural historians generally attribute the decline of cotton in the traditional Cotton Belt not to the boll weevil, but to four other factors: the Franklin Roosevelt administration's Agricultural Adjustment Administration (AAA), competition from low-cost western cotton producers, declining prices, and the introduction of other crops that brought higher overall cash returns. Cotton prices peaked during World War I at 35.2 cents per pound. Prices began dropping soon after the end of

slave labor. Ultimately, large holdings were subdivided and land rented to tenant and sharecropper cultivators, who either paid the owner a set fee or gave a share of the crop. Numerous abuses of both systems occurred during the late nineteenth century, especially during periods of economic stress brought on by periodic overproduction and falling prices. Tenant farming and large-scale land rental continued, however, until the 1930s when new forces appeared to

change the role of cotton in the southern economy forever. Low and unstable cotton prices lowered sales incomes occasionally below the cost of production. Western cultivators in Texas, California, and Arizona were able to produce the crop for as little as a third of the cost of producers in eastern states such as North Carolina.

Unrecognized at the time, but pivotal for southeastern farmers, was the report by a Corpus Christi, Texas, farmer of the

Cotton: 1840-1994

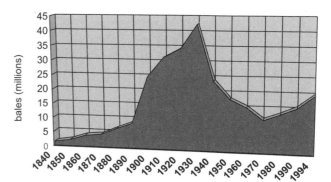

the war ultimately to bottom at 4.6 cents per pound in June 1932. Cotton producer earnings dropped from $1.5 billion in 1932 (representing sales from the 1931 season harvest) to $500 million in 1933, less than the cost of production for many Cotton Belt farmers. This lost income immediately triggered a round of farm bankruptcies throughout the region, especially among farmers with smaller holdings that had become overextended previously with the overall decline of prices that had been taking place over the past decade.

Insurance companies, the most common lenders in the region, acquired thousands of farms throughout the Depression, displacing thousands of farm families in the process. Depressed land prices encouraged the insurance companies to hold these lands until the late 1930s when land views had returned to higher levels. It is not known how many farms were affected, although records suggest that the twenty-six largest insurance companies owned $535 million of southern farmland in 1938, most thought to have been obtained after 1929. These companies efficiently utilized the AAA system to receive large government payments for their cooperation with the program. This, in turn, meant that even more of the AAA money, ostensibly targeted to assist the region's smallest farmers, never reached that group.

Western cultivators in Texas, California, and Arizona were able to produce the crop for a third of the cost of producers in eastern states. Larger operations, a higher degree of mechanization that lowered labor costs, and a drier climate that made pest control less costly all contributed to their cost advantages. As these cultivators thrived, southeastern farmers continued to

slip toward economic ruination, especially after the arrival of the boll weevil brought markedly increased production costs.

The AAA, though only in existence for a few years before being declared unconstitutional, generally is thought to have played the single most important role in destroying the Cotton Kingdom of the traditional South. The AAA attempted to protect farmers by setting acreage limits on production and paying farmers for unrealized harvests. Though altruistic in origin, the impact of this act on the traditional order of business was devastating. Large farm operators who agreed to decrease planted acreages needed to employ fewer tenants and sharecroppers to till their land. Thousands of tenant farmer contracts were not renewed and the tenants removed from their rented homes. The large landholders then used the leverage of excess labor to negotiate the remaining tenant and sharecropper contracts at lower rates. Simultaneously, tenants, sharecroppers, and owners of small holdings found it difficult to participate in the AAA program because a 20- to 30-percent reduction of their crop would reduce their net income below the level needed for basic family survival.

Finally, the AAA became the single most important force in changing southern agriculture for all time by underwriting the rapid mechanization of the region's agriculture. Once enrolled, a farmer was guaranteed a cash payment for crops not planted, which could, in turn, be used to purchase modern equipment. The cost of this transition was far less than might be imagined. In one example a farmer sold his mules and their associated equipment for $700, which was then applied toward the purchase of a tractor and equipment for $1,017. The net

transaction cost of $317 was put on credit with the new equipment as collateral. The increased efficiency of the new equipment allowed more acres to be planted (in other crops) and lowered labor costs for the cotton that was planted. The new equipment was often paid for with the increased profits in the first year. Unfortunately, only the larger farmers had the resources or credit to undertake the transition to mechanization in this way and the gap between the larger successful farmers and the tenants and owners with small holdings increased.

The mechanization of southern agriculture further reduced labor demand to make the plight of the sharecropper and tenant farmer even more tenuous. While the federal government attempted to convince farmers with larger holdings who participated in the AAA programs to allow their displaced tenants to remain in their homes, most became homeless during this period with no hope of obtaining either a job or shelter of any kind. Thousands of tenant and sharecropper families soon left the region for the northern cities to find work, forever banishing the traditional labor-intensive agriculture long identified with the region.

The newly mechanized large-scale farmer soon began searching for new uses for the lands that had been released from cotton production. Cotton cooperatives, such as Gold Kist, agricultural extension agents, and other agencies encouraged farmers to experiment with a host of new crops such as peanuts, soybeans, livestock, pecans, and broilers. Cotton production began declining in all but a few regions. Enterprise, Alabama, was so ecstatic about the new wealth brought by peanuts that a statue of the boll weevil was erected in the square. Thought dead for all time in the late 1980s, southern cotton began making a reappearance in traditional areas about 1990. North Carolina farmers began planting cotton as a rotation crop with soybeans on the Inner Coastal Plain with little thought of expanding production. Increasing cotton prices, coupled with increased water costs for many western farmers, soon altered the situation and a cotton boom began a few years later. While there is little danger of the Cotton Kingdom rising from the red hills of the South, farmers are once again meeting in the cafes around the square in Enterprise talking about cotton futures and new strategies of this new wonder crop.

Western Cotton Production

Texas and California are the largest cotton producing states in the nation today. Texas production began as an expansion of the Old South into eastern Texas in the midnineteenth century. Production spread into the Llano Estacado of the Texas panhandle and adjacent areas after the development of deep-well irrigation based on the Ogalala aquifer. This region soon became the largest producing area in the nation and continued until the decline of the aquifer began making cotton production more and more difficult. It appears, however, that cotton will ultimately be driven from the Llano Estacado and retreat to its original Texas production core on the black waxy prairies of east and central Texas.

Cotton was introduced into California by the Spanish, with experimentation still being carried on as late as 1808. The first successful crops were planted in the Palo Verde Valley (1895) and by 1910 15,000 acres were being harvested in the Imperial Valley. The Acala variety was introduced in 1916 to farmers in the San Joaquin Valley where it thrived. By 1925 cotton growers had successfully petitioned the state legislature to require that this variety (more resistant to blight brought on by the region's dry climate) would be exclusively cultivated in the state, except in the Imperial and Colorado River valleys. California cotton cultivation continues to be concentrated in the southern San Joaquin and often is the state's single most valuable field crop. Cotton's high water demands, however, have made it increasingly expensive to cultivate in this water deficit region and overall production has been slowly sliding downward over the past few years.

Notes

1. Sauer's discussion is the most accessible, detailed account of early cotton without the usual politics and economic theories. Anyone even casually interested should search out this source as a starting point.

Western dairy operators utilize low cost byproduct feeds to support large feedlot style operations as this one near Visalia, California. (RP/JF)

Dairy Cattle

Milk cows have been an integral part of the agrarian and dietary traditions of Western Europe throughout recorded history. Milch cows were introduced to colonial America during the early seventeenth century, though the animals themselves were often poorly bred. By the 1790 census, the New England states, New York, and Pennsylvania were exporting considerable amounts of butter and cheese, as well as serving home markets. The development of dairy centers away from urban areas was restricted until the development of rapid transport allowed the safe shipment of product with minimal spoilage. For example, the completion of the Erie Canal encouraged the expansion of milk production northward from New York City into the Lake Champlain lowland and westward across upstate New York. New York was the leading dairy state in 1839 with almost a third of the nation's produc-

tion by value, three times that of the less accessible Pennsylvania, the next most important state. The continuing importance of the New York metropolitan market for dairy products can be seen in the establishment of the first "milk trains" from Chester, Orange County, to New York City in 1842. Soon, milk runs became an impor-

tant part of railroad business across New York, southern New England, and Pennsylvania as these slow-moving trains dropped off empty cans and picked up full ones of cream and milk for urban markets from hundreds of stops too small for normal service.

New York dominated dairy production

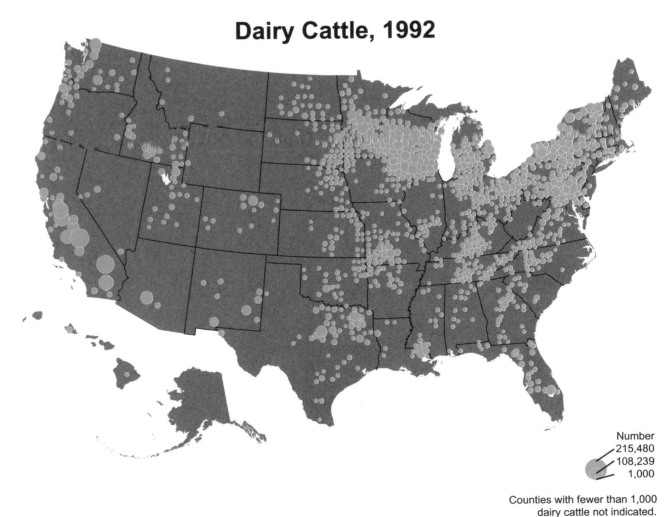

Dairy Cattle, 1992

Number
215,480
108,239
1,000

Counties with fewer than 1,000
dairy cattle not indicated.

U.S. Inventory: 9,491,818

Table III.8
Leading Dairy Cattle Counties, 1992

County	Inventory	Percent of Total
Tulare, CA	215,480	2.30
San Bernardino, CA	191,944	2.00
Merced, CA	144,771	1.50
Stanislaus, CA	137,351	1.50
Riverside, CA	116,799	1.20
Lancaster, PA	92,595	1.00
Kings, CA	86,235	0.90
Maricopa, AZ	79,075	0.80
San Joaquin, CA	76,003	0.80
Marathon, WI	65,892	0.70
Top ten counties	1,278,495	13.47
U.S. total	9,491,818	

Dairy Cows, 1860-1992

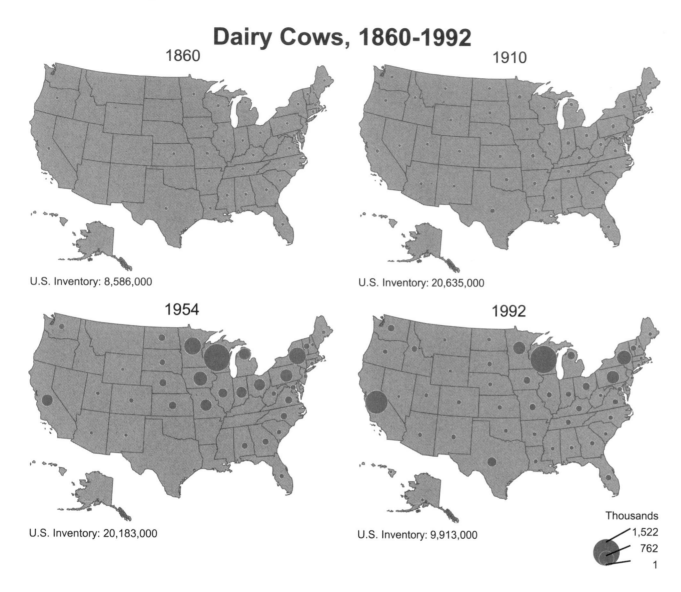

1860

U.S. Inventory: 8,586,000

1910

U.S. Inventory: 20,635,000

1954

U.S. Inventory: 20,183,000

1992

U.S. Inventory: 9,913,000

Thousands
1,522
762
1

profitable farming strategy. Wisconsin became the leading state in numbers of milk cows in 1920 to remain the "Dairy State," in appellation as well as fact, until 1992 when California became the nation's leading center of production. The total number of milk cows in the United States peaked at 27.8 million in 1945 and has since steadily declined. More productive milk cows, however, have continued to bring increases to overall production since that time. Changing demographics finally brought a peak in 1985 to per capita consumption of fluid milk and cream.

The recent growth of dairy herds in the West has been brought about by the phenomenal growth of the California market to more than 31 million people, its isolation from traditional centers of production, and the ability of western dairies to use low-cost agricultural byproducts for feed. Similarly, Oregon, Washington, and Idaho have also sustained growth in herds, while most eastern states have sustained declines. Northwestern producers have tended to concentrate on processed product because of their limited local markets. The growth of Idaho dairy production to seventh in the nation is especially astounding, though not

throughout the remainder of the nineteenth century, though the industry continued to sweep westward across the northern United States just north of what was then the Winter Wheat Belt in the East and the Corn Belt in the Midwest. Winter wheat usually did not survive in these more northerly areas; the corn varieties then available typically did not mature before the first frost; and spring wheat suffered a variety of problems, which significantly cut

returns. The area's cooler weather, however, was well suited for dairying before extensive refrigeration and a Dairy Belt within the Hay and Pasture Belt was largely formed by 1900.

Iowa passed New York in the number of milk cows according to the 1890 census, fell to second in 1900 as other crops became more valuable, and then dropped to third in 1910 as the corn/hog farming regime became established as a far more

Dairy Cattle: 1850-1992

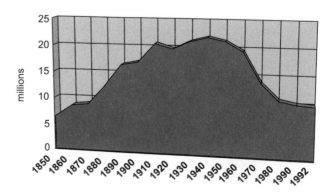

surprising considering the mountains of potato, sugar beet, and processed vegetable byproducts that are produced each year in that state. The vast majority of that state's milk output is converted into American cheese and dehydrated milk.

The transformation of the American dairy farm since World War II is nowhere more evident than among the California dairy operations around Visalia and in the Imperial Valley. The enormous two-floor dairy barn, the icon of eastern dairying for more than a century, has been replaced by a feedlot-style cluster of pens of one or more acres surrounding automated milking parlors. Each pen is provided with low galvanized roofed sheds for shade and automated feeding troughs with collars to hold the cattle's heads when desired. The central milking parlor, the only traditional looking building in the entire facility, is designed to handle a continuous stream of laden milk cows who exit a few minutes after entering, completely emptied by the high volume equipment. Unit costs of these automated facilities are minimized as thousands of dairy cattle are handled with a handful of personnel. Computers track feed prices, individual cattle production cycles, and meat prices as they constantly develop projections about costs, returns, and the handling of individual production units (the cows).

The typical American dairy herd has also been transformed over the past thirty years. While there are still 171,560 dairy farms in the United States (1992), the total number of producers has been declining while average herd sizes have been increasing. California typically has the largest dairy herds today with 97.9 percent of all milk cows being in herds of 100 or more.

Traditional Northlands farms have much smaller herds on the average. Only 17 percent of Minnesota dairy cattle are in herds of 100 or more and only 21 percent of those in Wisconsin.

Dairy Products

The development of an effective dairy products industry is an integral part of the evolution of the dairy farm in American life. Early farmers were hampered by inaccessibility, which meant that only those a few miles from a city could produce fluid milk for sale, while those more distant were forced to rely on the production of less perishable solid products. Commercial butter and cheese production began much earlier than fluid milk in most parts of the country as a result. Creameries were established in isolated areas where they serviced farmers within a fifteen- to twenty-mile radius. Traditionally, farmers provided gravity-separated cream to creameries, which normally was only collected every two to three days. This cream was often partially soured by the time it reached the creameries, lowering its quality. The development of the factory cream separator in 1885 allowed processors to collect milk daily, separate the cream, and return the skim milk to farmers for feeding livestock. The invention of the Babcock (butterfat) tester in 1890 encouraged increased cream production by allowing more accurate payments to farmers for their product, while the invention of the hand-powered cream separator in 1895 reduced spoilage of the skim milk byproduct by allowing farmers again to ship only the cream to processors.

Cheese

Wisconsin claims the nation's first commercial cheese factory. A dairyman from Ohio migrated to southeastern Wisconsin and began making cheese. In 1841, unable to meet the demand for his product, he contracted with neighbors to purchase excess milk to increase his volume. This basic model was elaborated and became the standard for the early industry after a skilled cheesemaker near Rome, New York, contracted with a dealer to deliver his superior product at a much higher price than was generally paid in the area. When demand increased he contracted first with his son, and then with other neighbors to process their milk into cheese for a fee.

Farm cheese virtually disappeared by the twentieth century because its manufacture required more skill and effort than manufacturing butter. While farm cheese production was concentrated in the Northeast, especially far western New York, the Lake Champlain lowland and adjacent New England, and central New York north of Schenectady, the rise of factory cheese production took place primarily in the more isolated northern Midwest. Almost two thirds of the entire national production of cheese was located in Wisconsin (246 million pounds) in 1921, especially along Lake Michigan, in the southwest, and in a broad band west of Green Bay. New York state (65 million pounds) was second with half the remaining national production, concentrated primarily along the St. Lawrence River from Watertown northeastward. Wisconsin remains the leading cheese state with more than 2 billion pounds of annual production (1992), but its market share has dropped to about one third of the nation's output.

Cheese Production, 1860-1992

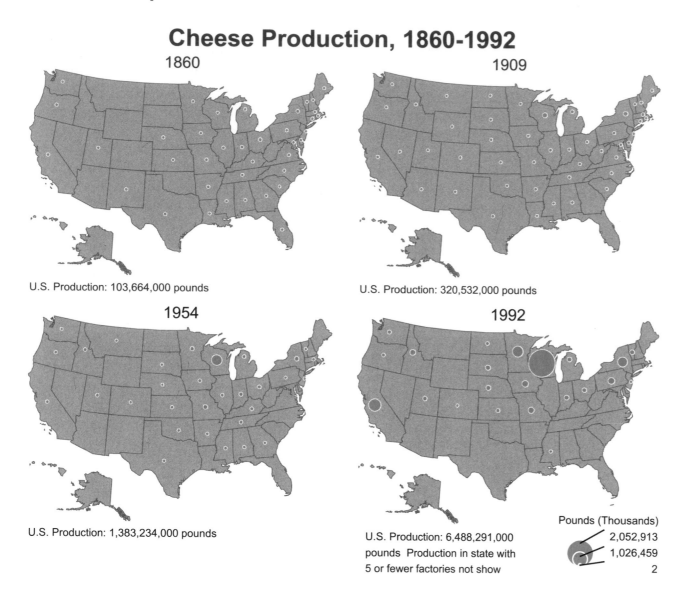

1860

U.S. Production: 103,664,000 pounds

1909

U.S. Production: 320,532,000 pounds

1954

U.S. Production: 1,383,234,000 pounds

1992

U.S. Production: 6,488,291,000 pounds Production in state with 5 or fewer factories not show

Pounds (Thousands)
2,052,913
1,026,459
2

Western states are growing in importance in the cheese industry because of their isolation from national markets and distance from large markets. California is the second most important cheese state with 789 million pounds; Idaho is seventh (primarily American cheese). New York is the third largest producing state, and the remaining large producers are concentrated in the northern Midwest.

Butter

Butter and cheese processing were combined until the late nineteenth century when they were separated for greater efficiencies. Butter production was long plagued by poor quality because of variations in the quality of cream, its pureness, and in problems ensuring its freshness when processed. All butter was sold by individual farmers to retailers until the establishment of the first commercial creamery (butter factory) in Orange County, New York, in 1861. Factory production increased slowly until the 1870s when increasing urbanization and better transportation fostered growth, and then exploded in the 1880s with the invention of the power cream separator. The higher quality factory product forced farm butter out of the urban markets within a decade or two. Factory production continued increasing at the rate of population growth until the Depression lowered demand and the development of better quality inexpensive butter substitutes (oleomargarine) gave the consumer a tolerable alternative. Postwar butter consumption has continued to decline as consumers have become less aware of taste differences, and concerns about saturated fats have spurred finding lower fat content alternatives. Many Americans rarely taste "real" butter today.

The geography of butter production has changed remarkably little over the years. While butter may be shipped farther distances than raw milk, farm production tended to be concentrated in the Northeast and northcentral regions until it disappeared. Only minor production for sale was found on the West Coast and in the South throughout the nineteenth century. The lower prices paid to farmers for butter over fluid milk favored the rise of commercial creameries in the less market-oriented northern Midwest. Minnesota was the leading state in 1921, followed by Wisconsin, Iowa, and Ohio. Commercial creameries were comparatively uncommon in the Northeast, and then only in the most isolated locations. Wisconsin is the leading producing state today (357 million pounds in 1992), followed by California (328 million pounds), and Washington (121 million pounds). The westward migration of butter production has paralleled the rise of the large-scale feedlot-style dairy operations throughout those areas of the West where feed costs are reduced by the availability of large quantities of agricultural byproducts.

Fluid Milk

Fluid milk has always been the core of the dairy industry. Urban milk delivery routes evolved in the nineteenth century with milk ladled out of large cans at each customer's door. The glass milk bottle (1886) introduced more sanitary conditions, while the Babcock method of testing for butterfat ensured quality standards. Pasteurization was introduced in 1895, but public acceptance was quite slow because of the subtle taste changes that accompanied the process. Only one third of the milk sold in Boston and New York was pasteurized as late as 1912, although half of that sold in Chicago and three-quarters sold in Milwaukee were treated.

Difficulties in shipping milk long distances has meant that producers and consumers have never been too distant from the source. The initial expansion of urbanization in the Northeast and later industrial and urban growth of the Midwest along the main east–west rail corridors during the nineteenth century meant that the commercial milk market was concentrated along a line connecting Chicago and New York and production was concentrated in the northern tier of states. The industrialization of the fluid milk industry expanded rapidly after World War I as the major dairy companies began to take shape. The ubiq-

Butter Production, 1859-1992

1859

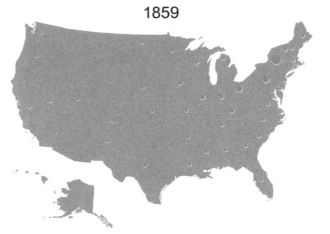

U.S. Production: 459,681,000 pounds

1909

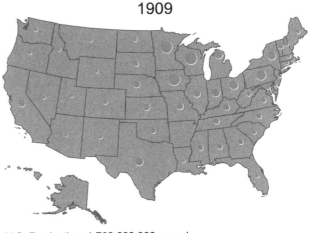

U.S. Production: 1,762,689,000 pounds

1954

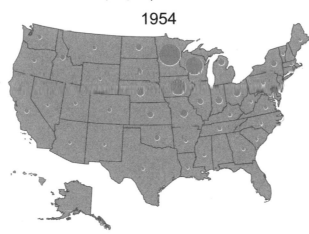

U.S. Production: 1,,327,862,000 pounds

1992

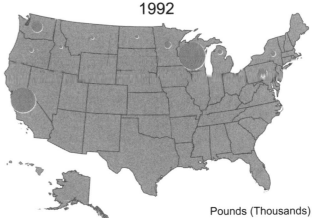

U.S. Production: 2,894,221,000 pounds

Pounds (Thousands)
357,452
178,896
340

uitous nature of the market, coupled with the relatively low entry costs, has meant that even today there are almost 700 fluid milk handlers, down almost a third from a decade ago.

Fluid milk production, distribution, and sales are almost entirely controlled by federal and state agencies today. After almost a decade of overproduction and prices below cost in the teens followed by underproduction during the 1920s, the

federal government attempted to control production and pricing in 1933. The Agricultural Administration Act and associated Commodity Credit Corporation set minimum support prices and attempted to control fluid milk production. Although the AAA was ruled unconstitutional in 1936, most of its provisions continued as parts of other legislation. The federal program works through a system of federal milk orders. The program regulates minimum prices dealers are allowed to pay producers. These prices are based on the average value of the finished product. Farmer payments thus are based on a pooled price for a region or processor rather than the value of the product actually sold. Farmers who enter this program are grouped into milk marketing areas and all product must be sold to handlers within their market area. The number of market areas has varied since the program's inception from a high of eighty-three to the current forty. State programs further control milk prices in an attempt to ensure local production of this perishable and psychologically volatile product. Though programs vary from state to state, more than half regulate prices in some manner. Many regulate retail prices, while others regulate wholesale or producer prices.

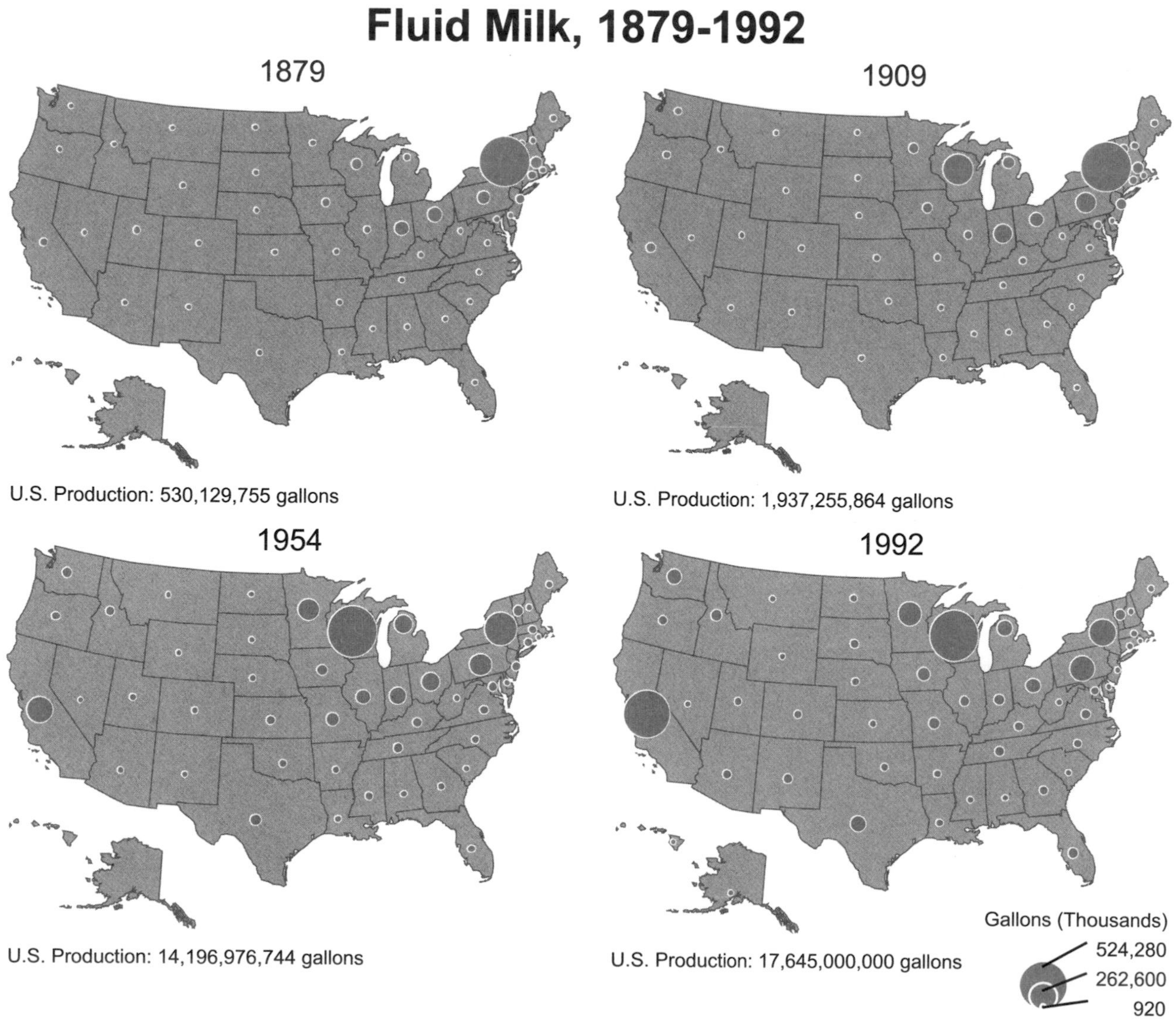

Fluid Milk, 1879-1992

1879

U.S. Production: 530,129,755 gallons

1909

U.S. Production: 1,937,255,864 gallons

1954

U.S. Production: 14,196,976,744 gallons

1992

U.S. Production: 17,645,000,000 gallons

Gallons (Thousands)

524,280
262,600
920

Dry Edible Beans

More than thirty varieties of dry edible beans evolved in the New World prior to European contact and more than a dozen were widely cultivated by Amerindians in North America by the time of first European settlement. The dietary advantages of beans were quickly recognized by the Europeans who spread their consumption and cultivation around the world. More than fifty varieties of beans are grown in the United States today, many the product of intensive breeding programs to make them easier to harvest. Beans generally cannot be grown continuously because of the danger of soil-borne diseases and usually are a part of a three- or four-year crop rotation system in humid areas. They are often rotated with winter wheat in the Far West where they are planted soon after harvest and harvested before the fall wheat planting. Beans can be grown continuously

Dry Edible Beans, 1909-1994

1909

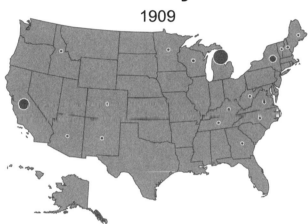

U.S. Acreage: 802,991
U.S. Production: 6,750,696 hundredweight

1954

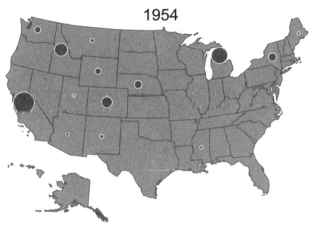

U.S. Acreage: 1,533,000
U.S. Production: 16,639,000 hundredweight

1994

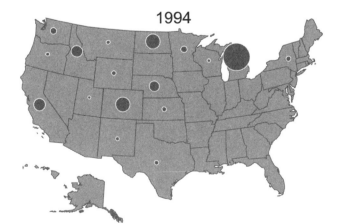

U.S. Acreage: 1,529,900
U.S. Production: 21,842,000 hundredweight

Dry Edible Beans by Variety: 1994

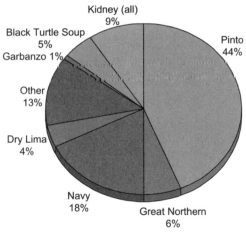

Kidney (all) 9%
Black Turtle Soup 5%
Garbanzo 1%
Pinto 44%
Other 13%
Dry Lima 4%
Navy 18%
Great Northern 6%

Thousand Hundredweight
6,080
3,045
10

Production less than 10,000 hundredweight not indicated.

Dry Edible Beans: 1870-1994

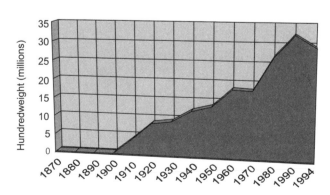

in some parts of the West, although soil erosion is a serious problem.

The distribution of bean varieties is quite distinctive as specific varieties require quite distinct local soil and climatic conditions. Michigan growers primarily concentrate on navy beans while the similar-in-appearance Great Northern is primarily produced in Nebraska. California is most noted for its red kidneys, pinks, blackeye peas, and dry limas. Red kidneys are also grown in New York and Michigan. Pintos are primarily from the Colorado Plateau of Colorado and New Mexico.

About one third of the dry edible bean crop is exported. Pinto beans (41 percent) are the most important export variety followed in importance by navy beans (24 percent) and Great Northerns (8 percent). Mexico imported 138,079 tons in 1990–91, primarily pintos and kidney beans. The United Kingdom is the second largest importer (81,254 tons) followed by the Netherlands (32,327 tons) and Japan (25,068 tons).

The center of dry bean cultivation has steadily moved westward throughout the twentieth century. Three states, Michigan (403,669 acres), California (157,987), and New York (115,698) produced 84 percent of all dry beans in 1909 with only minor production found in New Mexico, Wisconsin, Kentucky, and Maine. California was the leading producer in 1954 with about 27 percent of the entire harvest, primarily concentrated in the northern Central Valley. Michigan continued as the second largest state, with Idaho's Snake River Valley as the third largest production area and the high plains of Colorado coming in fourth. These four states accounted for almost 73 percent of the nation's output. Today, North Dakota is the leading state by acreage, followed by Michigan, New Mexico, Colorado, California, and Idaho. Michigan remains the leading state by value ($99 million), followed by North Dakota $94 million), and California ($88 million).

Table III.9

Leading Dry Edible Bean Producing Counties, 1992

Counties	Hundredweight (Thousands)	Percent of Total
Grand Forks, ND	1,243	5.4
Huron, MI	1,183	5.1
Twin Falls, ID	871	3.8
Tuscola, MI	747	3.2
Trail, ND	689	3.0
Scotts Bluff, NE	684	3.0
Weld, CO	622	2.7
Pembina, ND	600	2.6
Walsh, ND	597	2.6
Steele, ND	501	2.2
Total ten counties	7,736	33.4
U.S. production	23,100	

Dry Edible Beans, 1992

Hundredweight
1,242,725
621,863
1,000

U.S. Production: 23,059,577 hundredweight

Fruits, Berries, and Nuts

A small orchard, a vegetable patch, and a few chickens were all as much a part of the traditional American homestead as the privy and the well. While most fruit was produced on individual farms for home consumption through the Civil War, some American farmers on the urban fringe began to specialize in fruit production for sale as early as the late seventeenth century. Northern New Jersey farmers, for example, were producing more than 300,000 gallons of applejack for the New York market in 1810.

Two factors encouraged the development of concentrations of fruit production away from major urban centers during the midnineteenth century. The first was the invention of commercial scale canning processes, which allowed for large quantities of fruit to be grown without concern about shipping and marketing the produce before it spoiled. Several important fruit districts developed during this period to meet the increasing demands of the canneries. Two of the most interesting were Washington County (Maine) and the Chautauqua (New York/Pennsylvania). Washington County farmers produced small quantities of wild blueberries for fresh shipment and sale until a local entrepreneur devised a method of canning excess produce in 1863. Production escalated and this small area soon dominated the national wild blueberry industry. The Chautauqua district along the southern shore of Lake Erie was a locally important grape area through the midnineteenth century. In 1895, Dr. Thomas Welch left New Jersey in search of a

reliable source of Concord grape juice for his infant company. Stopping for a year at Seneca Falls, Welch moved to Westfield, New York, in 1896 and built the world's first large-scale grape juice plant. The plant processed 300,000 gallons of grape juice in 1897 and more than a million gallons a decade later. Chautauqua catapulted into the major grade district in the nation.

The second factor favoring concentrated production was the development of effective transportation technology for long-distance shipping of fresh fruit. Rail ship-

ments of strawberries from Norfolk, Virginia, to northern markets in 1860 demonstrated the feasibility of early season shipment of fruits. Production for rail shipment to the North rapidly increased in the South after the Civil War, but growers often sustained heavy losses because of delayed trains, spoilage, and overproduction until a more efficient system of rail scheduling and storage could be developed.

The cold storage warehouse was patented in 1858 and this innovation began appearing in most cities for meat and other

Fruits and Nuts, 1909

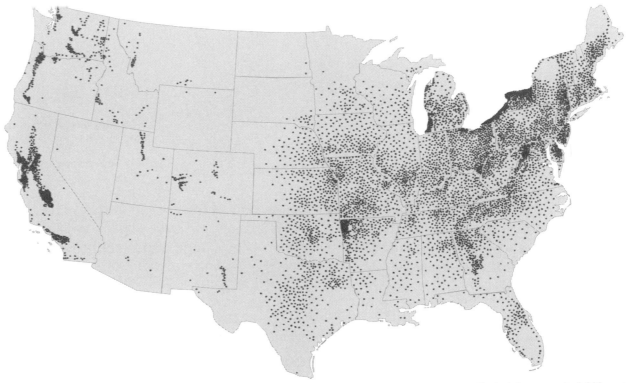

Each dot represents 1,000 acres.

Orchards, 1992

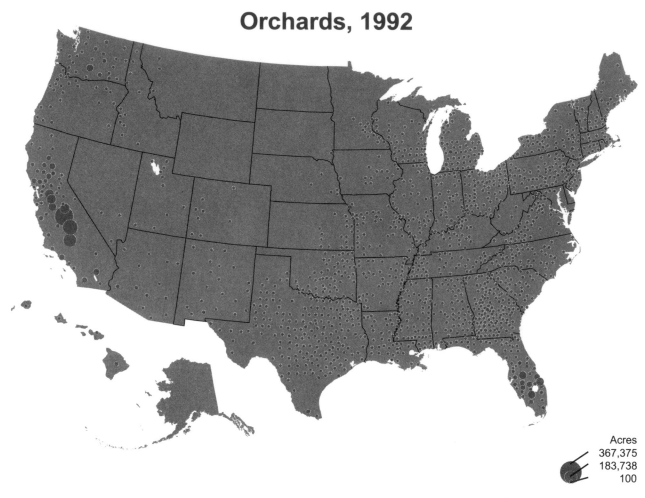

Acres
367,375
183,738
100

Counties with fewer than 100 acres in orchards not indicated.

U.S. Acreage: 4,770,778

Table III.10

Leading Orchard Counties, 1992

County	Acres in Orchards	Percent of Total
Fresno, CA	367,375	7.7
Tulare, CA	282,903	5.9
Kern, CA	236,526	5.0
Madera, CA	151,095	3.2
San Joaquin, CA	148,608	3.1
Stanislaus, CA	142,989	3.0
St. Lucie, FL	119,121	2.5
Merced, CA	117,902	2.5
Polk, FL	113,076	2.4
Hendry, FL	112,031	2.4
Ten leading counties	1,791,626	37.6
Total U.S. acreage	4,770,778	

produce from both the Far West and deep South to northern markets. It is estimated that there were 60,000 refrigerator cars operated by more than fifty companies in service in North America in 1901.

Even a cursory examination of the distribution of fruit production in 1909 suggests four basic geographic patterns of production at that time: (1) a very broad distribution of production, essentially for local sale; (2) the development of a few (sometimes isolated) areas that largely relied on local fruit processing; (3) the

perishables during the 1860s. Although mechanical refrigeration was invented in 1755, it was not until 1881 that Boston became the site of the nation's first mechanically refrigerated warehouse. It is estimated that there were about 150 million cubic feet of refrigerated warehouse space (for all purposes) in the United States by 1901. The development of a system of effective refrigerated rail shipment was more difficult. The first iced

shipments of fruit began in the 1860s, but were plagued with the effects of residual heat of the shipped fruit melting the ice, uneven cooling (freezing some portions of the car and not cooling others), misrouted railcars, and equipment breakdowns. Thomas and Earle of Chicago revolutionized the business in 1887 by developing a private-car line for the shipment of fruit. Starting with six cars, they were operating more than 600 by 1891 for shipment of

Apples, Peaches, Pears: 1890-1993

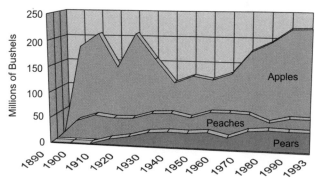

beginnings of some specialized off-season production areas for fresh sale; and (4) clumps of production around each of the larger urban centers. The most striking element of the map, however, is not the areas of concentration, but the continuing broad geography of production.

If decentralization characterized the geography of fruit production in 1909, intense concentration is the best description for today. Three states—California, Washington, and Florida—produced more than 80 percent of the nation's fruits and nuts (by value). California alone produced 54 percent of the nation's crop in 1993. Fruit production each year is increasingly concentrated as the number of processors and shippers, as well as the number of retail grocery distributors and chains, declines and the market share of the largest companies increases. Problems with handling these immense volumes, the need to reduce costs through amortization of fixed expenses over increased shipments, and the development of a worldwide marketplace have convinced distributors that production should be concentrated at fewer and fewer locations for them to remain competitive. Farmers in areas of expanding production and capacity are exhorted to increase their output; farmers in those locations not served by the new generation of processing plants and shippers find it ever more difficult to enter the national and international markets with their produce.

Both specialization and concentration dominate production (see separate crop discussions below). Central and south Florida, the Central Valley of California, the Yakima and associated valleys of central Washington, the Willamette Valley, the lee shores of Lakes Michigan, Ontario, and Erie, and the Shenandoah and Cumberland Valleys are the only major American orchard areas remaining today, and even some of these are in jeopardy. Apple Pie Ridge, near Winchester, Virginia, is increasingly a historic appellation; the "Peach State" (Georgia) has slipped to producing barely 5 percent of national output. Imports, environmental constraints, water problems, and increasing land costs join to make expansion of production in almost all of these areas difficult. Certainly some small niche areas, well located to service specialty markets, are thriving in this competitive world, but each year more of them are sliding into oblivion than are expanding. Environmental conditions may favor or disfavor production, but marketing skills and the rise and fall of the great fruit cooperatives and packers today play a far more important role in the contemporary geography of American fruit production.

Orchard crop production has exploded in the southern San Joaquin Valley as farmers shift to higher return crops in response to rising water costs. (RP/JF)

Apples, 1909-1994

1909

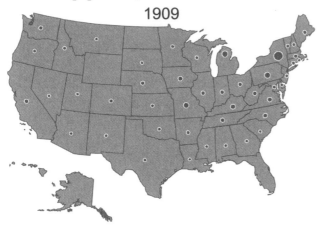

U.S. Production: 145,412,318 bushels

1954

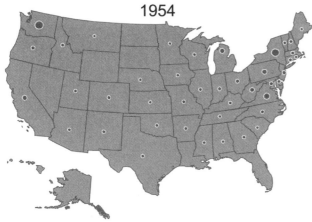

U.S. Production: 109,038,239 bushels

1994

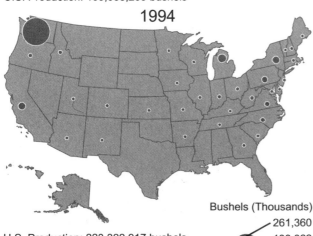

U.S. Production: 223,322,917 bushels

Bushels (Thousands)
- 261,360
- 130,862
- 364

Apples

The apple probably originated in central Asia and was widely distributed throughout central and northern Eurasia by the beginning of recorded history. Our earliest archeological evidence comes from a tenth century B.C. site in Judea, and the fruit was widely grown by the Phoenicians, Greeks, and Romans, as well as northern and central Europeans in later years. The traditional favorites among Americans today—the Northern Spy, Jonathan, and Winesap (all from New York about 1800), the Golden Delicious (West Virginia, *ca.* 1900), the Rome Beauty (Ohio, *ca.* 1850), and the beautiful looking Red Delicious (*ca.* 1880)—were created by plant breeders more than a century ago. The search for the perfect apple—one that not only looks good, but tastes good as well—still continues with some of the most interesting varieties coming into the marketplace during the past thirty years, especially the Fuji (Japan 1962), Granny Smith (created in 1868 in New South Wales, Australia,

Apples, 1992

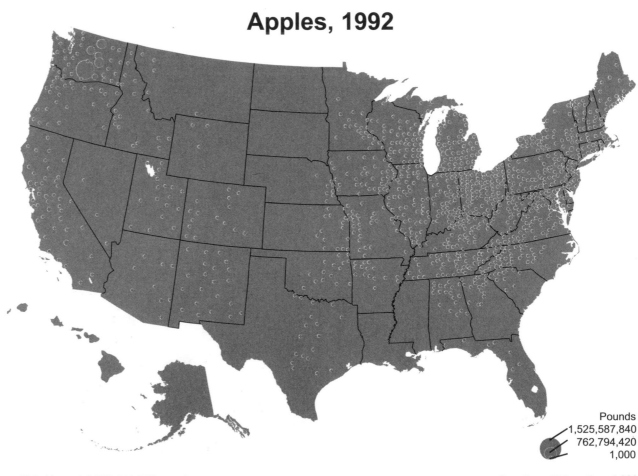

Pounds
- 1,525,587,840
- 762,794,420
- 1,000

Counties with less than 1,000 pounds production not indicated.

U.S. Harvest: 9,720,211,820 pounds
Value of Sales: $1,421,768,000

but not introduced into the United States until the 1960s), Gala (New Zealand 1965), and Braeburn (New Zealand 1952) varieties.

Like all members of the *Rosea* family, successful apple production requires a combination of large numbers of cold degree days (periods during the dormant season when the trees must endure continued cold temperatures to fruit), long summers while the fruit matures, and a bit of cold weather just at harvest time to trigger color mechanisms in the fruit. Production from individual districts, as a result, often varies widely as aberrant weather conditions do not produce sufficient cold, sun, moisture, and coolness from year to year. Typically, apple production is concentrated in those areas with the coolest weather possible without disastrous frosts during the flowering period, though California has become a major producer in recent years—concentrated in the cooler uplands where sufficient cold degree days can be found.

The geography of commercial apple production has changed significantly over the past century. Ohio was the most important state in 1890 with 12 percent of total production, followed by Michigan and Kentucky. The combined production of nine states was necessary to reach one half of all production, and eighteen states to equal 75 percent. The role of express trains is evident in the increasing concentration of production as early as 1909 with New York becoming the leading state, followed by Pennsylvania and Michigan. These three states accounted for just more than a third of total production. The center of production had again moved by 1932, with Virginia becoming the leading state and Washington second. The Northeast continued to dominate production after World War II, although Washington's future role was beginning to be evident. Today, Washington alone produces almost 45 percent of the nation's apples and Yakima County almost 16 percent by itself. Michigan's western lakeshore is the second largest production area (9 percent), followed by

Table III.11
Leading Apple Producing Counties, 1992

County	Pounds	Percent of Total
Yakima, WA	1,525,587,872	15.7
Grand, WA	617,849,692	6.4
Okanogan, WA	607,039,324	6.3
Chelan, WA	476,961,326	4.9
Wayne, NY	370,881,071	3.8
Douglas, WA	362,359,603	3.7
Kent, MI	274,421,724	2.8
Benton, WA	272,233,576	2.8
Adams, PA	271,084,252	2.8
Kern, CA	205,944,848	2.1
Ten leading counties	4,984,363,288	51.28
Total U.S. production	9,720,211,820	

Overhead wire trellis systems, such as this one installed over a Yakima Valley orchard of young Fuji apple trees, are increasingly common in the Far West as farmers attempt to increase yields and reduce harvest costs. (RP)

New York's Niagara Peninsula (8 percent), fast-rising California production concentrating on new varieties (8 percent), and eastern Pennsylvania (6 percent). Virginia's famous Shenandoah Valley has slid to less than 3 percent of the national crop. Granny Smith, Fuji, Gala, and Braeburn plantings dominate new acreages in the Yakima Valley (and California) as the poorly flavored Red Delicious variety, long the state's trademark fruit, begins to decline in importance. Eastern producers, in contrast, have generally continued production of the traditional favorites—McIntosh, Northern Spy, and Delicious—which bring lower prices in the fresh market.

Almost half of all apples are processed into canned (13 percent), dried (3 percent), frozen (3 percent), and cider and juice (24 percent) product. Michigan has the highest percentage of its crop processed (70 percent), followed by Virginia (68 percent), West Virginia (66 percent), Pennsylvania (62 percent), and New York (61 percent). Washington dominates juice and cider production (primarily with fruit below the region's self-imposed "beauty" standards), even though its processed product represents only 21 percent of the state's apple production. New York, Michigan, and Pennsylvania produce the most canned product and Michigan dominates the frozen market. Although California is now the fourth largest producer of apples, its strong local market and orientation toward organic natural product creates high prices and a strong demand for its fresh produce. Only a relatively small processed apple market has developed as a result, most notably the Martinelli brand of apple juice.

Exports are an increasingly important element of apple production, almost doubling between 1977 (142,360 tons fresh) and 1990 (371,309 tons fresh). Dried exports increased almost sevenfold during the same period. Canada, Taiwan, Hong Kong, and the United Kingdom were the principal destinations of the export crop. Extended storage technology has made fresh American apples available throughout the year, though the minor deterioration of crispness has underlain the development of the significant off-season market of imported fruit. Imports of fresh fruit have increased at about the same rate as exports (63,647 tons to 121,996 tons) during the

1977 to 1990 period as a result, primarily from Canada, Chile, and New Zealand. The recent end of the South African embargo will prompt renewed imports from that nation as well. Apple juice is imported in substantial quantities from Mexico. The passage of the NAFTA agreements promises to create even greater shifts in the off-season apple market.

Peaches

Peaches are the second most important of the deciduous fruits in the United States.

Peaches, 1992

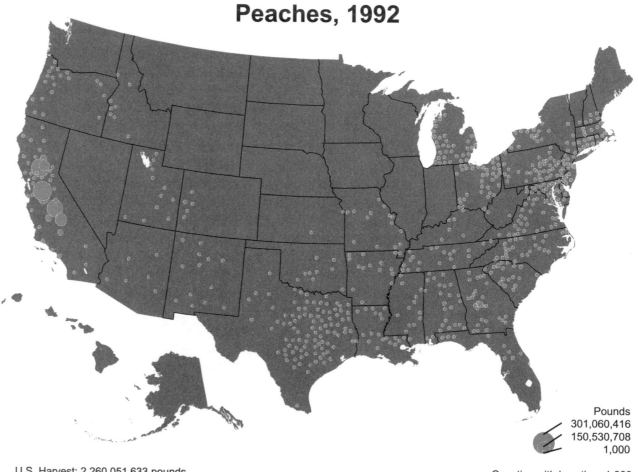

Pounds
301,060,416
150,530,708
1,000

U.S. Harvest: 2,260,051,633 pounds
Value of Sales: $372,787,000

Counties with less than 1,000
pounds production not indicated.

Their sweet flavor and geographic adaptability meant that peaches spread rapidly after their introduction; their fragility and poor shipability meant that fresh fruit production has traditionally been diversified.

Originating in China, the peach was brought to Europe about two thousand years ago. It was quickly adopted by the Amerindian population after it was brought to North America by early Spanish and French settlers. The fruit was so common among the Amerindians that mideighteenth-century naturalist John Bartram thought that the peach might be indigenous to the New World. The introduction in the 1850s of the Chinese cling peach from Britain brought a new era to American peach production. Elberta, J. H. Hale, and other cling varieties were quickly developed in Michigan and spawned a large peach canning industry.

The distribution of peach production has undergone even more change than commercial apple production over the past

Table III.12
Leading Peach Producing Counties, 1992

County	Pounds	Percent of Total
Stanislaus, CA	301,060,404	13.3
Sutter, CA	279,586,312	12.4
Fresno, CA	259,615,049	11.5
Tulare, CA	169,910,494	7.5
Yuba, CA	146,319,400	6.5
Merced, CA	130,202,543	5.8
San Joaquin, CA	73,656,130	3.3
Kings, CA	66,137,129	2.9
Butte, CA	60,090,412	2.7
Peach, GA	52,251,222	2.3
Ten leading counties	1,588,337,591	68.1
Total U.S. production	2,260,051,633	

century. Disastrous freezes in the Northeast virtually wiped out northern peach production through the 1890s, as production was just beginning to become important again by 1909. The dominance of the Northeast was never to return as southeastern producers dominated during the late nineteenth century. The combination of the submersible irrigation pump, the growth of the canned peach market at home and abroad, and rising urban markets brought revolutionary changes in California. By 1910, it had become the leading state with more than triple the production of Georgia, the second most important state. Arkansas, Michigan, and New York followed in importance. California's growing domination of peach production continued unabated, accounting for 56 percent of the nation's production in 1950 and 64 percent in 1993. Peach production in several California counties exceeds the production of every individual eastern state.

Processed fruit continues to play an important role in peach production. About half of the utilized commercial crop is sold fresh, the remainder canned (40 percent), dried (2 percent), frozen (3 percent), and used for such purposes as peach brandy, jams, and preserves. California's Central Valley grows 95 percent of the nation's canned crop, as well as most of the dried and frozen product. Peach preserves and brandy continue to be produced largely in the East.

Export of canned peaches has been an important part of the peach industry for years. Increasing world production, however, has had a significant impact on this trade, which rapidly declined in the early 1980s. Exports today are only about one-quarter of their 1977 levels. Japan, a large

Peaches, 1909-1994

1909

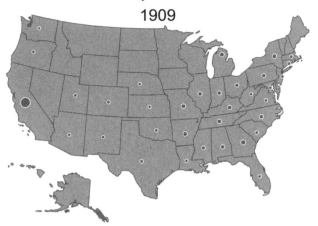

U.S. Production: 35,470,276 bushels

1954

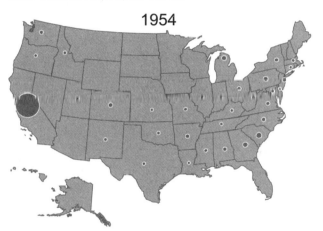

U.S. Production: 61,316,000 bushels

1994

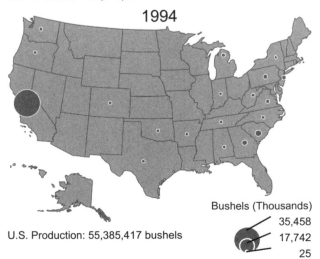

Bushels (Thousands)
35,458
17,742
25

U.S. Production: 55,385,417 bushels

California has recently become the leading pear producing state. (JF/RP)

Pears, 1909-1994

1909

U.S. Production: 8,840,733 bushels

1954

U.S. Production: 9,828,529 bushels

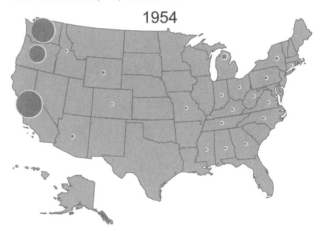

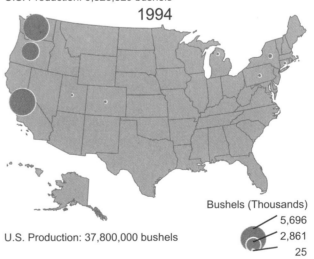

1994

U.S. Production: 37,800,000 bushels

Bushels (Thousands)

5,696
2,861
25

domestic producer of fresh peaches, is the largest importer of American canned peaches, followed by Taiwan and Canada. The fragility of ripe fruit has protected peaches from import competition, though in recent years Chile has shipped high quality, off-season fruit equaling about 4 percent of U.S. production (1993).

The nectarine is a smooth-skinned variant of the peach that first reached prominence in southwestern China and Turkestan. More than 95 percent of commercial production is concentrated in California and almost 85 percent of that in

Fresno and Tulare counties. About 205,000 tons were produced in 1993.

Other Deciduous Fruit

Pears are the third most important commercial deciduous tree fruit with 934,000 tons of production in 1993. The pear was introduced by English and Dutch settlers to the northeastern United States in the seventeenth century, and by the Spanish into California in the late eighteenth century. Contemporary commercial pear production is concentrated along the central

Columbia River and tributaries, most notably the Yakima, the upper Rogue River of southern Oregon, and in central California around Sacramento in the Central Valley and Ukiah in the Russian River Valley. Production has varied widely due to weather conditions, but with little change in overall acreages.

Prunes and plums have been concentrated in California throughout the twentieth century with smaller crops for fresh sale produced in Oregon, Washington, Michigan, and Idaho. Almost 90 percent of the California prune/plum crop is dried. The Santa Clara Valley, south of San Francisco,

Table III.13
Leading Pear Producing Counties, 1992

County	Pounds	Percent of Total
Hood River, OR	281,920,392	15.9
Yakima, WA	267,885,787	15.1
Sacramento, CA	247,529,217	13.9
Chelan, WA	167,920,987	9.4
Jackson, OR	147,397,986	8.3
Lake, CA	144,367,612	8.1
Mendocino, CA	110,400,892	6.2
Yuba, CA	44,973,070	2.5
Okanogan, WA	42,926,958	2.4
Solano, CA	33,405,178	1.9
Ten leading counties	1,488,728,079	83.7
Total U.S. production	1,778,224,532	

Plums and Prunes, 1992

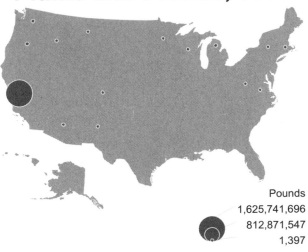

Pounds
1,625,741,696
812,871,547
1,397

U.S. Production: 1,709,000,879 pounds

Pears, 1992

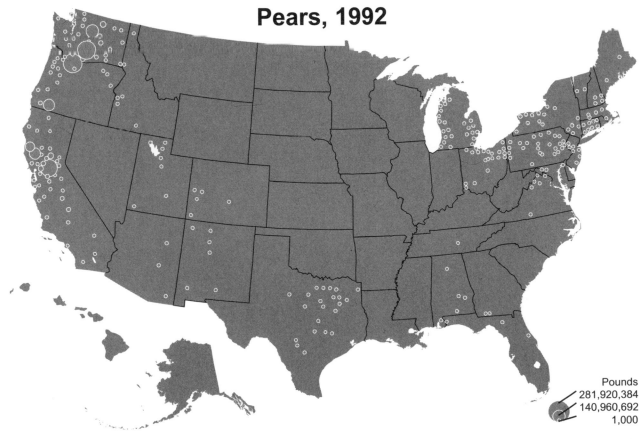

Pounds
281,920,384
140,960,692
1,000

U.S. Harvest: 1,778,229,532 pounds
Value of Sales: $273,188,000

Counties with less than 1,000
pounds production not indicated.

Cherries, 1992

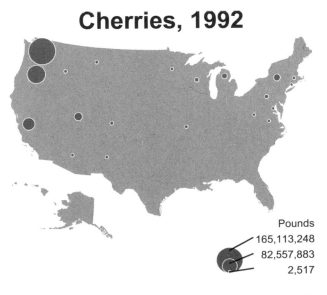

Pounds
165,113,248
82,557,883
2,517

U.S. Production: 669,040,111 pounds

has been the center of production and calls itself the "Prune Capital" of the world. Massive industrial and residential growth in the valley as the area became redubbed "Silicon Valley" sent most farmers northward into the Central Valley between Chico and Sacramento. Both prune and plum production have begun to increase in the southern San Joaquin Valley as well.

Cherries, like the other deciduous fruits, are highly concentrated in a handful of production areas. The western shore of Lake Michigan continues to be the largest center of cherry production in the nation, followed by the Yakima Valley (Washington), the northern Willamette Valley (Oregon), and the Stockton (California) area of the Central Valley. Minor levels of production are also found on the Pleistocene lake terraces north of Salt Lake City (Utah), the southern Lake Ontario shore (New York), the Cumberland Valley of Pennsylvania, and the Door Peninsula of Wisconsin.

Citrus Fruit

Citrus fruit originated in east Asia, yet little is known about this important crop in its early years. The sweet orange reached the Mediterranean by the early fifteenth century. Columbus collected seeds in the Canary Islands in 1493 for transport to Hispaniola. Oranges were introduced to Mexico in 1518 and thrived so well in Middle America and the Caribbean that they were virtual pests in some areas. All the important varieties currently grown in the United States owe their origins to late-nineteenth-century breeding research. Florida was the scene of the first commercial production, but frequent freezes virtu-

ally wiped out the industry several times. Southern California production became important in the late nineteenth century with the advent of economical rail service to the East. Various attempts to organize marketing eventually brought about the development of the Sunkist cooperative in 1909, which represented more than 6,000 California and Arizona growers at one time.

Citrus production has always been highly concentrated in a few states, and the relative position of those states within each of the individual citrus crops has changed little. Florida is the largest grower of oranges (110 million boxes), but a large percentage of the crop is processed into juice. Most of California's 71-million-box production is sold as fresh fruit. Arizona and Texas together account for slightly more than 1 percent of annual production. Grapefruit was first grown in Florida, and the state still dominates (77 percent), followed by California (18 percent), and Arizona (5 percent). Only 65,000 boxes were produced in Texas during the 1991–92 season. All domestic limes, temples, and tangelos are grown in Florida. Almost all lemons are grown in Arizona and California.

Oranges, 1909-1994

1909

U.S. Production: 963,587 tons

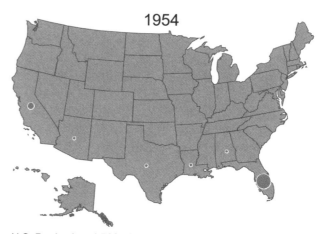

1954

U.S. Production: 1,766,437 tons

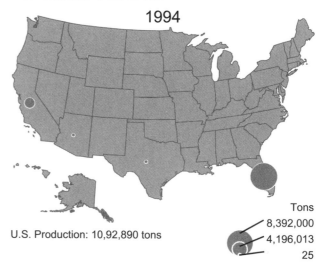

1994

U.S. Production: 10,92,890 tons

Oranges and Grapefruit: 1910-1993

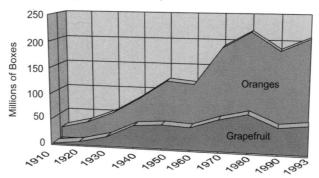

25

Oranges, 1992

Pounds

1,802,564,480
901,284,589
4,697

Counties with less than 4,500
pounds production not indicated.

Grapefruit, 1909-1994

1909

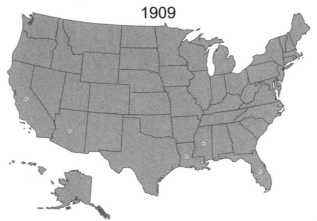

U.S. Production: 44,597 tons

1954

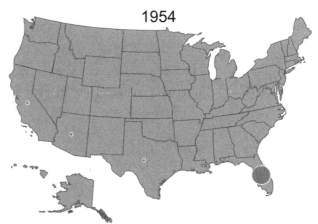

U.S. Production: 1,766,437 tons

1994

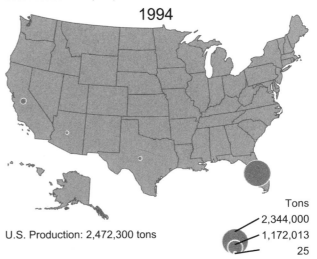

U.S. Production: 2,472,300 tons

Tons

2,344,000
1,172,013
25

Lemons, 1992

Pounds

1,150,602,880
588,492,148
26,381,416

Table III.14

Leading Orange Producing Counties, 1992

County	Pounds	Percent of Total
Tulare, CA	1,802,564,452	10.7
Hendry, FL	1,750,977,466	10.4
Polk, FL	1,426,741,884	8.5
Highlands, FL	1,298,673,238	7.7
De Soto, FL	1,183,806,090	7.0
Hardee, FL	1,038,244,094	6.2
Martin, FL	945,206,354	5.6
St. Lucie, FL	921,262,389	5.5
Kern, CA	693,458,307	4.1
Indian River, FL	589,574,382	3.5
Ten leading counties	11,650,508,656	69.2
Total U.S. production	16,827,363,194	

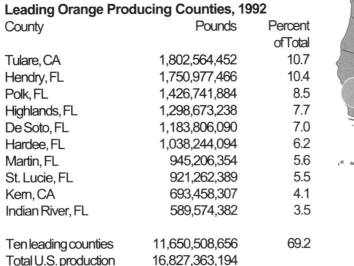

Although virtually all citrus production is found in only three states today, production within those states has undergone significant geographic adjustments over the past fifty years. A series of disastrous freezes has steadily driven Florida production southward. Suburbanization in the Los Angeles basin began driving that state's production into the southern San Joaquin Valley as early as the 1920s, where it has become extremely concentrated in only a few counties. Naval oranges have been grown for many years in northern California, but a combination of freezes and higher prices from specialty citrus fruit has hurt the production of the traditional crop.

The United States' position as the world's largest exporter of citrus products has been surpassed by Brazil, causing more stress in an industry already challenged by environmental problems. Brazil's 13-million-box (1992) annual production has assured its role as the world's largest exporter of both oranges and juice (including significant quantities of juice concentrate to the United States). Japan and Spain are the two largest producers of tangerines with each producing more than five times the American crop. The United States and Italy produce about equal tonnages of lemons, while Argentinian production is growing rapidly. The United States still produces about two thirds of the world's grapefruit, and that is unlikely to change. Orange and grapefruit growers exported more than 843 thousand tons of fresh fruit and 110 million gallons of concentrated juice during the 1990–91 season. The largest markets were in Canada, Japan, and Hong Kong. Significant declines have been experienced in this trade with expanding competition from Brazil and the Mediterranean.

Grapes

American grape production exceeded $2 billion in 1993. Grape production is especially difficult to characterize because the four principal grape products are derived from specific species with unrelated histories and geographies. The recent popularity of wine making has made this picture even more complex as new wineries have brought grape culture to states never before associated with commercial production such as Texas, Louisiana, and Idaho. Bonded wineries have appeared in forty-three states utilizing both local and imported juices from both traditional and nontraditional grape varieties.

The native range of the *Vitis sylvestris* grape stretches from Spain to western Asia. Early farmers modified this grape through cultivation into the *Vitis vinifera,* the species of all important European cultivated grapes. Juice, wine, and fruit have been widely consumed throughout the Western world since the Bronze Age. The Greeks and Romans spread production northward to Britain and westward to Iberia, the Portuguese and Spanish to the Atlantic islands. Columbus carried vines to Hispaniola in 1494.

Several indigenous wild grapes of the *Muscadina* subgenus were widely spread in the eastern United States prior to European contact, though were little cultivated by the native American farmers. The first wine produced in the United States probably was on Parris Island (South Carolina) in 1568. The Virginia Company imported French winemakers and had 10,000 Euro-

Grapefruit, 1992

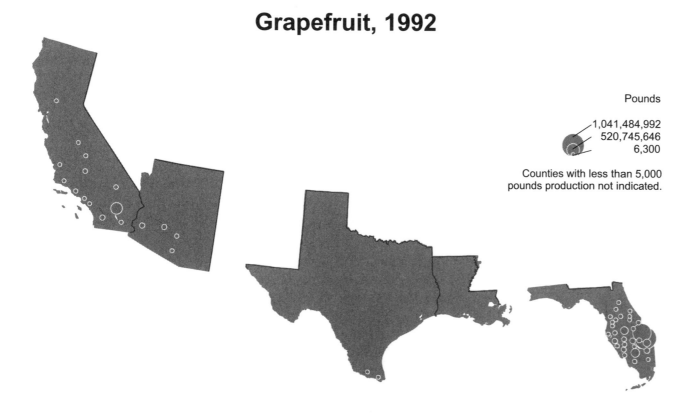

Pounds

1,041,484,992
520,745,646
6,300

Counties with less than 5,000 pounds production not indicated.

Grapes, 1909-1994

1909

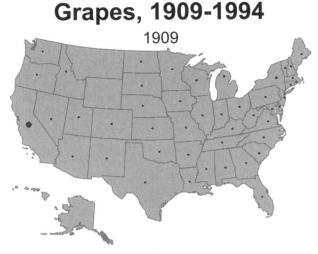

U.S. Production: 2,265,065,205 pounds

1954

U.S. Production: 4,835,850,681 pounds

1994

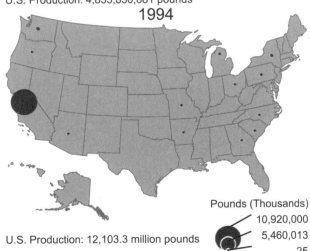

Pounds (Thousands)
10,920,000
5,460,013
25

U.S. Production: 12,103.3 million pounds

Grapes, 1992

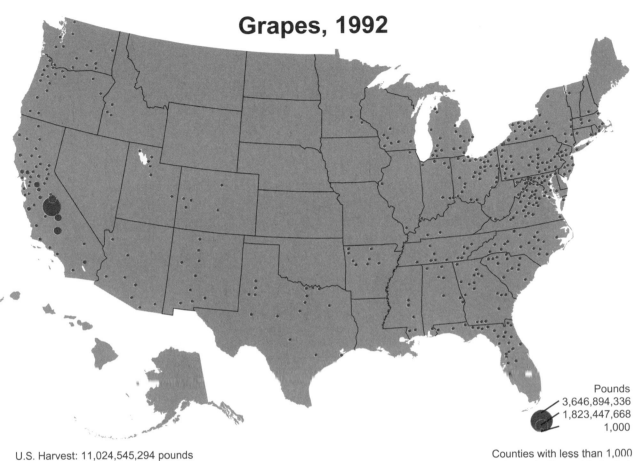

U.S. Harvest: 11,024,545,294 pounds
Value of Sales: $14,825,275

Pounds
3,646,894,336
1,823,447,668
1,000

Counties with less than 1,000
pounds production not indicated.

pean vines planted in 1620 to initiate a native wine industry. The project failed, as did all other experiments over the next 180 years because the European *Vitus vinifera* was susceptible to phylloxera and other diseases endemic to the United States. The development of the Alexander, an American hybrid, set the stage for wine production in this country. Indiana was the first successful wine center (1810), though it was not until the arrival of the Catawba a decade later that production increased significantly. Nicholas Longworth of Cincinnati was producing 75,000 bottles of Catawba wine

Table III.15
Leading Grape Producing Counties, 1992

County	Pounds	Percent of Total
Fresno, CA	3,646,894,341	33.1
Kern, CA	1,144,271,309	10.4
Madera, CA	1,139,562,189	10.3
Tulare, CA	976,561,051	8.9
San Joaquin, CA	675,927,464	6.1
Sonoma, CA	320,339,740	2.9
Napa, CA	313,467,405	2.8
Monterey, CA	313,467,405	2.6
Riverside, CA	242,848,080	2.2
Stanislaus, CA	223,793,653	2.0
Ten leading counties	8,968,060,082	81.4
Total U.S. production	11,024,545,294	

annually by 1852. Scientific grape breeding became popular in the 1850s and a spate of new American hybrids quickly renewed interest in grapes and wine making. The 1860 census noted an annual production of 1.6 million gallons, centered on Cincinnati and the Ohio River, although California and New York were beginning to emerge as challengers. Great wines, however, had to wait until the late nineteenth century when *Vitus vinifera* varieties began to be grafted on American root stocks to produce traditional European wines.

California's rise to preeminence as America's center of wine production was both slow and arduous. Wine production began at the Spanish missions in the eighteenth century, was elaborated by commercial entrepreneurs in the early nineteenth century, expanded by large-scale producers and charlatans in the late nineteenth century, and finally achieved some respect internationally in the twentieth century. Production began in southern California, but inexorably moved northward. The climate and soils of the Napa and Sonoma valleys led to that region's development as the center of high-quality wines in the state. Increasing competition for land and escalating prices ultimately forced expansion of production both northward into the Alexander and Russian River Valleys and southward beyond San Francisco. The intense heat of the Central Valley drew early attempts at vineyards and that area continues as the home of some of the nation's largest table wine producers (Table III.15). Central Valley wines, however, tend not to have the sophistication of those of the cooler coastal valleys.

American wine consumption burgeoned during the 1960s, spawning interest in wine production among those wealthy enough to undertake the enterprise. A rash of boutique wineries sprang up first in the Napa and Sonoma valleys and later throughout much of the nation, which has spread modern wine culture into many new areas. While quality of these early efforts ranged from excellent to awful, they spawned renewed interest in experimentation in vitaculture and wine making throughout the nation. Central Washington has become a small but important producer of high-quality wines in recent years, while the northern Willamette Valley in Oregon appears to be on the verge of a similar transition. Wineries are found and widely publicized in almost every state east of the Mississippi, most notably New York, Virginia, and North Carolina.

Virtually all table grapes, 773,000 tons annually, now come from California. Small amounts of native varieties are also produced in the Chautauqua district of New York/Pennsylvania (Concords and

The growth of the processed strawberry market stimulated the development of a rapid expansion of production and an almost year-around harvest in the Salinas Valley. Note the greenhouses producing cut roses in the background. (JF/RP)

Cranberries, 1992

Hundredweight
1,789,781
969,799
149,816

U.S. Production: 3,959,725 hundredweight

berry was created in France from two American species in about 1750. It is the most important berry in the United States with 712,000 tons harvested in 1993. About 30 percent of the crop is processed and the remainder sold fresh. The geography of production has undergone almost total reorganization in the past seventy-five years. In 1925, for example, the 193,000 acres under cultivation were predominantly centered in twelve areas with harvest timed so that as one area finished production a more northerly one took its place. The most southerly growing areas were in south Florida and southern California, followed by centers in Louisiana and the Carolinas, followed by those more northerly and so forth. Less than 10 percent of the nation's crop was from California. Currently, more than 90 percent of fresh (1993), and 83 percent of processed, fruit is grown in that state's cool coastal valleys, especially the Salinas. This dramatic concentration of production occurred with the introduction of rapid fresh-freezing technology and the subsequent market expansion for frozen strawberries. Leading California production areas, especially the Salinas Valley, were

Catawbas), Arkansas (muscadines), and Georgia (scuppernongs and muscadines). The Chautauqua district remains the largest producer of Concord grape juice, with additional Concord juice coming from the vineyards of western Michigan. Washington state is the largest grape juice producer, primarily from *Vitus vinifera*.

Raisin production is almost totally concentrated in the greater Fresno (California) area. Armenian settlers arriving in 1881 began working at menial jobs, saving to purchase lands for figs and grapes. Letters home brought more immigrants and ultimately this group dominated production of both figs and raisins.

Berries

Strawberries, blueberries, and cranberries are the three most important berry crops in America today. The common garden straw-

Blueberries, 1992

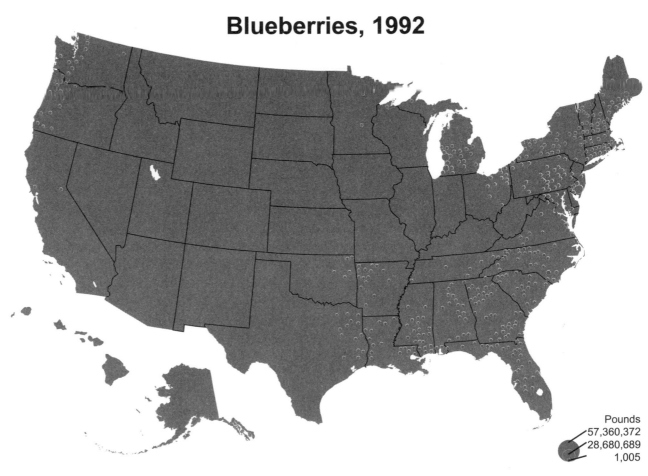

Pounds
57,360,372
28,680,689
1,005

U.S. Tame Blueberry Harvest: 117,941,093 pounds
U.S. Wild Blueberry Harvest: 72,752,596 pounds

Counties with less than 1,000 pounds production not indicated.

Strawberries, 1992

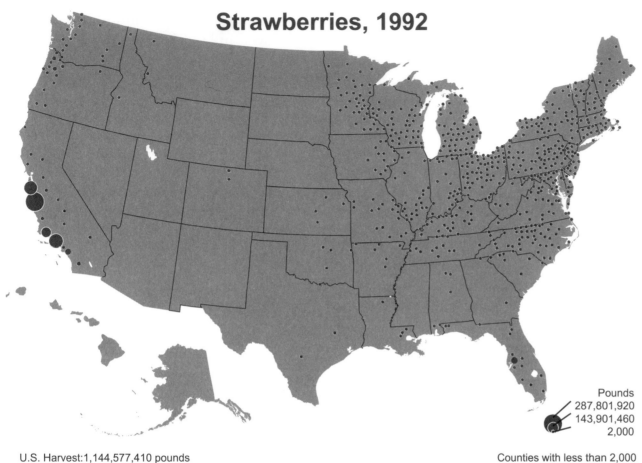

U.S. Harvest:1,144,577,410 pounds
Value of Sales: $684,754,000

Pounds
287,801,920
143,901,460
2,000

Counties with less than 2,000
pounds production not indicated.

Strawberries, 1909-1994

1909

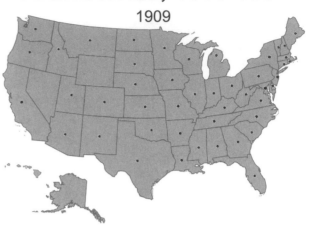

U.S. Production: 383,553,052 pounds

1954

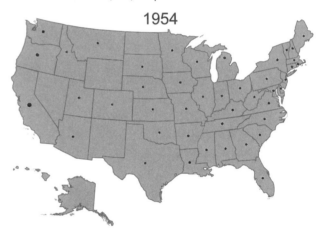

U.S. Production: 309,696,722 pounds

1994

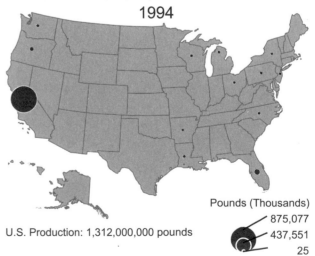

U.S. Production: 1,312,000,000 pounds

Pounds (Thousands)
875,077
437,551
25

Table III.16
Leading Strawberry Producing Counties, 1992

County	Pounds	Percent of Total
Monterey, CA	287,801,905	25.1
Ventura, CA	204,424,135	17.9
Santa Cruz, CA	193,369,485	16.9
Santa Barbara, CA	128,117,716	11.9
Hillsborough, FL	72,676,863	6.4
Orange, CA	56,017,050	4.9
Marion, OR	21,536,752	1.9
San Diego, CA	20,480,150	1.8
Washington, OR	19,350,187	1.7
Santa Clara, CA	9,281,105	0.8
Ten leading counties	1,013,055,346	88.5
Total U.S. production	1,144,577,410	

Almonds, 1992

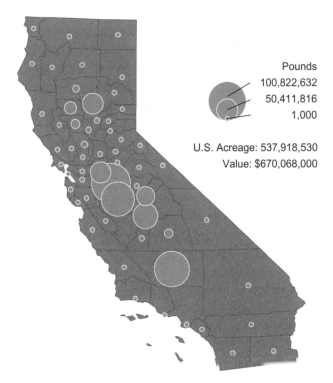

Pounds
100,822,632
50,411,816
1,000

U.S. Acreage: 537,918,530
Value: $670,068,000

Blueberry production (85,000 tons in 1993) is almost as concentrated, with more than 50 percent of the "tame" crop from the Michigan Fruit Belt, and smaller amounts grown in central New Jersey, the northern Willamette, and scattered across the southern Atlantic Coastal Plain. All significant wild blueberries (37,500 tons in 1993) are grown in Washington County, Maine. Cranberry acreages are about equally spread between Massachusetts, New Jersey, and Wisconsin with small amounts also produced near Grey's Harbor, Washington, and Coos Bay, Oregon. Com-

mercial bush berry cultivation, blackberries, raspberries, and the like, was once widely scattered throughout the eastern United States, especially in the northeastern agricultural region. Today production is heavily concentrated in the Pacific Northwest.

Nuts

The three most important nut varieties produced in the United States are almonds, walnuts, and pecans. A record $846 million almond crop was produced in California in

thereafter able to continue producing berries after market demand for their fresh market subsided. Continued production for processed fruit, however, ultimately expanded their fresh fruit "window" because of the reliability of their distribution network and the superior quality of their product. Fresh California strawberries have become a year-round product as a result, with Monterey County alone producing over a quarter of the nation's crop; coupled with adjacent Santa Cruz County and Ventura (near Los Angeles), the top three counties produce 60 percent of the nation's strawberries. Remaining centers are found in central Florida near Tampa, central New York, the Michigan Fruit Belt near the Indiana border, and the northern Willamette Valley.

Pecans, 1992

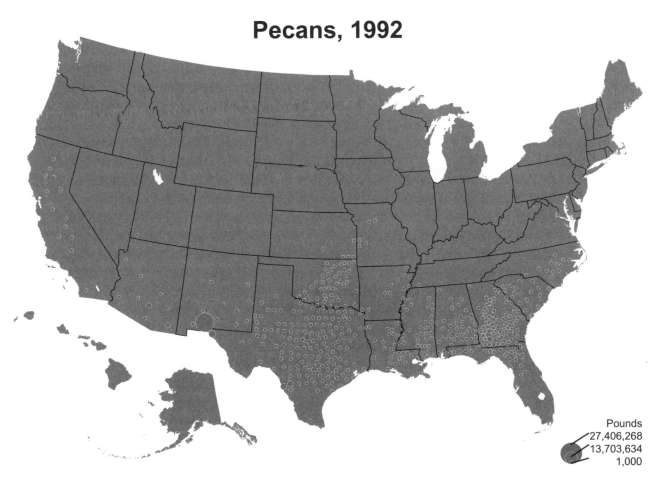

Pounds
27,406,268
13,703,634
1,000

U.S. Harvest: 133,492,969 pounds
Value of Sales: $240,362,000

Counties with less than 1,000 pounds production not indicated.

English Walnuts, 1992

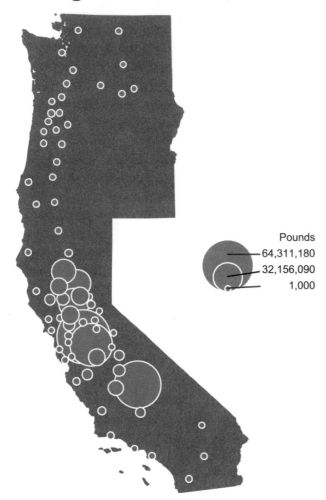

Pounds
—64,311,180
—32,156,090
1,000

tion was a 400 percent increase over 1992. Texas, for years the largest producer of pecans, harvested only about 70 million pounds in 1993, while Louisiana and Arkansas together produced about 50 million pounds.

1993 as a result of higher prices brought on by a smaller crop and increasing demands. Almond production is confined to California, but the center of production has shifted southward from the central Sacramento Valley to Merced and Stanislaus counties over the past thirty years. Virtually all of the nation's $288 million (1992) walnut harvest was from orchards scattered the length of California's Central Valley with minor production in the Willamette Valley. A record 1993 pecan crop lowered prices. Georgia's 130-million-pound produc-

Hay

America's second most important crop after corn is not wheat or soybeans, but the more prosaic hay. Farmers cut hay from nearly 60 million acres in 1992, compared with only about 50 million of wheat and 20 million of soybeans. Hay is any one of a number of grasses, legumes, and other types of plants grown and harvested for the stems and leaves, which are used for animal fodder, primarily for cattle. The most important hays in the United States are the legumes (alfalfa, lespedeza, and several of the clovers) and forage grasses (timothy and orchard grass). Alfalfa today is the most common and productive American hay crop with an annual harvest of 67 million tons. This crop is commonly used as a rotation element in many areas because it increases soil nitrogen levels by "fixing" free atmospheric nitrogen and storing it in nodules on the plant's roots. Alfalfa accounts for 40 percent of total hay acreage and 52 percent of the harvested crop. Not only does this crop have a higher per acre yield, but it also has a greater feed value per ton, which has made it the hay crop of choice where it is grown.

Cattle are ruminants with complex digestive tracts that allow them to digest roughages such as the stems and leaves of grasses and alfalfa. They also demand more protein, relative to energy, than hogs. Forage crops are generally a key component of that high protein diet. The distribution of alfalfa production, thus, is closely associated with the geography of cattle and uncommon elsewhere. As a result, hay dominates cropland acreages of the interior highlands of the Northeast from central Pennsylvania to central Maine; the Northlands dairy areas in northern Michigan and Wisconsin; portions of the Grasslands in Nebraska and South Dakota; and as a common rotation crop in much of the irrigated Ranching and Oasis region of the interior West and the Pacific Industrial regions.

Alfalfa was introduced to the New World by the Spanish in the seventeenth century, but was not introduced to the United States for another two centuries. The first seed apparently was brought around Cape Horn during the California Gold Rush in the 1850s. Ignored initially by California ranchers who grazed their animals on natural pasture, it was adopted

Alfalfa has become the most important hay crop in the United States, especially in the arid west. (RP)

Hay, 1840-1994

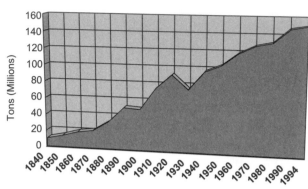

Hay, 1859-1994

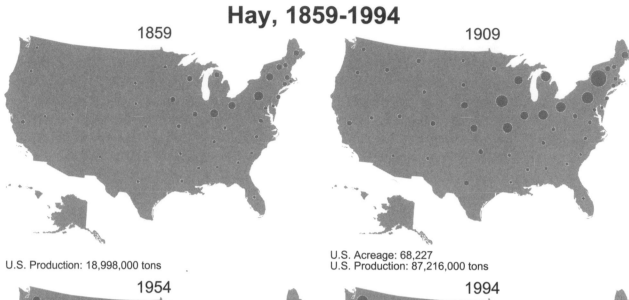

1859

U.S. Production: 18,998,000 tons

1909

U.S. Acreage: 68,227
U.S. Production: 87,216,000 tons

1954

U.S. Acreage: 69,940
U.S. Production: 103,597,000 tons

1994

U.S. Acreage: 60,398
U.S. Production: 148,854,000 tons

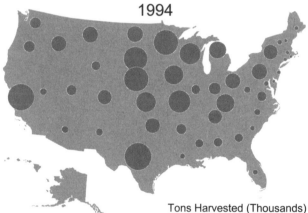

Tons Harvested (Thousands)
8,450
4,226
1

Hay, 1992

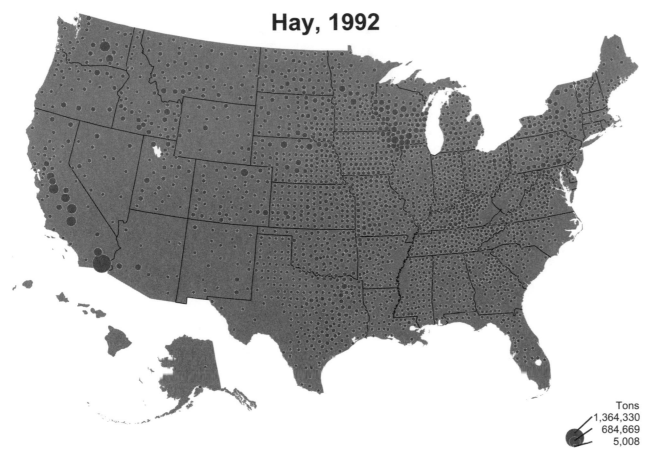

Tons
1,364,330
684,669
5,008

Counties with fewer than 5,000
harvested tons not indicated.

U.S. Harvested Acres: 56,596,466
Total Production: 126,981,302 tons

Table III.17
Leading Hay Producing Counties, 1992

County	Tons	Percent of Total
Imperial, CA	1,364,330	1.07
Grant, WA	686,254	0.54
Merced, CA	649,599	0.51
Kern, CA	611,321	0.48
Tulare, CA	593,890	0.47
Fresno, CA	525,656	0.41
Riverside, CA	470,994	0.37
San Joaquin, CA	448,558	0.35
Weld, CO	437,518	0.34
Franklin, WA	435,599	0.34
Ten county total	6,223,719	4.9
U.S. production	126,981,302	

by rancher and meat packer Henry Miller on his San Joaquin ranch. Once proven, the new crop spread through the irrigated sections of the West Coast and Intermontaine over the following decades. By the end of the century alfalfa was the most common hay crop throughout the West, although it was little known in the East.

Wendelin Grimm, a German migrant, brought a small quantity of hybrid alfalfa with him to Minnesota about the same time the crop was introduced in the West. Generations of careful selection eventually created the Grimm variety, which thrived in the cooler northern Middle West. The synthetic Ranger variety was created about the turn of the century by USDA plant geneticists crossbreeding Asian varieties with the Grimm alfalfa. Ranger allowed the rapid expansion of alfalfa across the northern Middle West and Northeast in the early twentieth century. More recently, hay (commonly alfalfa) cultivation has declined across the Corn Belt as the traditional corn/legume crop rotation was replaced by a corn/soybean cropping pattern. This decline has further been accelerated by the decline of beef cattle as an important component of the agricultural economy in many areas.

In colonial times hay traditionally was cut and stored loose, spurring the construction of the large, picturesque barns with great hay mows throughout the Northeast and Corn Belt. Rectangular hay bales became popular after the development of automated balers in the late nineteenth century, creating this more economical method of transporting and storing large quantities. While eastern farmers utilized their existing barns for bales as well, ranchers in the arid West tended to create open stacks of bales ten to twelve feet high and

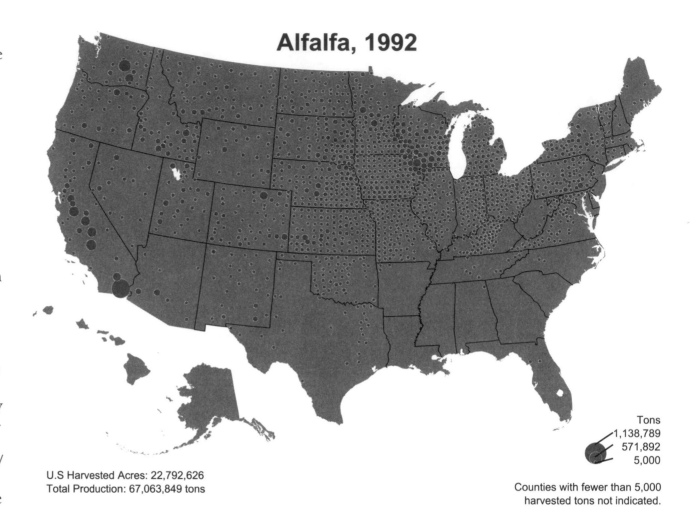

Alfalfa, 1992

U.S Harvested Acres: 22,792,626
Total Production: 67,063,849 tons

Tons
1,138,789
571,892
5,000

Counties with fewer than 5,000
harvested tons not indicated.

as long as fifty yards. Open shelters were provided in areas of moderate rain, though typically these stacks remained uncovered in most areas. Square bales continue to be popular in many areas, especially in the Ranching and Oasis region where flatbed trucks still clog the highways in late summer hauling baled hay hundreds of miles to where these widely traveled herds winter. Round bales began appearing in large numbers in more recent years as farm labor became increasingly expensive and difficult to find. These round bales could be easily

speared with a hydraulic lift mounted on a tractor moved by a single person. They tend to be most common in the eastern United States. Most recently alfalfa cubes have become popular, especially in areas where feedlots are common. Cattle tend to eat as much as 20 percent more of these one-inch pellets than traditional loose alfalfa, thus reaching market weight sooner.

Almost 20 percent of the hay crop today is cut wet, stored in glass-lined, airtight metal silos, and fed as silage to cattle in this moist form. The use of haylage is most common in dairy operations, and the distribution of this crop strongly parallels the number of milk cows. Though initially expensive to institute, this system provides a much more nutritious feed from the same amount of harvested crop. Green cut and chopped corn silage is also increasingly popular in the Northland. Corn silage acreage today is about two thirds that of all hay nationally.

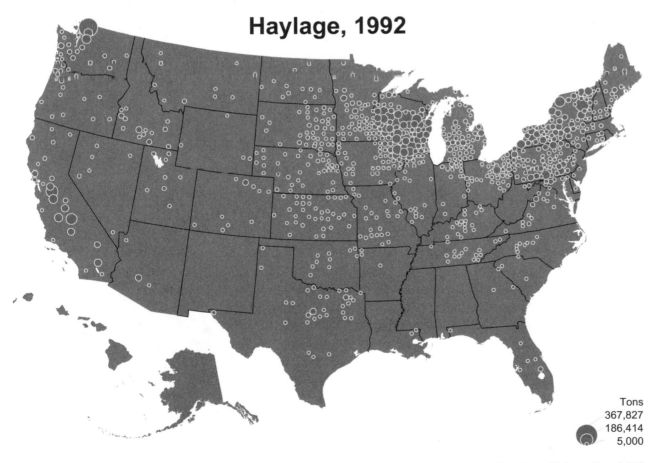

Haylage, 1992

Tons
367,827
186,414
5,000

Counties with fewer than 5,000 harvested tons not indicated.

U.S Harvested Acres: 56,596,466
Total Production: 126,981,302 tons

The adaptation of automated poultry production technology to swine has fostered the expansion of hog production in eastern North Carolina. These piglets will soon be transferred to growout barns holding 10,000 and more animals. (Karla Harvill/Gold Kist)

Hogs

Everything is used but the squeal.

Carl Sandberg

Hog production has long defined the Corn Belt more than any other product. While corn was the most common crop of the region, it was produced primarily to feed hogs. Simply put, the typical Corn Belt farmer raised corn, fed his crop to his hogs, and sold the hogs for cash income. That symbiotic relationship and traditional pattern, however, are coming to an end as the industry is undergoing one of the most massive, short-term transformations in the history of American agriculture. Structurally, the industry is moving from corn/hog farms that marketed from 100 to 130 pigs annually to automated production units capable of handling more than 20,000 animals a year. The overall lack of growth in the industry has meant that the total number of production units has shrunk from more than 750,000 in 1974 to less than 265,000 twenty years later. Geographically, the industry is also dramatically changing. Corn Belt hog production is increasingly centered in Iowa, while farmers in Illinois, Indiana, and Ohio turn to corn and soybean cultivation for sale. Simultaneously, Iowa's role as the nation's largest hog producer is being challenged from outside the region. North Carolina producers have increased production from 3.9 million in 1982, to 10.8 million in 1992, to more than 13 million in 1995. It is estimated that North Carolina sales will surpass those of Iowa by 1997. The factors underlying these changes are a vivid illustration of the basic changes in American agriculture, which are found in almost every section of this atlas.

Swine Breeds

Swine were first domesticated in Asia more than 8,000 years ago, and have been an

Table III.18

Leading Hog Producing Counties, 1992

County	Sales	Percent of Total
Sampson, NC	2,502,748	2.3
Duplin, NC	2,351,744	2.1
Sioux, IA	1,180,718	1.1
Lancaster, PA	828,599	0.7
Delaware, IA	813,394	0.7
Plymouth, IA	748,248	0.7
Greene, NC	728,661	0.7
Washington, IA	709,307	0.6
Dubuque, IA	642,169	0.6
Carroll, IA	609,937	0.6
Ten county total	11,115,552	10.0
U.S. sales	111,376,807	

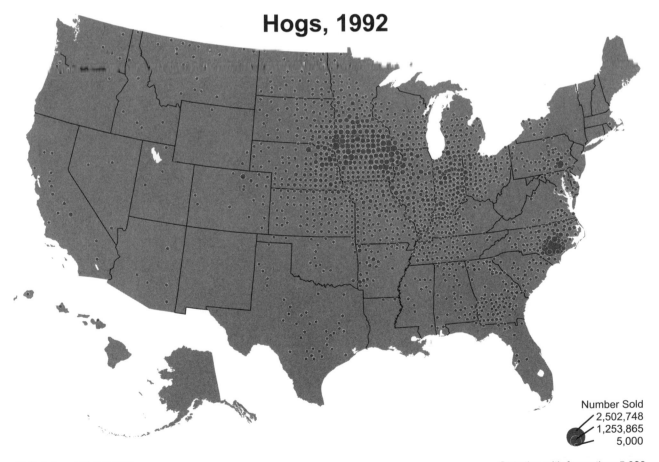

Hogs, 1992

Number Sold
2,502,748
1,253,865
5,000

U.S. Sales: 111,376,807
Value of Sales: $10,148,761

Counties with fewer than 5,000 sales not indicated.

important source of protein and fat in many Asian and European diets. Hogs typically were produced as a sideline to other activities. Little time was spent in attempting to increase production through improved stock until European farmers began crossbreeding the European wild boar, *Sus scrofa*, and the East Indian pig, *S. vittatus*, in the early nineteenth century. Nearly all North American hog breeds today result from these animals.

Most American farmers breed and raise crossbred hogs because these hybrids grow more rapidly and reproduce more prolifically than purebreds. The eight commonly raised breeds in this country are the American Landrace, Berkshire, Chester White, Duroc, Hampshire, Poland China, Spotted Swine, and Yorkshire. All were developed in the United States except for the Berkshire and Yorkshire, which were imported from England in the midnineteenth century. No one breed is greatly superior to another, although Landrace and Yorkshire sows generally produce larger litters. Traditionally, American farmers emphasized lard-type hogs (those with a high percentage of fat) well into the twentieth century. Leaner animals have been sought over the past several decades.

A History of Hogs in America

The first New World hogs were probably imported by Columbus, who kept them on his ships as a source of food during the voyages. Some of these animals escaped when allowed to forage after their New World arrival. Most Spanish land expeditions also drove herds of hogs to provide a supply of food and many of these escaped as well. De Soto's expedition introduced

hogs to the American Southeast where they were quickly adopted for food by local American Indian tribes. By the mideighteenth century "Spanish" hogs were common throughout the woodlands of eastern North America where they thrived on the mast of roots and nuts in the southeastern woodlands. Later European settlers similarly allowed their marked hogs to roam the woods until late fall when the excess were rounded up and butchered as a winter meat supply.

Feral hogs normally revert to the general body configuration of their wild ancestors within a few generations. Thus, the "Razorback" hog of the American frontier woodlands, with its long snout, long legs, sharp back, and roaming disposition, bore a close resemblance to its wild European predecessor. This hog demanded little of its owner, but gave little in return. The animal was big and tough, and could be driven great distances to market, but it also produced little meat.

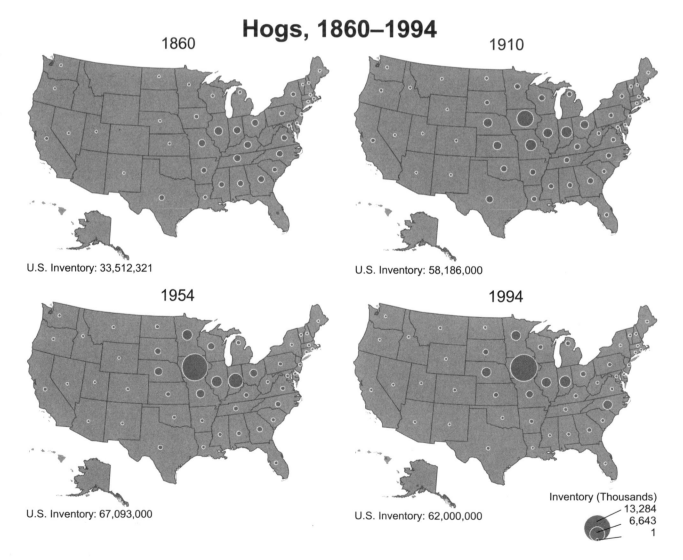

Hogs, 1860–1994

1860
U.S. Inventory: 33,512,321

1910
U.S. Inventory: 58,186,000

1954
U.S. Inventory: 67,093,000

1994
U.S. Inventory: 62,000,000

Inventory (Thousands)
13,284
6,643
1

The earliest interest in breeding improved hogs in the United States was concentrated in the German populations of the lower Delaware River Valley. The Chester White, named after Chester County, Pennsylvania, was one of the first (1818) American breeds. Farmers across the river in New Jersey soon bred the Jersey Red, from which came the Duroc Jersey. Farmers settling the valley lands of southern Ohio brought their hog-raising tradition with them. Union Village Shaker Colony farmers created the Warren County (Ohio) hog by crossbreeding the Big China, Russia, and Byfield breeds in 1816. This new animal gained weight rapidly and could walk great distances. The larger Berkshire was introduced from England in the 1820s, but was a poor walker. Farmers in the Miami Valley (Ohio) crossbred the Warren County hog and the European Berkshire and Irish Grazier breeds to create a better traveling hog in the early 1840s. This new animal soon became the standard of western Ohio, and set the stage for the emergence of the Corn/Hog Belt.

Cincinnati evolved into the nation's hog processing center in the 1830s and soon took great pride in calling itself

Hogs: 1850-1992

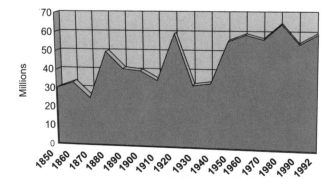

"Porkopolis." The Cincinnati hog packers concentrated on processing lard, rather than meat, from hogs driven to market from across much of western Ohio, eastern Indiana, and neighboring Kentucky, because hog oil was replacing whale oil in American lamps. The city had thirteen factories making lard oil in 1843, with a total output of over 100,000 gallons. Cincinnati's lard industry was the foundation of the city's soap and cosmetic industries in later years.

Hog production shifted westward during the last half of the nineteenth century. By century's end Corn Belt production was concentrated in a zone from western Illinois across Iowa into eastern Nebraska with a smaller outlier in Ohio and Indiana. The concentration of rail lines fanning out across the central and western Corn Belt from Chicago catapulted that city into the nation's leading hog processing center in the 1860s, although other Corn Belt cities closer to the center of hog farming, like St. Paul and Omaha, also developed important processing industries.

Modern Hog Farming

The classic Corn Belt corn/hog farm for more than a century was a relatively small, family establishment with several hundred acres of cropped land providing feed for a hundred or so pigs that were bred and brought to market weight once or twice a year. Change began in the 1970s, when a few growers, initially on the Inner Coastal Plain of eastern North Carolina, began applying poultry production technology and strategies to hog production. The early role of North Carolina in this phase, rather than in other major poultry areas, probably was due to a combination of lax environmental

laws, cheap labor, cheap feed, and large numbers of small tobacco farmers who were facing a declining market. North Carolina hog production nearly quadrupled between 1982 and 1994, when the state leapfrogged from seventh to second among swine producing states. Total income from the state's hog industry more than doubled just during the first half of the 1990s to almost $1.2 billion. In 1995 hogs jumped ahead of both tobacco and poultry to become the state's leading farm industry in annual value of the product. Two neighboring North Carolina counties, Sampson and Duplin, now each have hog sales of more than 1 million animals. Only one other county in the nation, Sioux County, Iowa, produced more than a million hogs for sale in 1992.

The modern hog farm bears a striking resemblance to its poultry predecessor. The process begins in breeding barns where sows are kept in small pens to produce young. Sows are mated or artificially inseminated under close supervision, with computers monitoring the reproductive rates and gestation cycles of each animal. When reproductive performance declines, usually after about two years, the sow is slaughtered. Piglets are weaned after about twenty-one days, and trucked to nursery farms where their initial growth can be more carefully monitored. When the piglets reach about fifty pounds they are taken to finishing farms, where they are fed a carefully controlled diet of corn, soybeans, and supplements. Their ability to develop muscle, rather than fat, peaks in about six months when they are about 240 pounds. The animals are then slaughtered.

The entire operation takes remarkably

little labor. Sow breeding operations are the most labor intensive, and may require eight to ten workers for 2,000 sows and their offspring. The finishing or operation may not need any full-time workers. Although these football-field-length barns hold a thousand or more animals each, the entire operation is almost totally automated. Like poultry, these animals live on a metal grating suspended above a concrete floor. Automated water and feed distribution systems precisely control the amount of food and water delivered. Waste passes through the floor grating and is regularly washed out of the building into a holding pit, or lagoon. This animal waste slurry is then pumped into holding tanks until it can be sprayed onto nearby fields as fertilizer.

Producing units often operate several of these "growout" barns, each costing as much as $200,000 to build. Total farm investment often tops $1 million. In 1984 most North Carolina hogs were raised on farms with inventories of between 500 and 1,999 animals, and a quarter were on farms with fewer than 500 animals. By 1994, 86 percent were on farms with more than 2,000 animals, and fewer than 3 percent were on units with less than 500. The small producer had virtually disappeared in North Carolina.

Contract farming and corporate involvement have become major components of this new industrial structure, although the two largest farms in North Carolina in 1994 were both family operations. Murphy Family Farms, headquartered in Rose Hill, North Carolina, owned 180,000 sows, the largest inventory of a single farm in the nation. Murphy Family Farms also contracted with several hundred independent farmers to produce 3 million hogs annually.

The fact that Murphy Family Farms and Carroll's Foods, in nearby Warsaw, are the nation's two leading hog producers suggests that the distinction between family and corporate farms is becoming increasingly difficult to determine. But even here they stand as the exceptions rather than the rule. Large "protein" corporations are rapidly increasing their share of the industry. The third and fourth largest sow owners in 1994, Tyson Foods and Cargill, are two of the world's largest food conglomerates.

The Future

Per capita pork consumption in the United States declined steadily from just over 75 to about 60 pounds during the 1980s. Concerns about fat consumption and an effective marketing campaign extolling the health virtues of the less expensive poultry protein pushed fowl consumption ahead of pork on the American dinner plate. The retaliatory marketing of pork as "the other white meat" and effective cost containment through more efficient production has not only kept prices stable, but generated a modest increase in consumption during the early 1990s.

The shift to large-scale hog farming, however, brings real costs to both the regions undergoing growth and those left behind. Hogs produce more organic waste daily than humans. North Carolina's Inner Coastal Plain's added waste tonnage has risen in less than a decade to become equivalent to that produced by the entire population of New York City. Many decry both the odor and potential groundwater damage that these large-scale farms introduced into the region.

The Corn Belt family hog farm faces a threat of a different sort. The traditional family farm is a mere vestige of its former self, rapidly dwindling in numbers in the face of corporate competition. Corn Belt farms represent the last substantial region of potentially economically structured family farms in the United States. Contract farming in Iowa accounted for 15 percent of all hogs raised in 1994, compared to 82 percent in North Carolina. While contract farming superficially brings great benefits to the small farmer by guaranteeing a market and often a price for his finished goods, psychologically it also makes him a laborer rather than an entrepreneur. If the contractors decide to shift operations to another area, as they already have done with broilers in North Carolina, the seemingly independent farmer is destroyed with a simple swipe of a pen. The reality of the threat, whether to bring contract farming and its evils to Iowa corn/hog farms or to crush the traditional, independent producer, is seemingly inevitable. Murphy Farms is already a major hog producer in Iowa and has brought its hog raising strategies with it. Some upper middle western states have enacted laws to limit corporate ownership of farms, but their effect can only slow the pace of corporate farm expansion. And if it does grow in Iowa, then it will move farther west. Four North Carolina hog producers recently opened a new 2-million-hog-per-year processing facility in Utah. Change is inevitable; only the geographic focus is in question.

Horses

The horse was the most important mode of land transport at the beginning of the colonial era with a population that peaked at 21,431,000 in 1915.[1] Horse ownership has been declining since that time with fewer 2 million animals remaining in 1993. The nation's transition from horse power to the internal combustion engine had a dramatic influence on the agricultural landscape. More than 20 million acres of farmland devoted to the production of feed for these animals went out of cultivation or changed crops between 1915 and 1925. Production of oats, used primarily as horse fodder, plummeted.

The transition in the role of horses, however, was far from over in 1925, and it was not until the 1960s that the horse finally disappeared from America's farms in Appalachia and other isolated areas with the introduction of maneuverable small tractors and large tillers that could economically work steep hillsides and small plots. Today Amish and some Mennonite farmers are virtually the only farmers who continue to utilize horses for plowing and other draft work. Horses continue to be widely used by western ranchers, but today even the trusty cow pony is beginning to be replaced by helicopters and all-terrain vehicles.

Most horses are now kept either for pleasure riding or show competition. The thoroughbred has become the most important breed, with the fastest used for racing and the remainder for polo, steeplechase, and dressage and other show events. While these equestrian activities may seem relatively minor, thoroughbred sales in the ten leading counties alone totalled almost a quarter of a billion dollars in 1992. Though a relatively unimportant farm product

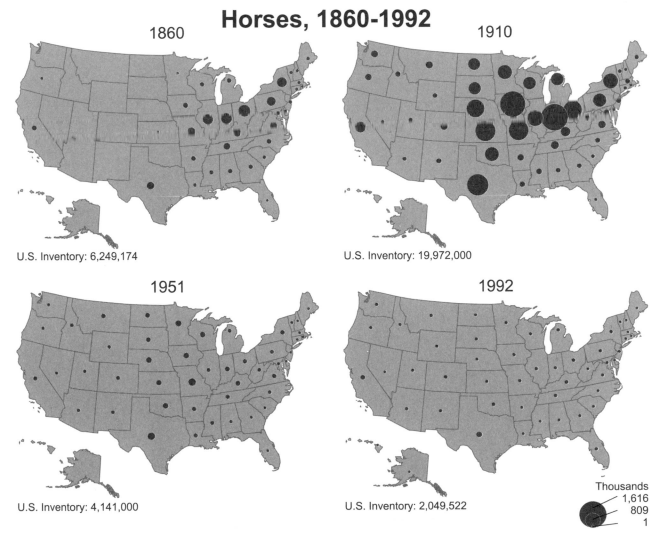

Horses, 1860-1992

1860 — U.S. Inventory: 6,249,174

1910 — U.S. Inventory: 19,972,000

1951 — U.S. Inventory: 4,141,000

1992 — U.S. Inventory: 2,049,522

Thousands
1,616
809
1

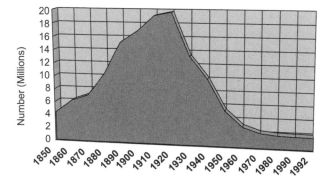

Horses and Ponies: 1850-1992

Number (Millions)

1850 1860 1870 1880 1890 1900 1910 1920 1930 1940 1950 1960 1970 1980 1990 1992

Horses and Ponies, 1992

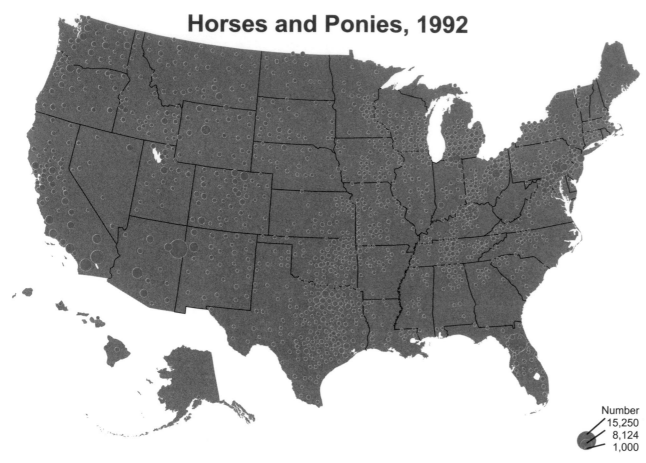

Number
15,250
8,124
1,000

Counties with fewer than 1,000
horses or ponies not indicated.

U.S. Inventory: 2,049,522

all transactions by value. This concentration by value is almost entirely due to the role of the relatively valuable thoroughbred sales, especially those used in racing.

The West today has the most horses. Most of those in the Ranching and Oasis and Grasslands regions are primarily agricultural work animals, although there is a surprising number of nonfarm animals in these areas as well, especially those used in rodeo competitions. The large number enumerated in Apache County, Colorado, is largely associated with local Indian reservations, while those in nearby Maricopa (Phoenix, Arizona) are more likely to be pleasure animals. The large group in central eastern Texas is primarily associated with cattle ranching.

The West Coast distribution of horse ownership is more complex. The bulk of the animals found along the southern California coastal counties are pleasure animals. The coast ranges from Santa Maria southward are home to hundreds of estate horse

perceptually, numerically, and spatially, horse sales in the United States in 1992 reached $647 million. Unlike almost every other agricultural crop discussed here, however, these sales represent not a transfer of an agricultural commodity to a processor, but usually to the beginning of a working career either in agriculture or pleasure competition. About 259,000 horses are slaughtered each year under federal inspection, primarily for dog food. An increasing amount of American horse meat is also exported annually to Europe for human consumption.

The distribution of horse ownership in the United States is interesting in that it simultaneously exhibits one of the most diffuse patterns of any crop mapped here, with virtually all counties having at least some horses raised within their confines; yet the most notable activities are concentrated in a handful of counties. The top ten counties account for only 4.8 percent of all horses counted, one of the lowest concentration indices in contemporary American agriculture. Similarly, the numbers of horses sold are also widely spread, but transactions by value are highly concentrated. The ten top counties in sales, including four in Kentucky, account for more than a third of

Table III.19

Leading Horse and Pony Counties, 1992

Counties	Number	Percent of Total
Marion County, FL	15,250	0.74
Apache, AZ	14,996	0.73
Fayette, KY	11,347	0.55
Riverside, CA	10,316	0.50
Lancaster, PA	9,901	0.48
San Diego, CA	8,768	0.43
Maricopa, AZ	7,529	0.37
LaGrange, IN	6,955	0.34
Los Angeles, CA	6,703	0.33
Woodford, KY	6,559	0.32
Ten county total	98,324	4.80
Total U.S. inventory	2,049,522	

The traditional cowboy and his horse are rapidly being replaced by all-terrain vehicles, pick-up trucks, and aircraft. (RP)

Maria southward are home to hundreds of estate horse ranches, as the region has become one of the nation's leading centers for polo, fox hunts, and competition horse shows. The Palm Desert centering on Palm Springs similarly is the center of western polo. The horses surrounding the San Francisco Bay area also largely represent urban spillover. In contrast, most found in the Central Valley are work animals that winter there (the inventory is taken on January 1) and are moved to upland pastures in conjunction with transhumance

sheep and cattle grazers who similarly winter their herds there. Basically the same pattern is repeated in the areas east of the Cascade Mountains in the Northwest.

Virtually all the horses in the East are pleasure animals, except those kept by the Amish in southeastern Pennsylvania, western Pennsylvania/eastern Ohio, and northern Indiana. While Amish influences may seem small, Lancaster County, Pennsylvania, and LaGrange County, Indiana, had the fifth and eighth highest horse inventories in 1992. The largest center of horse owner-

ship in the eastern states is in a narrow zone extending across the piedmont from Charlottesville, Virginia, northward through Maryland, Pennsylvania, and New Jersey to the southern margins of metropolitan New York City. This is the largest center of polo, steeplechase, equestrian events, and horse shows in the United States with some areas exhibiting estate horse farm landscapes as concentrated as anywhere in the world. Fox hunts are most common in Virginia, but the other equestrian competitions are an integral part of rural life throughout the remainder of this zone. Middleburg, Virginia, continues as the single most prestigious center of this horse culture, but other places are of increasing importance.

Thoroughbred competition and production diffused westward from Virginia to the Kentucky Bluegrass during the nineteenth century after the Virginia legislature banned horse racing. Kentucky became the center of thoroughbred racing and the breeding of race horses during this period and has remained so, psychologically, if not in fact. Several of the nation's most important tracks are found in the region, and many of the nation's most famous thoroughbred stables dot this rolling landscape of green pastures, white or black wooden fences, and white stone and wood barns. The spring sales of thoroughbreds still attract buyers ranging from the royalty of Europe to Middle Eastern sheiks to California film stars. The Nashville Basin became a secondary center soon after the Bluegrass, while the central Hudson Valley (around Saratoga, New York) developed a few years later.

Early industrial entrepreneurs often maintained winter homes in the South and

Horse and Pony Sales, 1992

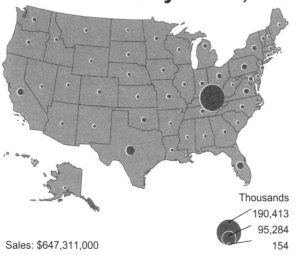

Sales: $647,311,000

Thousands
190,413
95,284
154

through the auction centers. Ocala is the second largest center of sales. While more scattered, the sales in the Virginia/Maryland/Pennsylvania/New Jersey thoroughbred zone actually surpass the Kentucky Bluegrass and Florida in number of animals sold. A zone of sales stretching from Dallas northward into Oklahoma is the largest sales center in the Grasslands, primarily of rodeo stock and working herding animals. The increasing importance of the southern California thoroughbred horse culture is seen in the appearance of both Santa Barbara and Riverside (Palm Desert) counties among the top ten counties in value of sales.

Notes

1. Thousands of horses are kept off farms as pets and are not included in any of the statistics presented here.

several areas became the center of thoroughbred activities as a result. Ocala (Marion County), Florida, is the largest southern center of thoroughbred breeding and was tabulated with the largest inventory of horses on January 1, 1992. Many of these animals move to Kentucky and other northern centers in the summer racing and competition season. The Florida Gold Coast is a secondary racing center, but West Palm Beach is home to several of the nation's top polo teams. Aiken, South Carolina, is a smaller southern center numerically, but only slightly less prestigious than the others. Eastman, Georgia, is the winter training center for a large number of pacers and trotters.

Horse sales further amplify these patterns. The Kentucky Bluegrass spring sales are the center of race horse sales and dominate sales in value because of the young racing thoroughbred hopefuls that pass

Nursery and Greenhouse Crops

Production within the nursery and greenhouse group of crops is growing so quickly that significant data collection did not begin almost until these products reached the billion-dollar annual production level. Known as the horticulture/environmental agricultural group in the 1987 census, the 1992 census used the term nursery and greenhouse crops to apply to these products, which include cut flowers, cut cultivated greens, bedding and garden plants, potted flowering plants, potted foliage plants, bulbs, nursery plants, turf grass (sod), flower and vegetable garden seeds, cut Christmas trees, and unfinished plant materials. This ubiquitous and little noticed complex of activities had sales of $7.7 billion in 1992, making it the nation's ninth most important "crop" behind wheat and ahead of cotton. Production has increased fourfold since 1976, yet this industry has received only minimal interest, research, and financial support from government agencies.

Growth of Interest

Modern environmental gardening had its beginnings in the early nineteenth century when European and American social reformers theorized that gardening was morally uplifting and would help curb the nation's growing materialism. Andrew Jackson Downing, architect and writer, of Newburgh, New York, became one of the gardening movement's most persuasive supporters through his numerous books, articles, and speeches. Gardening was initially promoted by social theorists during the nineteenth century because they believed that it would help bring peace and serenity to the rapacious capitalists of the time. Social reformers soon determined that digging in the dirt would benefit the lower economic classes as well. Numerous public projects, such as the window-gardening program of the Massachusetts Horticultural Society (1876), were promoted to counter vice among the children of the poor. School gardening programs began appearing in the 1890s for much the same reason. The middle class simultaneously was bombarded by speeches, magazine articles in such popular outlets as Godey's *Lady's Book*, and books such as F. J. Scott's *Art of Beautifying Suburban Home Grounds.*

Home landscaping was also made easier during this period by the invention of a number of crucial tools. The all-important lawn mower was invented in England in 1830 by Edwin Budding and exported to the United States in large numbers before the Civil War. The first internal combustion "power" mower was introduced in 1902, interestingly the same year the internal combustion engine tractor went into factory production. Other crucial inventions of the time included "scientific" fertilization, hoses and sprinklers for home irrigation, and the increasing spread of seed catalogs to bring ever more exotic plants to the attention of the home gardener.

The post–World War II era was the fastest period of expansion of market demand. The creation of massive housing developments, the exodus of the urban middle class in search of means to separate themselves from their past, and the evolution of an image that these houses should be set in the center of a perfect meadow all contributed to the industry's phenomenal growth. In many ways the environmental horticulture industry has been a microcosm

Table III.20

Leading Nursery and Greenhouse Crop States, 1992

State	Sales ($ Thousands)	Percent of Total
California	1,661,762	22
Florida	1,024,315	13
Pennsylvania	532,465	7
Oregon	364,343	5
Texas	358,770	5
Michigan	309,521	4
Ohio	288,731	4
Illinois	221,264	3
New York	218,241	3
North Carolina	183,777	2
Ten state total	5,163,189	68
U.S. production	7,634,924,000	

Nursery and Greenhouse Crops, 1992

Sales ($ Thousands)
1,661,762
834,201
6,639

Total Number of Farms: 47,425
U.S. Sales: $7,634,924,000

185

California's central coast is the nation's leading center of nursery and floriculture production. (JF/RP)

Cut Flowers, 1992

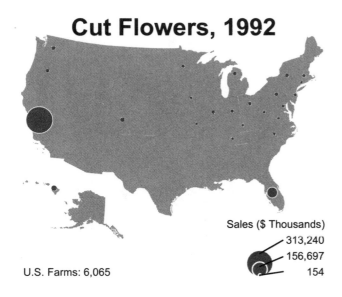

Sales ($ Thousands)
313,240
156,697
154

U.S. Farms: 6,065

Table III.21
Leading Cut Flower Producing States, 1992

	Sales ($ Thousands)	Percent of Total
California	313,240	49
Florida	102,634	16
Hawaii	36,371	6
Colorado	24,365	4
Washington	16,929	3
Michigan	15,999	2
New York	13,989	2
Ohio	12,688	2
Indiana	12,543	2
Pennsylvania	12,315	2
Total ten states	561,073	88
U.S. production	645,104,000	

of American life as suburban America has moved from the manicured "yard of the month" of the 1960s to the "back to nature" movement of the 1980s to automation and beyond today.

The rapid growth of the industry, the increasing dominance of nursery and greenhouse retailing by larger and larger companies with national, rather than local, roots, and the growing importance of the "home and garden" magazine as a source of ideas have brought decreasing tolerance of individuality and regionality of landscaping taste and ideals. This, coupled with the rise of Kmart, Wal-Mart, and other national mass merchandisers as the primary retailers of garden plants, vegetable and flower flats, and indoor plants, has brought an interesting dichotomy of uniformity of landscaping design, accompanied by a remarkable increase in the availability of exotic flora.

The geography of nursery and greenhouse horticulture is becoming increasingly bimodal each year with the industry simultaneously being dominated by massive producers while thousands of small specialty niche producers pop up each year. Farms with annual sales of $5 million accounted for 40 percent of all sales in 1992, while farms with $1 million sales accounted for 59 percent. These farms averaged $160,989 sales in 1992. Despite this increasing concentration, every state has commercial production of nursery and greenhouse products, often showing strong profits through aggressive marketing of highly specialized products. Interestingly, this is the most important agricultural activity in four unlikely states—Alaska, Connecticut, Massachusetts, and Rhode Island—and is the second most important commodity group in California, Florida,

Bedding Plants, 1992

Sales ($ Thousands)
313,240
156,697
154

U.S. Farms: 6,065

Sod, 1992

Sales ($ Thousands)
79,357
43,016
6,674

U.S. Acreage: 218,161
U.S. Sales: $471,640,000

Fields of flowers cultivated for their seeds and corms dot California's south coast. (RP/JF)

Potted Flowering Plants, 1992

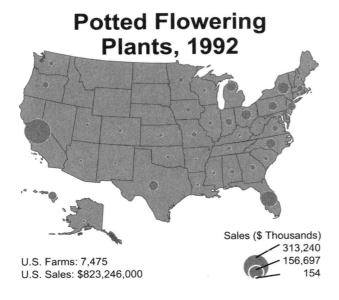

U.S. Farms: 7,475
U.S. Sales: $823,246,000

Sales ($ Thousands)
313,240
156,697
154

Maryland, New Hampshire, New York, and Oregon. While California is the nation's largest producer (22 percent of total production by value), and Florida the second largest state (about one eighth of national production), these levels of geographic concentration are far lower than for most other crops. While some products are highly concentrated, cut flowers and bulbs, bedding plants, nursery plants, and other less specialized products seemingly can be successfully produced commercially anywhere. High transportation costs and the relative fragility and the need for reasonably similar environmental conditions for many plants are the most important factors in maintaining a wide distribution of production. It should be noted, however, that the cut flower market has long been a global one. The Aalsmeer (Netherlands) flower auction, for example, handles more than a billion blooms a year as the world's largest flower market. More than $500 million of nursery and greenhouse crops are imported annually, largely cut flowers,

tropical potted plants, and cut Christmas trees (from Canada). Competition from Latin American growers is rapidly increasing with most cut carnations sold in this country already being imported. Rose production in Columbia is beginning to have a significant impact on U.S. sales and production of cut roses as well.

Distribution

Cut flowers are the most visible and one of the most concentrated elements of the nursery and greenhouse product group. More than one half of the nation's cut flowers are grown in California's coastal valleys between San Francisco and San Diego. The two largest concentrations are found near Watsonville on the edge of Monterey Bay (especially roses) and around Santa Barbara. The area south of Lake Okeechobee in Florida is the second largest center of production, while the third largest concentration surrounds the nation's largest market in New York City.

Traditionally, virtually all bedding and

Foliage Plants, 1992

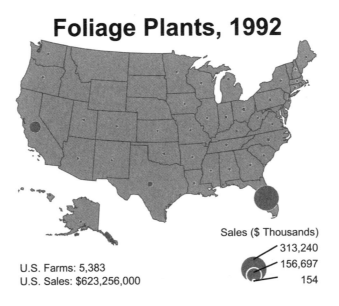

U.S. Farms: 5,383
U.S. Sales: $623,256,000

Sales ($ Thousands)
313,240
156,697
154

Christmas Trees, 1988

Total U.S. Sales: $45,228,052

Units
6,520,303
3,280,502
40,700

garden plants were cultivated in local greenhouses for regional distribution with little attempt at national distribution. The rise and domination of Wal-Mart, Kmart, Home Depot, and other mass retailers as the primary distributors of these products in the last several years, however, has tended to alter that pattern. Larger urban retailers often sell several truckloads a day of these products during the critical planting season. Lacking the time and expertise to develop local contacts, these chains have tended to purchase from national or regional producers or distributors who could guarantee the delivery of large quantities of product at the required times. Several large regional and national distribution networks have evolved, as well as some large producers, such as southern California's Monrovia Nurseries, which attempt to service the entire nation. This process has tended to concentrate production, and increasingly even larger quantities of prosaic bedding plants are highly concentrated in their production for national distribution. Cali fornia and Florida are the leading states in the production of bedding and garden plants and their dominance will most likely increase. The Megalopolis agricultural region currently has the largest area of concentration of these products, because of the size of their local market. It is likely that the most entrenched of the producers in this area will continue through the midterm future, but ultimately the production of "commodity" bedding and ornamental plants will shift to the large-scale producers outside the region. The eastern shore of Lake Michigan is the fourth most important area. These producers often are small family operations interspersed among the apple orchards and grape vineyards along the southern lakeshore near the Indiana border. The growth of bedding and garden plants in Texas partially stems from the state's rapid urban expansion, although several Texas growers are developing national distribution systems for their products.

The weight of sod makes long-distance shipment prohibitive. Most production is concentrated near large urban areas, except for a few exceptionally large producers whose economies of scale make bulk multistate marketing possible. California is the largest producer by value ($79.3 million), followed by Florida ($64 million). All other states have relatively smaller sales and acreages. Florida has almost five times the acreage of any other state, however, but the sale of the vast majority of this output to commercial users keeps the values of sales comparatively low.

Potted flowers and foliage plants are also susceptible to the problems of shipment and traditionally have been characterized by very decentralized production. This pattern is likely to change in the future, and predictably California is already the largest producer of this product. Christmas tree production demonstrates the greatest departure in geography from the other products in this group. Pennsylvania is the largest producing state, followed by Michigan, Oregon, New York, and Washington. North Carolina has been increasingly important in recent years.

Oats

Oats, peas, beans, and barley grow;
Do you, or I, or anyone know
How oats, peas, beans and barley grow
Anonymous

This classic English ditty suggests the long-term centrality of oats in the cropping system of the British Isles and much of the rest of northern Europe. Oats were a standard element of the classic three-field crop rotation of food, fodder, and fallow that characterized British farming and maintained the quality of their agricultural lands of centuries. While they could be ground into oatmeal for human consumption, this occurred primarily in Scotland, according to Dr. Samuel Johnson. More commonly, oats were used as horse feed throughout England.

Oat Origins

Oats are of the genus *Avena* of the grass family (*Gramineae*), which grows best in cool climates. The common cultivated oat (*A. sativa*), thought to be derived from the common wild oat, is the dominant variety of the "Oat Belt" across the northern margin of the Corn-Soy Belt. The Tartarian (side) oat, *A. orientalis*, is second in importance. It is most commonly found at the northern edge of oat range in the United States and in Canada and Scandinavia. The common red oat, *A. byzantina*, grown largely in the Southeast, is usually planted in the fall as a cover crop.

Oat Cultivation in America

Oats were brought to the American colonies primarily as a source for horse feed and were cultivated in every state by the

Oats: 1850-1993

[Graph showing bushels (thousands) on the y-axis ranging from 0 to 1600, and years from 1850 to 1993 on the x-axis]

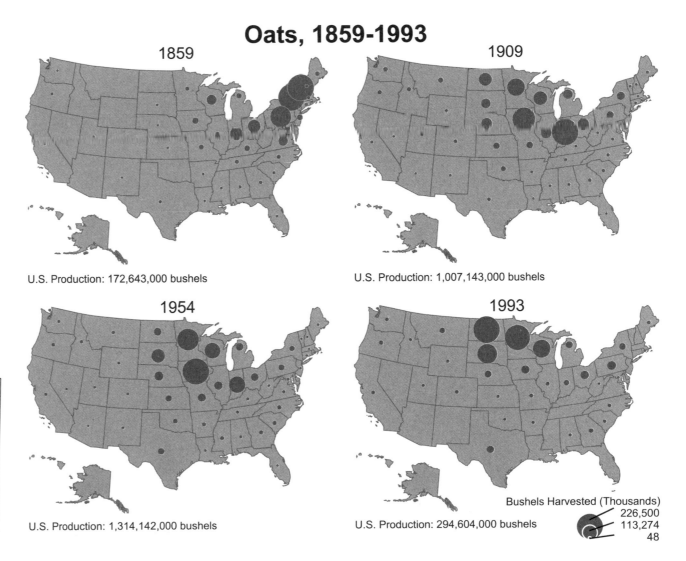

Oats, 1859-1993

1859
U.S. Production: 172,643,000 bushels

1909
U.S. Production: 1,007,143,000 bushels

1954
U.S. Production: 1,314,142,000 bushels

1993
U.S. Production: 294,604,000 bushels

Bushels Harvested (Thousands)
226,500
113,274
48

midnineteenth century with volume increases paralleling the expanding human and horse populations. All that has changed. Acreages planted in oats peaked about 1950 at 45 million acres and by 1970 had declined by half. The continuing decline has been less dramatic, but nevertheless overwhelming. Fewer than 8 million acres of oats were planted in 1993, a sixth of the level of forty years earlier. Output has slid to 300 million bushels annually. In this period of concentrated high-volume production the marginalization of a crop like oats ultimately tolls a death knell for continued cultivation of the crop. The large agricultural seed and machinery companies are unwilling to invest in the research necessary to increase yields and quality of crops that are grown only in small quantities. Already some available chemicals have not received Environmental Protection Agency approval for use because the needed research has not been undertaken. A single floor broker now handles all oat trading at the Chicago Board of Trade.

Much of the decline has been unavoidable. Oats have traditionally been used as horse feed in the United States, and the decreasing horse population has reduced demand for the crop. Oats are high in fiber and protein, but relatively low in energy content, making them unsuitable for hogs, which prefer the higher sugar content of corn. In contrast, the horse's sensitive digestive system is able to extract more completely the energy content of its food, making oats ideal.

Other changes in American agriculture have also contributed to the declining importance of oats. Ideally, many corn farmers could devote some land to oats because they can be planted very early in the spring and harvested by midsummer, well before the corn crop. In the past the farmer would use the midsummer hiatus to harvest oats to maximize machine and labor input. New varieties of corn with staggered maturation periods and eight-row corn picker/combines today allow the farmer to harvest vast quantities of corn over a longer period, which, in turn, has

Oats, 1992

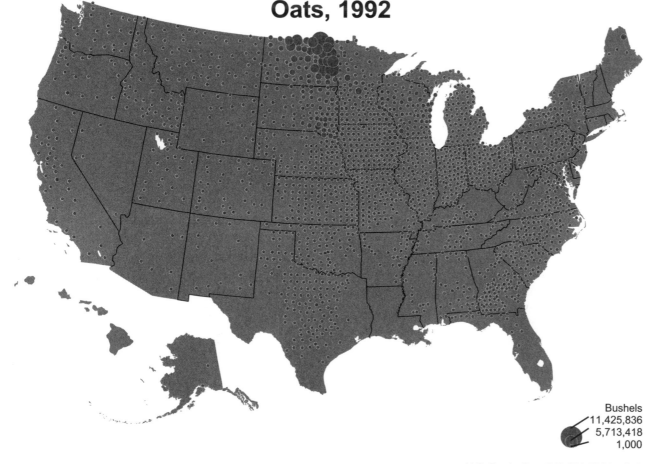

Bushels
11,425,836
5,713,418
1,000

U.S. Production: 249,874,036 bushels

Table III.22
Leading Oats Producing Counties, 1992

Counties	Bushels	Percent of Total
Cavalier, ND	11,425,836	4.6
Barnes, ND	8,467,054	3.4
Grand Forks, ND	8,167,996	3.3
Cass, ND	8,150,557	3.3
Trail, ND	6,223,758	2.5
Bottineau, ND	6,024,854	2.4
Ramsey, ND	5,876,826	2.4
Towner, ND	5,352,859	2.1
Walsh, ND	4,879,592	2.0
Steele, ND	4,834,352	1.9
Total ten counties	69,403,684	27.8
Total U.S. production	249,874,036	

virtually wiped out the traditional summer lull period and obviated the need for oat cultivation.

Oats were also a desirable rotation crop in the past. They represented an undemanding crop that was able to utilize residual nutrition from the previous corn crop. Today crop rotation has largely been replaced in the Corn Belt by increased usage of agricultural chemicals, and overall rotation does not play the role that it once did in the region. Soybeans have largely replaced oats in the rotation cycle, except in areas too cool for soybean cultivation. Oats thus remain primarily a rotation crop in the nation's colder regions, such as upstate New York, and in the Maine potato fields. It is also cultivated as a winter cover crop in parts of the South.

Finally, the structure of the price support system for feed grains does not favor oats. The price support level is established for each grain as a ratio of its energy potential as an animal feed in comparison to corn as a feed. As oats have a low-energy level, their price supports are low. In 1993 only about half of all farmers cultivating oats chose enrollment in the program, compared to almost 90 percent of all corn cultivators.

The oat bran health food phenomenon of the 1980s increased market demand, but had little impact on production. While American consumption increased from approximately 50 million bushels in 1987 to 100 million in 1990 to 109 million in 1993, American farmers were slow to make the shift back to the crop and the increased demand was met primarily with increased imports, primarily from Canada and Scandinavia. The United States has shifted from responsibility for 20 percent of total global exports to 60 percent of all imports in a relatively short amount of time.

The Geography of Production

The earlier widespread distribution of oat production in the country stemmed from the ubiquitous distribution of the horse, the several different oat varieties with different growing habits available, and the ease with which the crop can be planted and harvested. Nevertheless, it was always most especially a northern, cool environment crop, where it was best adapted and most alternatives were least well suited. There were more acres of oats than wheat harvested in the United States in 1850. The large northeastern states dominated oat production at the beginning of the Industrial Revolution. New York production had climbed to over 35 million bushels in 1860, followed by Vermont and Pennsylvania. Midwestern production was significantly lower, though Ohio, Illinois, and Wisconsin followed in size of output.

The evolution of the Corn Belt in the second half of the nineteenth century also promoted the production of oats, an important crop in the five-year rotation cycle widely practiced in the region prior to the introduction of soybeans. In 1910 Illinois was the leading producer with more than 150 million bushels of grain, followed by Iowa (128 million bushels), Minnesota (94 million bushels), and Wisconsin (66 million bushels). The disappearance of the horse as a draft animal drastically lowered the importance of oats in the nation, and the introduction of the soybean restructured the rotation cycle in the Corn Belt during the twentieth century. Today North Dakota is the leading state with only 37 million bushels, followed by Minnesota, South Dakota, and Wisconsin. Pennsylvania, the fourth largest state, has barely one-quarter the production of North Dakota.

The future of oats in American agriculture is unclear, but not bright. The conditions that led to production decline have not changed nor are changes likely. The purchase of oats on the international market is really only a modest issue given the small quantity involved. Oats, with an average yield of about 60 bushels per acre and a low market price, offer little enticement to the farmer. They seem destined to disappear in our national perception as a significant crop just as they already have in reality.

South Georgia is the leading center for cultivation of peanuts for peanut butter. (RP)

Peanuts

The peanut originated in Latin America where four principal varieties flourished at the time of Spanish colonization. Seeds were carried to Africa where several varieties were quickly adapted into the existing agricultural and dietary systems. Even today it continues as one of the region's most distinctive dietary staples. The peanut in Europe was primarily cultivated as a curiosity and an ornamental during the Age of Exploration. Of the varieties important in America, the Virginia variety was commonly used as a provision on seventeenth-century slave ships to the United States. Once in the United States, it spread through the lower South as slaves cultivated peanuts to supplement other food supplies. The Spanish variety was not introduced to the United States until 1871. Initially taken to Africa from Brazil in the seventeenth century, this variety was then carried to Spain in the late eighteenth century. It was introduced to America from France, where it was known as the "Spanish" variety. The Valencia peanut was taken from Argentina to Spain about 1900 and from there to the United States about 1910. Within the United States it is grown only in New Mexico today.

The peanut is the fourth most important source of cooking oil in the world behind soy, cotton, and canola. Americans, in contrast, use the peanut almost entirely as an edible nut. About one half of U.S. production is consumed as peanut butter. Another quarter of annual production is used in salted nuts, about a fifth in candy, and the remainder for peanut butter crackers (2 percent) or crushed for oil. Most peanut butter is made from peanuts grown primarily in the Georgia-Alabama district.

Peanuts, 1992

Pounds (Thousands)
126,198
63,101
5

U.S. Production: 4,085,296,395 pounds

Table III.23
Leading Peanut Producing Counties, 1992

Counties	Pounds (Thousands)	Percent of Total
Worth, GA	126,198	3.1
Houston, AL	106,370	2.6
Henry, AL	106,357	2.6
Gaines, TX	94,732	2.3
Southampton, VA	88,905	2.2
Caddo, OK	86,571	2.1
Early, GA	85,618	2.1
Jackson, FL	85,343	2.1
Comanche, TX	82,978	2.0
Mitchell, GA	77,849	1.9
Ten county total	940,921	23.0
U.S. production	4,085,296	

Peanuts, 1909-1994

1909

U.S. Production: 388,316,000 pounds

1954

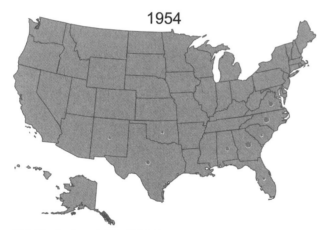

U.S. Production: 1,008,495,000 pounds

1994

U.S. Production: 4,264,550,000 pounds

Pounds (Thousands)
1,864,050
932,040
29

Salted nuts are primarily from the Virginia-North Carolina district.

Peanut production has long been supported by government subsidies, and the geography of production was fossilized with the assignment of acreage allotments. While allotments may be rented, they may not be moved far and the overall distribution has stayed stagnant. Nine states account for 98 percent of American production, though seven additional states receive very small allotments. The Southeast district produces almost two thirds of the annual crop. This growing district spreads across the coastal plain of Georgia as a broad wedge beginning near the South Carolina border and ending in southeastern Alabama, the third largest producing state. Additional acreage is also found in contiguous Florida. The second major zone (18 percent of the annual crop) concentrates on the production of the Virginia variety and is found in a narrow band extending from the Chesapeake Bay into central North Carolina. A loose zone of production extends from near Dallas, Texas, into central Oklahoma, accounting for about 17 percent of national production, mostly in Texas. National peanut production may vary by as much as 20 percent and regional output even more from year to year because of variations in growing conditions.

Peanut Uses, 1993

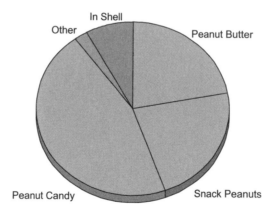

In Shell
Other
Peanut Butter
Peanut Candy
Snack Peanuts

Peanuts: 1910-1994

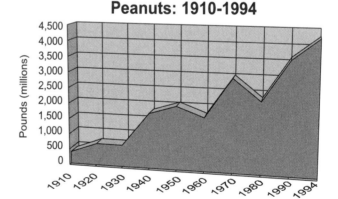

Pounds (millions)

4,500
4,000
3,500
3,000
2,500
2,000
1,500
1,000
500
0

1910 1920 1930 1940 1950 1960 1970 1980 1990 1994

Potatoes

The Irish potato (*Solanum tuberosum*) is grown for market in every state in the Union, although 95 percent of the commercial crop is produced by only 3 percent of the growers. Almost 20 million tons worth more than $2.4 billion were harvested in 1990. Per capita consumption of the potato in the United States has increased tremendously over the past 150 years, especially after the arrival of large numbers of emigrants from Ireland, Germany, and eastern Europe where consumption traditionally was high.

The potato is indigenous to the Andean highlands of Peru and Bolivia. Spanish explorers to Middle America and the Andes encountered potato cultivation and sent samples back to Spain where it was grown as a curiosity (Sauer, 1993). Several attempts were made to introduce the potato as a food crop in the Mediterranean, but these largely failed. The crop also met resistance when it was introduced into northern Europe, but its champions successfully coerced farmers in Germany, and later in eastern and northern Europe, to try the new crop with startling results. The potato was such an important staple in the Irish tenant farmer diet by the early nineteenth century that a series of devastating invasions of potato blight brought famine, death, and the first great out-migration of Irish to the Americas.

The first "European" potatoes in colonial America were introduced to Virginia farmers from Bermuda in 1621. The new crop brought little serious attention and remained largely a curiosity. Scots-Irish immigrants introduced the potato to the climatically better suited Londonderry, New Hampshire, area in 1719 with quite different results. The arrival in eastern

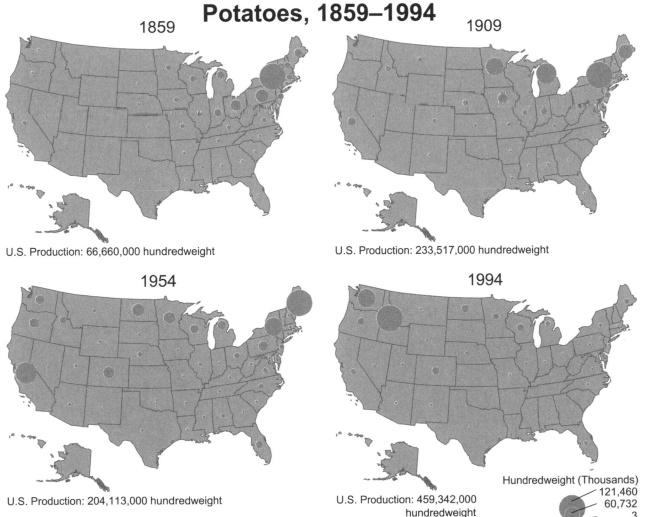

Potatoes, 1859–1994

1859
U.S. Production: 66,660,000 hundredweight

1909
U.S. Production: 233,517,000 hundredweight

1954
U.S. Production: 204,113,000 hundredweight

1994
U.S. Production: 459,342,000 hundredweight

Hundredweight (Thousands)
121,460
60,732
3

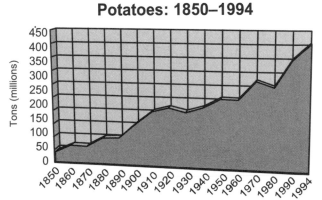

Potatoes: 1850–1994

Tons (millions)

cities of tens of thousands of emigrants who were familiar with the potato over the next century or two secured its future in the nation's basic diet.

The potato's role in American culinary life has been transformed over the past century. Served almost always as a filler in stews or meat pies in the past, the potato is an important visual ingredient of today's menu and is most often served baked, French fried, or as potato chips. The transformation of the American diet since World War II has changed the preparation and role of many foods, but few as much as the potato.

Cultivation

The potato is a cool-weather plant, which generally requires a low soil temperature. Young sprouts do best with a soil temperature of about 75°Fahrenheit. Tuber production is retarded at soil temperatures above 68 degrees Fahrenheit and totally halted at 84 degrees Fahrenheit. Production in the South is only possible when the crop is planted in the fall, winter, or early spring. Mature plants can withstand light frosts, but not heavy freezes. The interrelationship between soil character, temperature, and moisture significantly alters the texture of

the growing potato. Much of the changing geography of potato cultivation over the past fifty years is a direct result of the changing American dietary preferences. The best French fried and baked potatoes are prepared from the more mealy textured western potato to the detriment of the market for eastern potato market.

Four principal varieties of potatoes are grown in the United States. The russet, most associated with Idaho, but actually grown in many areas, is the most important variety today and is generally perceived to be the best for baked potatoes and frozen French fries.[1] The red-skinned triumph is most often harvested when the tuber is quite young and served as "new" potatoes—once only in the early spring, but today all year around. The Katahdin is popular because it is adaptable to a wide range of climates and once was the most widely grown variety. The Green Mountain is a general purpose potato also grown

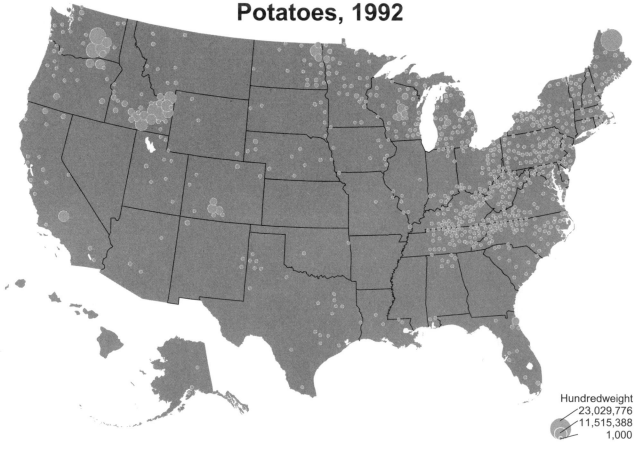

Potatoes, 1992

Hundredweight
23,029,776
11,515,388
1,000

Counties with less than 1,000 hundredweight harvested not indicated.

U.S. Production: 410,508,931 hundredweight

Table III.24
Leading Potato Producing Counties, 1992

Counties	Hundred-weight	Percent of Total
Aroostook, ME	23,029,776	5.6
Bingham, ID	21,296,484	5.2
Grant, WA	16,801,601	4.1
Benton, WA	14,536,037	3.5
Franklin, WA	14.536.037	3.4
Cassia, ID	11,442,781	2.8
Fremont, ID	11,341,076	2.8
Walsh, ND	10,783,951	2.6
Madison, ID	10,682,805	2.6
Bonneville, ID	9,415,125	2.3
Total ten counties	143,345,104	34.9
U.S. production	410,508,921	

throughout much of the nation. Literally hundreds of variations from these four basic varieties are planted each year, as well as some lesser known varieties. For example, many European potatoes have a yellow cast and a sweeter flavor than the traditional American favorites. First introduced to American consumers in the 1940s, these varieties have received little attention until the past decade when the Yukon and Finn varieties started appearing in larger numbers in specialty stores. South American purple varieties are also starting to appear on American produce counters today.

Potato cultivation must be rotated on a three- to five-year schedule for continued success. The rotation varies from region to region, but typically a year of potatoes is followed by one of small grains, which is followed by a year(s) of clover and/or corn. Western farmers often rotate potatoes with sugar beets, followed by wheat or barley, and finally with alfalfa, in which the last cutting is crowned in the fall and replowed in the spring.

Distribution

Potato production was widely distributed throughout the nineteenth century in the United States, except in the Southeast, where strategies of fall and late winter plantings had not yet been devised. The largest areas of cultivation were concentrated in the Middle Atlantic and Corn Belt states at the end of the century, the areas of the highest concentrations of Irish, German, and eastern European immigrants in the nation.

Declining relative shipping costs and higher return alternative crops in the Mid-

west began altering this basic pattern in the twentieth century. By the 1930s there had been significant decline of potato production in the Corn Belt, where pork and beef production had risen to dominance, with commensurate increases in production in the northern Midwest and Maine. Farmers in western states had also discovered the potential of the crop by this time with important areas of production centering in Colorado, Idaho, and California. Maine's Aroostook Valley was the preeminent potato area in the nation, but with only 11 percent of the nation's total production.

Possibly of more interest is the general lack of concentration seen in the low percentages in other states. Only six other states produced more than 5 percent.

Postwar America brought massive changes to American life, especially in suburbanization and frequent dining out. The idealized meal of Middle Americans in the 1950s was a steak, baked potato, and (for those who ate rabbit food), a salad, a striking change from the pot roast, mashed or boiled potatoes, and cooked carrots, cabbage, and onions of the preceding generation. The favorite meal of American

This Idaho farmer is having his truck loaded with seed potatoes for spring planting. (RP)

teenagers was a hamburger, French fries, and a milkshake—all served preferably at a drive-in.

The result of these food preference changes is only too obvious in the geography of potato production in the United States. The mealy textured potatoes favored for baking (and French fries) grew best in the irrigated far West. Maine remained the largest producer (65 million bushels) in 1950, but California's Tulare and Klamath Lake basins (48 million bushels) and Idaho's Snake River plain (47 million bushels) were already increasing in total importance. The perfection of the frozen French fry by an Idaho entrepreneur a decade later made the ascension of Idaho as the "Potato State" inevitable.

Idaho is now the nation's largest producer of potatoes with a total of 6.7 million tons in 1994. Just under half of the entire nation's crop is harvested in the combined Snake River production area of Idaho, Washington, and Oregon. Aroostook County, Maine, continues to lead the nation in potato production, but production in virtually all other areas has declined relatively and absolutely. The fame of Long Island new potatoes remains; the fields largely do not since the suburbanization of New York City onto the island. Much of the crop from Pennsylvania is for potato chips.

A potentially devastating new problem is now beginning to threaten potato production in the United States. The *Phytophthora infestans*, a blight fungus related to the same blight that struck Ireland a hundred and fifty years ago, is sweeping East Coast potato fields. Apparently originating in Mexico, the blight was first identified in U.S. potato fields in the late 1980s. The blight's ability to reproduce sexually with the cells of other plants has so far made it resistant to chemical treatments. The blight has now diffused westward and growers in Oregon and Idaho encountered it during the 1995 season with unknown results.

Notes

1. The modern restaurant French fry is almost always from a frozen product. Developed during the late nineteenth century, the French fry was a common item on restaurant menus throughout the country. Restaurateurs found it difficult to provide a quality fry during the summer, however, because of deterioration of the stored potatoes. J. R. Simplot developed a frozen fry during the 1950s, but it was not until he was approached by McDonald's a few years later to improve the product that the fry we know today was created. Utilizing a blanched product that was then quick-frozen, the new potato product was so good that it replaced the fresh potato entirely in this market. Virtually all commercial French fries today are based on Simplot's concept.

Poultry

Poultry has been a part of the Western European diet for centuries, but did not take on its current role in the American diet until after World War II. The following discussion examines the evolution of production of individual varieties of fowl; while related, their production continues to be responsive to different forces.

Chickens

Not many years ago preparing a Sunday chicken dinner meant walking out into the backyard, catching a prime bird, and proceeding along the preparation process. The transformation of this repast from a Sunday treat to a fast food favorite took an unusual path. The role of chicken in the American diet changed little in the first three decades of the twentieth century with apparent consumption declining slightly

from 14.9 pounds per capita in 1912 to 14.6 pounds in 1928. Consumption increased during World War II to over 25 pounds per capita (chicken and turkey), but began declining again until the early 1950s. Three factors seem to have coalesced at that time to bring new popularity to chicken consumption. An obscure, failed railroad engineer running a motel/gas station with a kitchen and a couple of tables in his repair bay in Corbin, Kentucky,

developed a method to pressure cook "fried" chicken. Prior to this time an order of restaurant fried chicken had taken as much as an hour. Harland Sanders, a born entrepreneur, began marketing his secret-recipe coating mix and cooking equipment to midwestern restaurants in the early 1950s with great success. In 1954 he developed the Kentucky Fried Chicken concept as a freestanding restaurant and began selling franchises internationally to become

Table III.25
Leading Broiler Counties (Sales), 1992

Counties	Sales	Percent of Total
Sussex, DE	194,185,730	3.60
Cullman, AL	121,253,358	2.20
Benton, AR	93,596,018	1.70
Washington, AR	93,108,511	1.70
Wilkes, NC	81,219,669	1.50
De Kalb, AL	79,349,113	1.50
Wicomico, MD	76,497,668	1.40
Rockingham, VA	76,248,729	1.40
Scott, MS	73,923,508	1.40
Fresno, CA	68,433,240	1.30
Ten county total	957,815,544	17.64
U.S. sales	5,428,589,485	

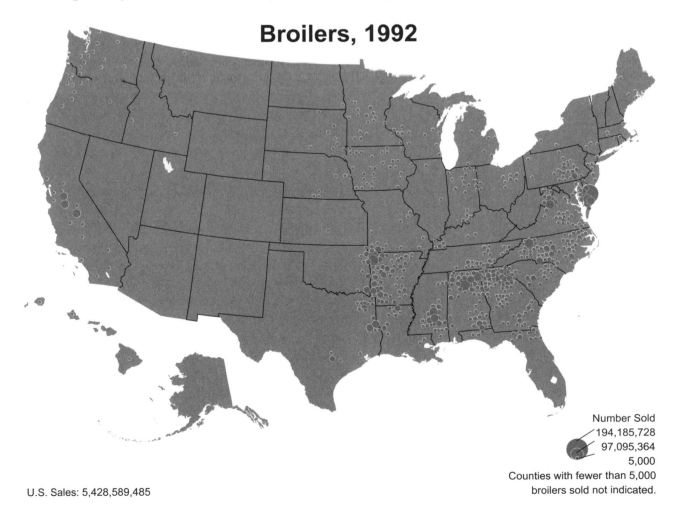

Broilers, 1992

U.S. Sales: 5,428,589,485

Number Sold
194,185,728
97,095,364
5,000
Counties with fewer than 5,000 broilers sold not indicated.

201

Chicken Consumption: 1910–1992

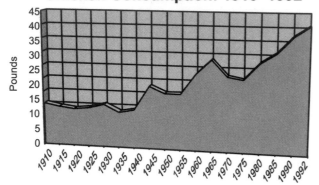

Broilers, 1954–1994

1954

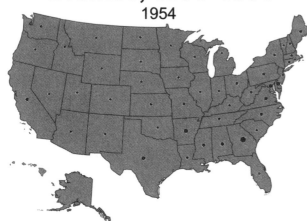

U.S. Inventory: 765,575,000

1994

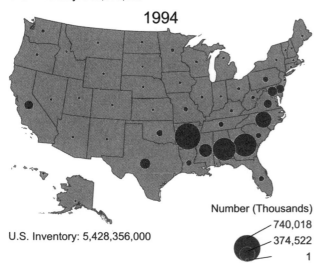

U.S. Inventory: 5,428,356,000

Number (Thousands)
740,018
374,522
1

the most recognizable entrepreneur of the early fast food revolution.

The new focus on fast food chicken, in conjunction with increased production to meet the enlarged demand, lowered unit production costs, and the relative decline of chicken costs, increased supermarket sales as well. Finally, the discovery that a major factor in heart disease was cholesterol obtained from animal fats further focused attention on the relatively low-fat content of the chicken, though certainly not as a fried entree. Consumption has been rising ever since. Apparent per capita consumption today is almost seventy pounds annually.

Patterns of Production

Chickens were generally transported to market alive until after World War II and the geography of their production reflected their limited capacity to survive the long distance travel of the day. The Midwest was the lowest cost production area throughout the nineteenth century, but most rural and many small-town families there and elsewhere raised their own birds. Many chickens were raised in the region, but commercial production concentrated on eggs, rather than on meat. The largest meat chicken production and distribution centers developed within a few hours' transport of the nation's major cities. While producers generally operated on a small scale, their accumulated sales began to mount. Perth Amboy (New Jersey) distributors, for example, shipped 320,000 fowl and 1.8 million pounds of dressed poultry a year as early as 1856.

Larger-scale commercial chicken producers began to appear during the 1870s with

the development of effective incubators, faster rail shipment, and the beginnings of refrigeration. The American Poultry Association began sponsoring exhibits to encourage the development of higher quality stock during this period and published their first "Standard of Perfection" in 1874. Central hatcheries also developed using the newly introduced parcel post system to distribute the newest breeds of chicks to even the most isolated farm home or commercial flock.

The first stages of industrial production started to appear in the 1920s. State-funded agricultural experiment stations began research on all aspects of the chicken industry. The first growth studies were launched at this time to determine the rate of maturation, feed use, and net return from different combinations of feed and chicken breeds. It took a typical Leghorn twenty-four weeks to reach 3.3 pounds, consuming 22 pounds of grain in this time. The goal was to reduce both the time and amount of food while increasing the amount of meat in the desirable body areas of the chicken. Today's forty-four-day growth cycle generates about a third of a pound of chicken for every pound of feed.

Chicken consumption declined during the Depression, but increased during World War II because neither chickens nor eggs were rationed. Chicken farms increased in number and flock size after the war ultimately to create overcapacity. The cyclical nature of the industry tended to drive small producers out of business, and eventually the industry became concentrated in the hands of a few giant vertically integrated companies. Tyson Foods, the largest, began inauspiciously in 1935 when John Tyson bought a load of chickens and hauled them to Chicago in the hope of receiving better prices than were available in Arkansas. He parlayed the profits of this and subsequent runs into feed processing, hatcheries, and processing plants, to become the world's largest chicken processor. The company today has more than sixty processing plants and markets more than 25 million chickens per week. The bulk of Tyson's business is with restaurant companies, but the acquisition of Holly Farms in 1989 increased capacity and sales, and added a well-known brand name for supermarket sales. The company has now relabeled these products with its own brand name. The second largest processor, ConAgra of Omaha, Nebraska, began as a grain milling company, but today is a diversified international food company. Gold Kist, starting as a Georgia cotton cooperative in the 1920s, has grown to become the third largest chicken marketer, while Frank Perdue's marketing genius has made his company number four in the marketplace. Foster Farms of California completes the big five that together now market more than two thirds of all chickens sold in the United States.

Today's chicken industry is almost totally vertically integrated. Profit margins are narrow, making efficient, large-scale operations necessary to maintain reasonable returns on investments. The processing companies have spent millions of dollars developing better breeds of birds, better feeds, and more efficient feeding systems, and on controlling disease. Processors typically provide the chicks, feed, and technical assistance and guarantee purchase of the finished birds at a given price. The new flock often arrives at one end of the chicken house while the finished birds are still being loaded at the other end. Most processors also have construction units to build chicken houses to their specifications for their operators.

The Pattern of Distribution

The evolution of the geography of chickens is a reflection of the growth of the dominant chicken processors. Chicken production was quite diffuse as late as the beginning of World War II. The concentration of the industry into the hands of fewer and fewer major processors has tended to concentrate production in their traditionally favorite areas. The vast majority of the nation's chickens were raised in only fourteen areas in 1954. In 1992 they were concentrated in only five: Arkansas and adjacent Oklahoma and Missouri (Tyson and ConAgra), northern Georgia and Alabama and adjacent Tennessee (initially Gold Kist, but Tyson and ConAgra as well today), North Carolina (Gold Kist and Holly Farms, now Tyson), the Delmarva Peninsula with outliers in southeastern

Poultry has moved from the farmyard to become a billion-dollar industry controlled by a handful of producers. (JF/RP)

Pennsylvania (Frank Perdue), and central California (Foster Farms). A host of smaller producers and centers still thrives, but the narrow profit margins and cost advantages of the largest processors make it difficult for smaller competitors to challenge them in major markets.

The dominance of the processor in the chicken industry is complete. The birds must be processed in USDA approved dressing plants. There is virtually no market for birds not grown on contract, with the possible exception of the growing "free-range chicken" niche market. If the owner of the local processing plant determines that its continued operation no longer meets company needs, the plant and growing region can disappear overnight (or at least at the end of the current forty-four-day growing cycle). Many processing plants today were built in the 1950s and only marginally pass government health regulations. The incentive to close poorly located facilities and reopen at new state-of-the-art

plants at more convenient sites is high. As many poultry processing plants are staffed by newly arrived immigrants, the need to find inexpensive local labor is not as important as it once was. The power of this potential on production location (and economic dislocation for the providers) was demonstrated by Holly Farms Foods in North Carolina in the early 1990s. The company decided that shipping chickens from the traditional production center in Appalachia was too difficult and that a coastal plain location (nearer to their hog growout operations) was more convenient.

Within a single year the center of North Carolina chicken production moved more than a hundred miles eastward.

Chicken Eggs

Egg production has long been separate from the raising of chickens for meat. Nine-teenth-century distributors discovered that dipping eggs in a lime or oil solution pre-served them for shipping to distant mar-kets. Thousands of barrels of eggs moved along the Ohio River system before the Civil War and overland to eastern markets

Table III.26
Leading Laying Chicken Counties, 1992

Counties	Sales	Percent of Total
Lancaster, PA	12,577,143	4.2
Riverside, CA	9,270,221	3.1
Darke, OH	5,701,904	1.9
Mercer, OH	3,967,791	1.3
San Diego, CA	3,929,964	1.3
Stanislaus, CA	3,901,643	1.3
Washington, AR	3,634,141	1.2
San Joaquin, CA	3,553,314	1.2
Fayette, TX	3,526,320	1.2
New London, CT	3,163,293	1.2
Ten county total	53,225,794	17.7
Total U.S. sales	301,467,288	

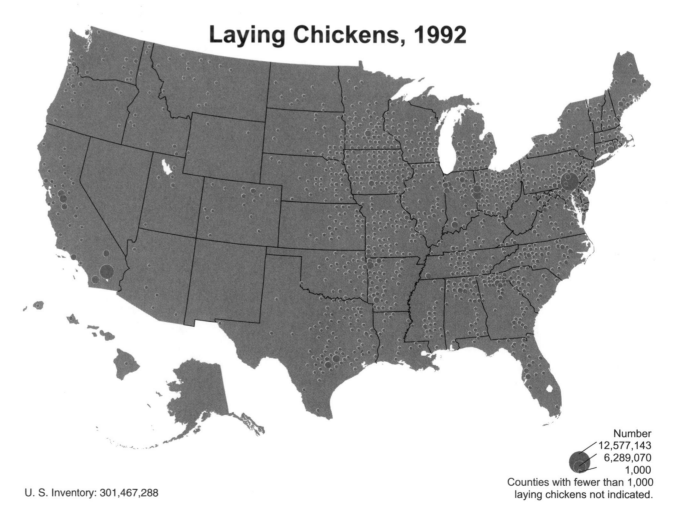

Laying Chickens, 1992

Number
12,577,143
6,289,070
1,000
Counties with fewer than 1,000 laying chickens not indicated.

U. S. Inventory: 301,467,288

Laying Chickens, 1954–1994

1954

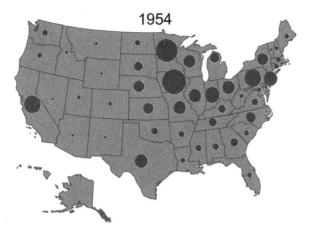

U.S. Inventory: 375,800,447

1994

U.S. Inventory: 301,406,728

Numbers Sold (Thousands)

31,231
15,689
2

Egg Production: 1899–1992

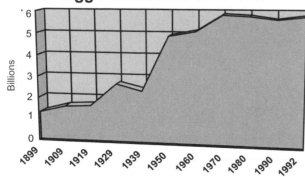

after the completion of the rail network. The transition from home to commercial production began in earnest in many midwestern areas after the Civil War. The relative cost of eggs remained low throughout the period. Consumption reached almost 300 per person by 1900 and temporarily peaked at 342 per year in 1927 before declining during the Depression. Consumption increased again during the war years to peak at 403 in 1945. It has been steadily dropping ever since. Egg consumption today is 235 per person, declining by almost one percent per year for more than a decade.

The geography of egg production has never been as concentrated as that of broilers. The distribution of laying chickens in 1919 was quite diffuse with slightly heavier than average in the Corn Belt and near the larger urban markets. Additional consolidation continued during the interwar years. Seven major production areas are seen in the 1954 Census of Agriculture, three on the East Coast, two in the Corn Belt, and two on the West Coast. Even today the inability of egg producers to create a branded identity for their product continues to both reduce profits and make significant production consolidation difficult. The pattern of production continues to be relatively diffuse with the seven identifiable regions of 1954 more or less in place, although the East Coast concentrations have moved westward and the southern California zones have been partially replaced by two larger areas in the Central and the Willamette (Oregon) Valleys.

Turkeys

The turkey has not played an important role on American dinner tables until the past thirty years. Poultry generally increased in number throughout the nineteenth century, although at very small numbers. Peaking in 1880, the number of turkeys, ducks, and geese began to drop dramatically, from about 11, 4.4 and 5.7 million respectively, over the next three decades to less than their 1880 totals. Turkey flocks stabilized around 3.5 million in the 1920s and 1930s and began rising dramatically in the late 1930s and during the war years because they were not rationed. The development of new, meatier "heavy" breeds after World War II spurred turkey consumption. Rising public concerns about cholesterol contributed to this increase and the comparatively low-cost turkey burger began appearing in West Coast supermarket meat cases during the 1960s. Turkey franks, processed meats, and sausage began appearing soon after. Acceptance of these products has been much slower in the more conservative East where the turkey burger is rare even today. Turkey-substitute processed meat products, however, are universally available.

Turkey production has increased dramatically in the past few years to about 290 million birds per year. While this is still minuscule in comparison to the 6.4 billion broiler chickens consumed each year, it has created a market sufficiently large to attract larger and larger competitors. Turkeys have always had much more concentrated production patterns than chickens. Commercial turkey varieties are very susceptible to disease and more care must be taken to protect them from infection. An outbreak of Newcastle's disease on the West Coast virtually wiped out the industry during the 1980s. As a result,

Turkeys, 1954–1994

1954

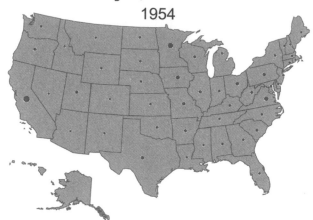

U.S. Sales: 67,507,507

1994

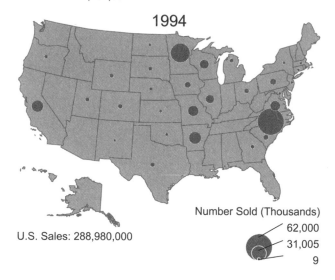

U.S. Sales: 288,980,000

Number Sold (Thousands)
62,000
31,005
9

Turkeys: 1930–1992

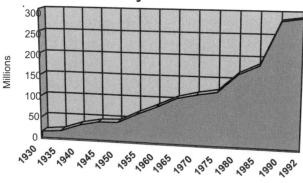

U.S. Sales: 279,230,136

many modern West Coast turkey farms could easily be mistaken for prison work farms with their barbed wire fences and other protective measures. The largest center of turkey production for many years was Rockingham County, Virginia, but it has declined in recent years. North Carolina is the turkey state today with 62 million head raised in 1992, followed by Minnesota with 43.5 million. Arkansas, California, and Missouri each produced slightly more than 20 million birds in that year. Like chicken, turkey processing is vertically integrated and virtually all birds are raised under contract. Louis Rich of Madison, Wisconsin, is the largest company focusing on turkey production. Most poultry and some meat packers also produce turkeys.

Other Fowl

Ducks are the most important of the other fowl. Both ducks and geese have not fared well in the American diet, which has consistently moved toward lighter-tasting foods for almost a century. Long Island was the most noted early center of duck production, targeting the large "live" New York

Turkeys, 1992

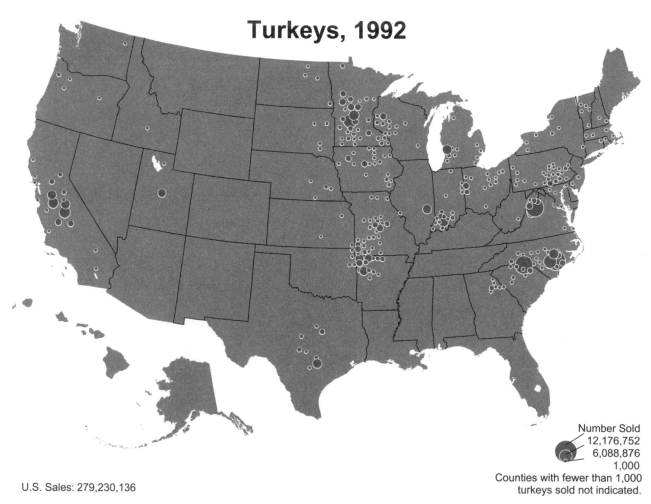

Number Sold
12,176,752
6,088,876
1,000
Counties with fewer than 1,000
turkeys sold not indicated.

Ducks, 1992

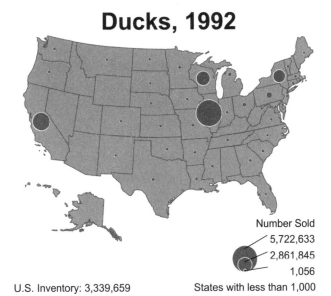

Number Sold
5,722,633
2,861,845
1,056
States with less than 1,000
sales not indicated.

U.S. Inventory: 3,339,659
U.S. Sales: 16,361,031

Geese, 1992

Number Sold
73,752
37,126
500
States with less than 500
sales not indicated

U.S. Inventory: 183,217
U. S. Sales: 324,472

City market. The relative decline of agriculture on Long Island moved production westward, although more than 2 million ducks a year are still raised there. Some of this production moved to Lancaster County, Pennsylvania, but northern Indiana, especially in and around Amish-dominated LaGrange County, soon became the nation's leading duck production center. California is the second largest producing state with almost 4 million sold in 1992. The large Asian population in the state seems to be the primary stimulus for this growth in production. Goose consumption, originally concentrated among eastern and central European immigrants, was at first concentrated along the eastern seaboard. The Midwest has become the center of production in more recent years, especially Indiana. The remainder of production is scattered through the northern Midwest.

California Central Valley rice cultivation is heavily concentrated on the poorly drained clay soils north of Sacramento. (RP)

Rice

Rice is one of the most widely consumed cereal grains in the world with a total harvest of 348 million metric tons in 1991. More than 90 percent of this harvest was grown in Asia; slightly less than 2 percent was produced in the United States. The grain was not an important dietary staple in colonial America, except among the African slave population along the South Atlantic Coast and in Acadian Louisiana. Consumption has increased in recent years with the arrival of millions of immigrants from traditional rice areas and the general rise in interest in ethnic foods in this country.

Rice is unique among the widely used domesticates in that the plants are cultivated in a partially submerged state through much of the growth cycle. Growers control weeds by keeping the roots and stems under water all or most of the time. The exposed rice leaves absorb oxygen and transmit it to the roots through ducts in the plant body, which respirate the carbon dioxide through the roots. The plants remain in these submerged fields until just prior to harvest. The most well-known system of rice cultivation is the paddy rice system of Asia, which was not introduced into the United States until 1912. All previous planters used versions of the African systems brought to America by the slaves who worked these fields.

Early Rice Cultivation in America

Rice culture was introduced to the Carolinas from West Africa in 1672 where its cultivation was taught to white planters by their African slaves. It was recognized as an

Rice, 1860–1994

1860

U.S. Production: 1,872,000 hundredweight

1909

U.S. Production: 9,827,000 hundredweight

1954

U.S. Production: 65,284,000 hundredweight

1994

U.S. Production: 156,110,000 hundredweight

Hundredweight
62,904
31,459
2

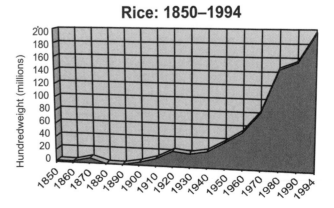

Rice: 1850–1994

Hundredweight (millions)

200 180 160 140 120 100 80 60 40 20 0

1850 1860 1870 1880 1890 1900 1910 1920 1930 1940 1950 1960 1970 1980 1990 1994

209

Rice, 1992

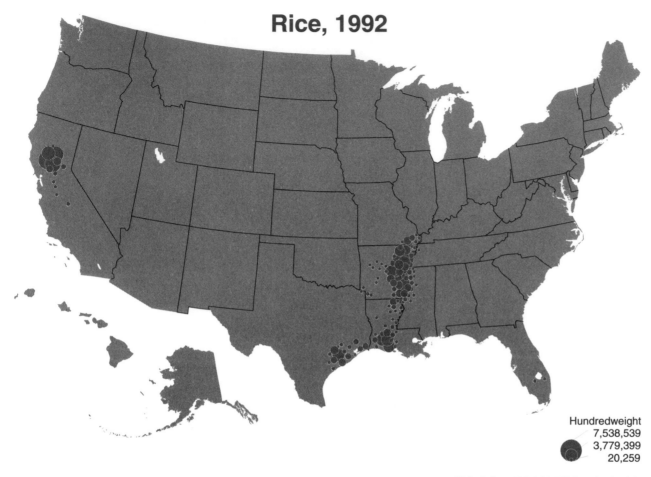

Hundredweight
7,538,539
3,779,399
20,259

U.S. Sales: 175,941,723 hundredweight

Table III.27
Leading Rice Producing Counties, 1992

Counties	Hundred-weight	Percent of Total
Colusa, CA	7,538,539	4.3
Arkansas, AR	7,320,152	4.2
Poinsett, AR	7,230,152	4.1
Sutter, CA	6,265,614	3.6
Butte, CA	5,615,509	3.2
Glenn, CA	4,987,132	2.8
Jackson, AR	4,708,028	2.7
Bolivar, MS	4,690,220	2.7
Cross, AR	4,677,613	2.7
Acadia, LA	4,463,276	2.5
Ten county total	57,496,711	32.7
U.S. production	175,941,723	

important export crop by 1690 and production expanded northward toward Hatteras and southward almost to Florida over the next fifty years. New varieties from the Caribbean, Southeast Asia, China, and Egypt were constantly introduced into the East Coast cultivation areas to increase production throughout the eighteenth century. Carolina planters dominated the American rice industry until the late nineteenth century when a combination of hurricanes, field degradation, labor shortages, and competition from expanding production in the Mississippi basin ultimately brought a halt to East Coast cultiva-

tion in the early twentieth century.

The earliest crops in the Carolinas were cultivated under the West African pluvial system, a form of upland rice that utilized the region's heavy rainfall to maintain proper water levels during the crucial growth periods. African slaves adapted their traditional method of flooding lowland areas by utilizing the dry Carolina Bays (geomorphic features) as basins to hold rain and spring water for the rice. Carolina planters generally allowed these fields to dry out after use and grazed cattle in them when they were uncultivated to increase soil fertility. The restrictive distribution of

these features encouraged the introduction of the phreatic system, also developed in West Africa.

The phreatic system utilized natural marshlands and other floodable areas near fresh-flowing streams. It utilized a combination of springs and fresh tidal water from the adjacent rivers to keep the growing rice plants partially submerged during the growing period. Sluice gates were alternately opened and closed to retain freshwater ponds at the correct height to promote rice plant growth behind embankments. Like the African system, the Carolina fields also had ridges to raise the rice plants slightly above the bottom of the fields to increase water flow and help ensure that they were kept above any saltwater that might enter the fields while they were being flooded by the daily tidal movements. This system began replacing the pluvial system in the early eighteenth century. Like the pluvial period, however, the acreages for this type of cultivation were quite limited and the expanding rice industry rapidly utilized all available areas.

Market expansion during the mid-eighteenth century forced the introduction of the much more expensive, but more widely adaptable, fluxial system. Though appearing similar to the phreatic system, the fluxial system is totally dependent on tidal movements to bring fresh water in and out of enclosed unridged fields. These fields can be built in marginally brackish water environments by carefully monitoring tidal movements so that sluice gates are open when the tides bring freshwater flow and closed when saltwater is present. But the fields must be enclosed on all sides and the construction of proper sluices, canals, and dikes is far more labor intensive. The returns, however, are greater because much of the coastal Carolinas and Georgia could be utilized with this system. Most of the 150,000 acres of rice that were planted at the crop's zenith in the Carolinas used this system.

The Beginnings of the Modern Era

Sugar planters facing insurmountable labor, capital, and disease (cane mosaic disease) problems began experimenting with rice cultivation along the distributaries of the Mississippi River in the 1870s. Water was pumped or siphoned over the levees of the major rivers where it flooded the fields throughout most of the growing season. The crop continued to be labor intensive in this area because the equipment used by southwest Louisiana farmers could not work these muddy fields. The crop prospered on the Mississippi Delta through the 1880s until an improving sugar market combined with lower cost production in southwest Louisiana shifted the center of rice cultivation westward.

Small quantities of rice had long been cultivated by Cajun farmers in the marshes of southern Louisiana utilizing a form of upland rice cultivation. The Southern Pacific Railroad acquired millions of acres of land in southwestern Louisiana in 1881 and dispatched Sylvester L. Cary of Manchester, Iowa, to promote them. Cary mailed thousands of advertisements to farmers throughout the Midwest and Northeast from his office in Jennings, Louisiana, with great initial interest and few sales. He was soon joined by the North American Land and Timber Company. Hiring Seaman A. Knapp, the North American Land and Timber Company began experimental farms, which ultimately included rice. Early fields emulated the Cajun method of dependence on rainfall, usually with great success, but ultimately these waters were augmented by wells and canals. The distinctive elements of the Louisiana rice were a combination of the development of large irrigation systems from both near surface well water and extensive canals, and the adaptation of midwestern wheat technology to the harvest. The Louisiana rice farms, quickly followed by those in Texas, became the most mechanized in the world with the widespread use of gang plows and binders. Tractors and steam-powered threshers soon followed. The size of the harvest exploded from 834,112 bushels in 1879 to 2,721,059 bushels in 1889 to 10,839,973 bushels in 1909. Texas farmers joined the bonanza in 1886, and almost 9 million bushels were harvested around Beaumont in 1909.

W. H. Fuller moved to Arkansas's Grand Prairie from Ohio in 1896 and soon decided to participate in Louisiana's rice bonanza. He planted his first crop in 1897, but it failed. Undaunted, he rented a farm near Jennings, Louisiana, and began rice cultivation there the following year to learn the business. Returning to Arkansas in 1904, Fuller began promoting rice cultivation on the Grand Prairie with increasing success. Stuttgart (Arkansas) investors spurred production in 1906 with the construction of a local rice mill to process the exciting new crop. The Arkansas harvest reached 1.2 million bushels in 1909 and 6.7 million bushels in 1919.

Rice in the Grand Prairie proved to be such a success that farmers in the adjacent northern Mississippi Delta and northeastern Arkansas soon joined in the bounty. Problems continued, especially with the continuing conflict between the rice millers and farmers, but even these were eventually ameliorated with the creation of the Arkansas Rice Growers Cooperative in 1932. Riceland, Inc., the modern-day progeny of these cooperative efforts, has become the largest processor and rice marketer in the nation.

The California Rice Industry

Cultivation of rice was introduced into California from Asia in 1912 and it has been grown continuously using a version of the Asian paddy rice system in the southern Sacramento Valley from Woodland to Chico since that time. The technology of the period was quickly adapted to the new crop, including continuous track tractors, combines, and rice dryers. Aerial application of seed and fertilizers followed in the 1930s and helicopters in the 1980s. A

California rice farmer today hardly needs actually to set foot on paddy soil.

The California annual cycle begins in the early spring when road graders and earth-moving equipment are used to create the paddies. Land "planes" are then used to level fields to within millimeters of the desired flatness. Levees are constructed around the field perimeter, and sluice gates control water levels so that the rice is protected both from weeds and drowning. The fields are flooded with a few inches of water in mid-April and fungicide-treated, germinated seed is aerially applied to the shallow paddies. Water levels are increased as the plants grow. Fertilization and any other chemical treatments are also aerially applied through the growth cycle. Paddies are drained three to four weeks before harvest to allow the fields to dry. Most fields are left dry until the early spring when the stubble is burned to add nutrients to the soil and to kill the millions of insects that live in the paddies.

Contemporary Rice Cultivation

California rice technology has been adopted in virtually all areas today. Combines and their attendant dryers were adopted in Louisiana and Arkansas during World War II when the combination of unquenchable demand, escalating market prices, and labor shortages made the new equipment investments feasible even during the wartime shortages. Rice demand after the war did not wane, as had been feared, but continued, especially during the Korean War years. Rice production today is at levels inconceivable even a few years ago.

The Texas Gulf Coast and southwest Louisiana, the Arkansas Grand Prairie, the Central Valley of California, and the Mississippi Delta remain the nation's principal rice producing areas, though significant differences in yield, varieties, and technologies continue. Arkansas is the leading rice state with 1.4 million acres harvested (1993). Louisiana is the second largest state in acreage, but third in output. The almost 50 percent higher yields achieved by California growers make that state second in output, although its dependence on short and medium grain rice because of environmental constraints means that little of its crop is sold in the United States. The United States exports more than 2 million tons of rice annually with almost half going to Asia. The remainder of the export market is almost equally split among the Western Hemisphere, especially Mexico, Europe, and Africa.

Sheep

Sheep were imported into the American colonies within the first years of settlement, but generally were less important than other large domestic animals. Most early farmers maintained some sheep as a source of wool for clothing, though the animals generally fared poorly because of their susceptibility to attack by wolves, inability to survive the winters without shelter, and the unsuitable natural forage. Most early flocks were small, as shown in a 1635 inventory of a large farm in the Massachusetts Bay Company that listed fifty-eight cattle, ninety-two sheep and lambs, twenty-seven goats, sixty-four swine, and twenty-two horses.

New York was the leading state for sheep production in 1840 with more than 5 million head, primarily concentrated between Rochester and Schenectady. Vermont had the highest density of sheep, partially due to a lack of viable agricultural alternatives and its proximity to the woolen mills in Massachusetts. The center of sheep production moved westward after the 1840s as New England and New York farmers found higher returns from dairying.

Sheep fared well during initial settlement of the Corn Belt prairies, but were again replaced as decreasing transportation costs and increasing markets for wheat and hog production proved to be more profitable. The center of sheep production shifted westward again a decade later to the western Corn Belt, and ultimately to the far West, although many eastern farmers continued to maintain small flocks to utilize otherwise wasteland. The numbers of sheep in the Corn Belt began decreasing at an even higher rate over the past few decades as farmers shifted to a two-crop rotation of corn and soybeans and tore

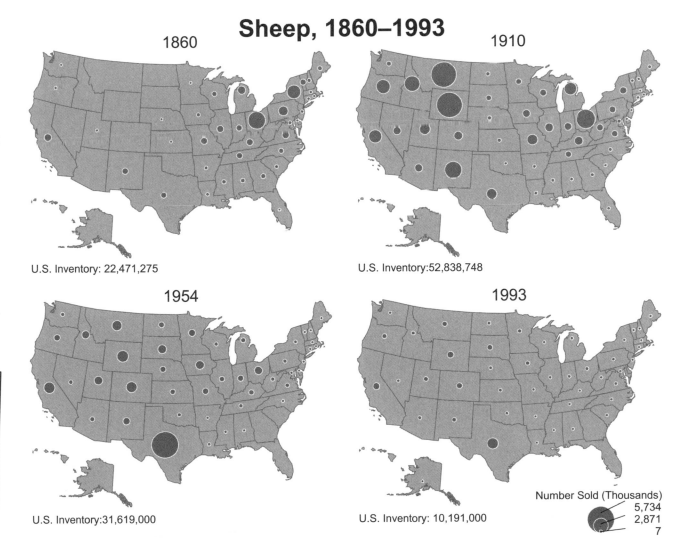

Sheep: 1840–1993

Sheep, 1860–1993

1860
U.S. Inventory: 22,471,275

1910
U.S. Inventory: 52,838,748

1954
U.S. Inventory: 31,619,000

1993
U.S. Inventory: 10,191,000

Number Sold (Thousands)
5,734
2,871
7

down the fences separating the fields to ease the movement of their farm equipment.

Problems Facing the Sheep Industry Today

Sheep production increased through the nineteenth century, temporarily peaking in 1905. Production wavered through World War I and the 1920s and the number of stock sheep peaked in 1935. The total number (including lambs) over the entire year peaked at 56 million animals in 1942.[1] Sheep inventories have been declining since that time largely because of decreasing demand for lamb and mutton, inefficiencies in the slaughter and meat distribution systems, low woolen prices, increasing problems with predators, and problems in obtaining reliable labor. About two thirds of cash revenues from sheep sales today come from meat, and only about one third from woolen sales.

The rapid decline of per capita lamb consumption from 4.08 pounds in 1955 to 1.4 pounds in 1988 hurt the industry badly. The aging traditional American lamb consumer, most often an emigrant from southern and eastern Europe, has not been replaced by new consumers, although the increasing number of immigrants from the former Soviet bloc and the Middle East may alter this in the future. There is little likelihood that the American dietary mainstream will embrace lamb in the near future, as the vast majority of new emigrants are from Asia, Africa, and Latin America where lamb is rarely consumed. Declining consumption has also meant that packing companies are reluctant to modernize plants and distribution systems, increasing the costs of moving fresh meat from carcass to consumer, lowering demand even further with higher prices and frequent unavailability.

Wool consumption was hurt by the introduction of nylon and other man-made fibers during the 1930s. Wool consumption for textiles peaked in 1946 at 737 million pounds. Domestic consumption dropped to less than 358 million pounds by 1957 and to 133 million pounds in 1990. Woolens are currently undergoing a resurgence in moderate to expensive fashion clothing, but

Table III.28
Leading Sheep Counties, 1992

County	Number in Inventory	Percent of Total
Weld, CO	289,605	2.7
Val Verde, TX	197,655	1.8
Cockett, TX	185,138	1.7
Kern, CA	177,874	1.7
Chaves, NM	140,041	1.3
Tom Green, TX	130,885	1.2
Pecos, TX	119,434	1.1
Butte, SD	113,687	1.1
Johnson, WY	107,946	1.0
Converse, WY	106,661	1.0
Ten county total	1,568,926	14.6
U.S. inventory	10,519,343	

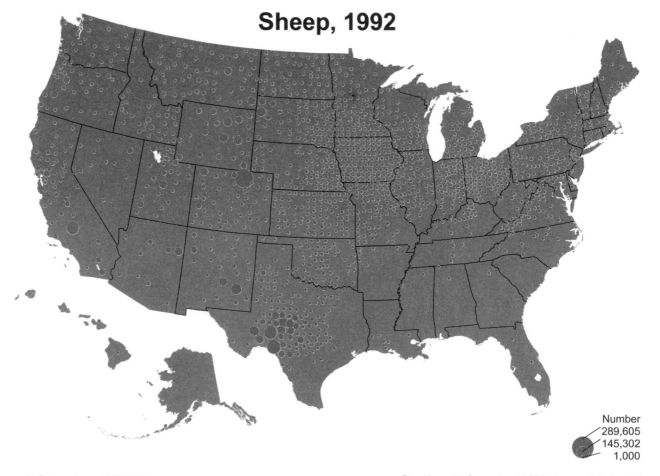

Sheep, 1992

Number
289,605
145,302
1,000

U.S. Inventory: 10,770,391

Counties with fewer than 1,000 sheep not indicated.

there is little likelihood that the outerwear market will follow suit as man-made fabrics and fillings are much more water- and wind-repellent than traditional woolens. The demand for woolen carpets, once consuming between 20 percent and 30 percent of the domestic woolen market, collapsed with the increasing popularity of nylon, orlon, and other fibers in the 1960s, though there is increasing interest in fine woolen carpets in recent years (14.3 million pounds, 1991).

Sheep are especially susceptible to predators, and unlike range cattle, must be constantly tended. Traditional predators, coyotes, eagles, wolves, and other large carnivores, once were virtually extinct in the Far West, but are currently increasing in numbers in response to the efforts of environmental interest groups to protect them. Predator sheep kills have increased dramatically. The small profit margin of this business makes even small losses significant.

A continuing problem for large sheep ranchers has been finding trained shepherds. Differing solutions have evolved in various areas. For many Amerindians in the arid West, sheep and goats have been the only source of income, and they have tended their own flocks. Sheep and goats served as the primary income for many Hispanic farmers as well, and these farmers also tended their comparatively small flocks with their sons. The Mormon ranchers of Utah and Idaho tended to have large families and strong kinship ties, which meant that there were always sons, nephews, and others to watch the herds. The remaining ranchers were forced to hire outsiders to tend their flocks. Anglo ranchers in the Southwest primarily hired Mexican emi-grants searching for work, but the ranchers in Nevada and to the North looked to the Basque shepherds entering the region from California.

Basques are a culturally distinct group of Spain and France where their homeland lies astride the Pyrenees. The first Basques in the Far West emigrated to California in 1848 to seek their fortunes in the gold fields. Some soon discovered that it was more profitable to provide meat to the miners than to mine themselves. Ranching initially in southern California, the Basque ranchers moved to the San Joaquin Valley in the 1860s to be closer to their markets. Their cattle were soon replaced by sheep in this new, drier environment. Competition from farming in the San Joaquin forced increasing numbers of sheep ranchers to summer their herds in Nevada's Great Basin by 1870. The core of the contemporary Basque community stretches from Boise in the north to Winnemucca and Elko in the east to Bakersfield on the west.

Distribution Today

Commercial sheep production takes place in virtually every state outside the Southeast. Texas has the largest sheep population (19 percent of the breeding ewe inventory), followed by California (8 percent), Wyoming (7 percent), and Montana and South Dakota (6 percent each). Declines in flocks are about equally distributed throughout the nation. The average flock nationally is comparatively small, averaging sixty-five breeding ewes, though the number increases to 143 in western flocks. The ability of sheep to consume weeds, shrubs, and other fodder unusable by other domesticates, their small water needs, and their ability to thrive in difficult environments have made them popular among ranchers with parcels of land with little other economic potential.

The typical far western flock is well traveled. Gone are the picturesque canvas wagons of the past. Travel trailers and three-decker tractor-trailers move the herds from their protected winter pastures to increasingly higher mountain meadows as the snows melt. Almost a million sheep are grazed on National Forest and federal grazing lands annually. A typical Basque sheep outfit, for example, might winter a flock of 8,000 sheep around Bakersfield with a smaller flock in Ely or Elko, Nevada, in what become alfalfa fields in summer. The California sheep are shipped back to Nevada in April where they graze in the lowland grasses waiting for the snows to melt. Typically, the herd is broken into several units as summer arrives and transported to permitted National Forest lands throughout the Intermontaine West. Herds are moved more frequently today to reduce permanent damage to the range. The herd will grow to over 20,000 in the early sum-

Sheep and Lambs Sold, 1992

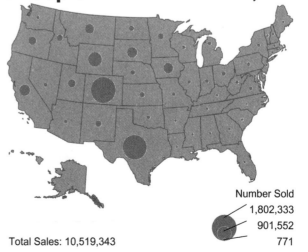

Number Sold
1,802,333
901,552
771

Total Sales: 10,519,343

mer as the ewes lamb. By fall the herd will begin to be brought back into the lowlands as the owner checks meat, wool, and stock prices to determine how many of the flock will be wintered and how many will be sold. The flock is finally shipped to its winter base in Elko, Ely, or Bakersfield where additional flock cutting may be made depending on forage and hay prices during the winter.

Lamb accounted for 94.2 percent of all sheep and lamb meat sales in 1993. Ranchers began shipping lightweight lambs to feed on alfalfa a few decades ago to enhance their weight prior to sale. Lambs may gain as much as one-half pound per day in a feedlot environment. Lamb-targeted feedlots first appeared in Colorado and California in the 1980s, but have spread to other feedlot areas as well. Weld County, Colorado, had an inventory of almost 300,000 sheep and sales of 1.2 million in 1992. Other counties with important sheep feedlot operations include Laramie, Wyoming; Umatilla, Oregon; Menard, Texas; and Washakie, Wyoming.

Notes

1. The number of sheep and most other farm animals varies widely during a production year with the smallest number in the winter, the time when most census inventories are taken. Stock animals (those carried over the winter) are the number usually tabulated, but number sold is often more meaningful, especially for animals with harvest times less than a year. Thus a January 1 broiler inventory is virtually meaningless as commercial broilers pass from chick to dressed chicken in less than two months.

This well-traveled flock of sheep is being loaded onto trucks to summer at several alpine locations in northern Nevada before returning to its winter home outside Bakersfield, California. (RP)

Sorghum

Grain sorghums evolved in sub-Saharan Africa and spread eastward into Asia. Archeological evidence dates cultivation in Oman by 2500 B.C. and in Pakistan a millennium later. Sorghums were widely cultivated in India soon after, especially in the south around Madras where they became an important food source. Evidence of early cultivation in southeast Asia and China is also present, but the date of introduction is unknown. The kaoliang race evolved in northern China, probably through intermingling with local wild species. Sorghums were introduced into the New World directly from Africa, probably on ships carrying slaves to the Caribbean. Sorghums were successfully cultivated in Central and South America where they continue to be an important source of food, especially in the highlands.

While sorghums were almost certainly introduced into the United States in conjunction with the slave trade as early as the seventeenth century, they were not widely cultivated in this country until much later. The species used today were introduced in the late nineteenth century through two unrelated incidents. Two races of durra sorghum were introduced to California farmers from Algeria in 1874. These plants were well adapted to the high temperatures and semidrought conditions of California's interior valleys where they were often cultivated as interrow crops in young orchards. "Kafir corn," a South African variety more adapted to eastern climates, was displayed two years later at the Philadelphia Centennial Exposition. Samples were given to two visitors who independently took them to Georgia where experimental plantings were made. Selective breeding soon created stock that thrived,

Sorghum, 1992

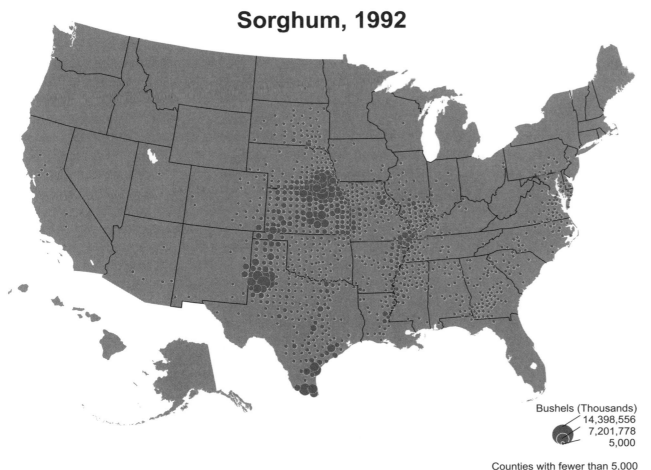

Bushels (Thousands)
14,398,556
7,201,778
5,000

Counties with fewer than 5,000 bushels harvested not indicated.

U.S. Production: 733,312,667 bushels

Table III.29
Leading Sorghum Producing Counties, 1992

Counties	Bushels	Percent of Total
Gage, NE	14,398,556	2.0
Lancaster, NE	12,200,594	1.7
Lubbock, NE	10,160,344	1.4
Marshall, KS	9,955,678	1.4
Hockley, TX	9,804,397	1.3
Nemaha, KS	9,393,841	1.3
Nueces, TX	9,351,166	1.3
Hidalgo, TX	8,842,146	1.2
Washington, KS	8,531,981	1.2
Hale, TX	7,882,592	1.1
Ten county total	100,521,295	13.7
U.S. production	733,312,667	

and seeds were widely distributed in the Southeast as far west as Texas and Oklahoma by the mid-1880s. Milo, an ancient type of sorghum, also appeared about this time in South Carolina, but it is not known whether its introduction dated to the slave trade or from later contacts with Africa and the Caribbean.

The expanding agricultural frontier in Kansas suffered serious crop losses in the 1880s as drought repeatedly destroyed the corn crop. Kafir corn seed was obtained from the Georgia State Department of Agriculture about 1886. It was found to be highly adaptable both as silage and forage grain for the growing cattle industry. Kansas farmers planted 47,000 acres of kafir corn in 1893 and more than 700,000 acres by 1902. Kansas continues to be the largest center of grain sorghum production in the United States with more than 3 million acres in 1992, followed by Texas (2.95 million acres) and Nebraska (1.6 million acres). No other states exceeded a half

Sorghum gained popularity on the arid margins of the Grasslands as farmers expanded feed cattle operations in the early twentieth century. (RP)

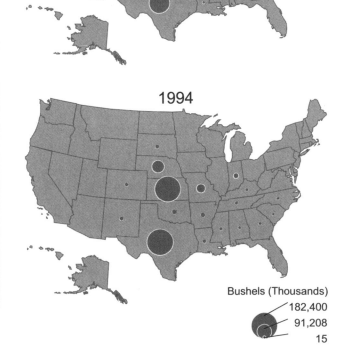

Sorghum, 1909–1994

1909

1954

1994

Bushels (Thousands)
182,400
91,208
15

Sorghum: 1899–1994

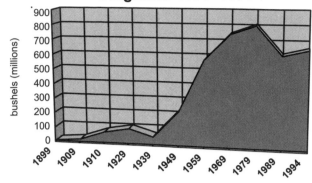

million acres that year. Sorghum is also widely used as a silage crop, especially in the southern Corn Belt. Dried sorghum hay is used in some areas, but is much less common.

Two other varieties of sorghum also have historic importance in the United States. A cane variety of grain sorghum was introduced in the nineteenth century and was widely used as a sweetener, especially during the Civil War. Today it is most common on the Gulf Coast and Appalachia where it is often used in exhibits of folk culture (see also Sweeteners). Another variety of sorghum constitutes almost the only source of natural broomstraw grown in the United States. Despite numerous attempts to promote a market for household brooms from artificial materials, the traditional "corn" broom remains the nation's favorite material.

The adaptability of existing equipment to soybean cultivation and harvest aided in the rapid spread of this crop throughout the Corn Belt. (RP/JF)

Soybeans

The ubiquitous soybean has become one of the most important elements of the American diet over the past twenty years, yet it is rarely recognized, tasted, or noticed. Used as a protein additive, binder, oil, and indirectly as animal fodder, there are few complex processed foods that do not directly or indirectly benefit from this bean. The soybean evolved from a wild perennial, native to Southeast and East Asia (Sauer, 1993). Little is known of its early cultivation history, except that it was one of the two most important staples (millet was the second) of the northern Chinese diet by 200 B.C. Its early spread was fostered by the diffusion of Buddhism, which requires a vegetarian diet of its followers. Soybeans thus became a common building block of many regional and national diets within the Asian Buddhist realm.

Development of Soybean Cultivation in the United States

Soybeans were introduced to the United States by Samuel Bower in 1765. He manufactured a variety of foods for export at his Georgia farm for a few years, but ultimately his business failed. Despite this early introduction and support from such luminaries as Benjamin Franklin, the soybean found little early favor in this country. The period of modern cultivation began after the United States Department of Agriculture initiated a research project in 1890 popularizing the crop as a "green manure," which would increase soil fertility by fixing nitrogen in the soil. These early attempts to popularize the crop primarily stressed cultivation of the crop as silage for animal fodder. Cottonseed oil shortages during World War I spurred planting and cultivation of the crop for its oil. The plant's beans did not become an important product until after 1924 when the federal government began stressing their value as a source of oil. Soybeans continued to be grown primarily as silage, however, until World War II vegetable oil shortages could not be met from traditional sources. Soybean acreages harvested as beans, for example, thus rose from slightly more than one quarter of total acreage in 1925, to about one third in 1930, to 95 percent of all harvested acres by 1949.

The collection of 4,000 seed lots by USDA scientists in China, Korea, and elsewhere in the late 1920s apparently spurred cultivation for both silage and beans as the nation moved into the Depression. The demand for inexpensive butter substitutes, the collapse of the cotton market, and improved technology all played a role in creating this expanding

Soybeans: 1910–1994

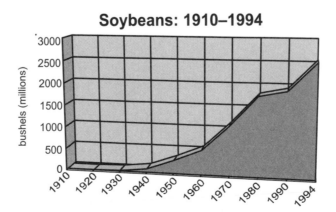

Soybeans, 1909–1994

1909

U.S. Acreage: 114,457
U.S. Production: 1,135,141 bushels

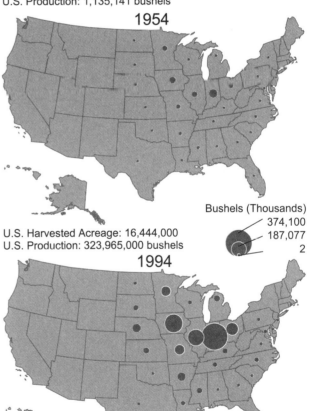

1954

U.S. Harvested Acreage: 16,444,000
U.S. Production: 323,965,000 bushels

Bushels (Thousands)
374,100
187,077
2

1994

U.S. Harvested Acreage: 56,447,000
U.S. Production: 1,808,538,000 bushels

Production less than 1,000 bushels not indicated.

soybean market. Acreages harvested for beans had increased to almost one half of total production by 1941 with the increasing use of soybean oil in margarine and other processed foods. Total bean production had climbed over 100 million bushels.

Wartime shortages of processed edible oils such as soy, traditionally imported from Asia, coupled with increased demand for margarine and other oil-based products during World War II, spurred domestic production, which had doubled by 1945. Escalating demand and prices linked to new food technologies rapidly brought soybean production to new levels.

Soybean acreages peaked about 1980, although total production has continued to grow because of rising yields. Value of production has also continued its upward climb, keeping soybeans as the second most important crop by value in the nation. The United States has long been the world's largest producer (59.8 million metric tons), followed by Brazil (22.3 million metric tons), Argentina (11.4 million metric tons in 1992–93), and China (10.3 million metric tons). Japan, the Netherlands, Mexico, and Taiwan are the largest export markets for U.S. soybeans, Morocco and India the largest importers of soybean oil, and the former Soviet Union, Canada, Mexico, and the Netherlands of cake and meal. Brazilian production has been increasing rapidly in the past few years.

Distribution

Soybeans have played an instrumental role in reshaping the geography of agriculture of the Corn Belt over the past several decades. Corn and soybeans have similar climatic demands and can be harvested by the same

machinery with minor modifications. Rising soybean prices supported increasing production at a time when traditional livestock activities were coming under severe competition from large cattle dry feedlot operations in the Grasslands. The continuing strong international market for soybean products has meant that bean prices have been far less volatile than corn prices. Today the Corn Belt is the center of soybean cultivation and its landscape has become monolithic in its domination by the two-crop rotation system of corn and

soybeans. The flat, rich lands of central Illinois's Grand Prairie, long focused on a cash grain economy, rather than on livestock because of its highly productive soils, was the early core of Corn Belt soybean cultivation. The heaviest concentrations are still located there, although it is rare to be out of sight of soybeans anywhere in the Corn Belt for more than a few minutes while driving country roads. More broadly, the Middle West geography of soybean production appears as a broad Y, with the lower portion extending down the Missis-

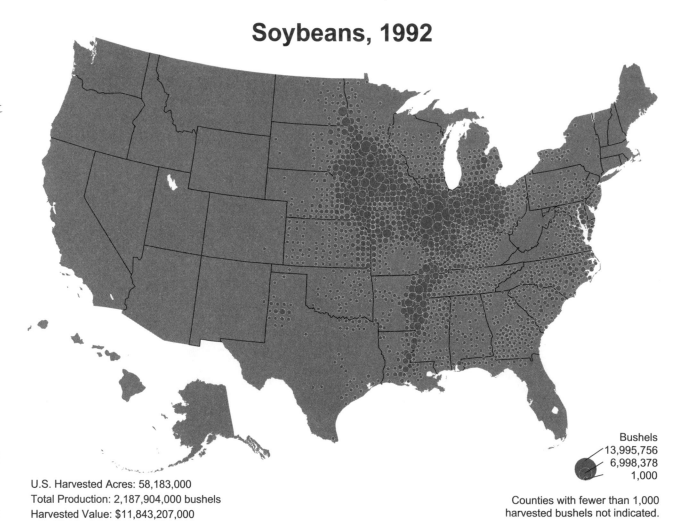

Soybeans, 1992

U.S. Harvested Acres: 58,183,000
Total Production: 2,187,904,000 bushels
Harvested Value: $11,843,207,000

Bushels
13,995,756
6,998,378
1,000

Counties with fewer than 1,000 harvested bushels not indicated.

sippi Valley from broad arms, one stretching northeast toward Ohio's Maumee Valley, and the other across the northwestern margins of the Corn Belt into the eastern Dakotas. The Mississippi Valley is the second largest center of production, the southward spreading soybean almost totally obliterating any clear demarcation between the Lowland South and the Corn Belt along the river margins today. The fall line stands out sharply along the eastern seaboard, as Lowland farmers have increasingly relied on soybeans as a low labor alternative to many traditional crops on the Atlantic Coastal Plain from Georgia to Virginia and across the Gulf South. Rising cotton prices, however, have caused some retrenchment in recent years with the explosion of cotton production in the same sections of the Lowland South. Finally, Llano Estacado farmers cultivate a small but highly visible crop of soybeans in the Texas panhandle.

Table III.30
Leading Soybean Producing Counties, 1992

County	Bushels (Thousands)	Percent
McLean, IL	13,996	0.64
Livingston, IL	12,557	0.57
Iroquois, IL	11,706	0.53
Champaign, IL	11,434	0.52
La Salle, IL	10,689	0.49
Kossuth, IA	9,445	0.43
Vermillion, IL	9,112	0.41
Sangamon, IL	8,892	0.40
Christian, IL	8,037	0.37
Webster, IA	7,664	0.35
Ten county total	103,532	4.71
U.S. production	2,196,504	

Heavy government subsidies support the cultivation of the sugarcane processed in this Belle Glade, Florida, sugar mill.

Sugar and Other Sweeteners

The average American consumes 140 pounds of caloric sweeteners and the equivalent of an additional 24.3 pounds of artificial low-calorie sweeteners each year. This represents an almost 20 percent increase of calorie sweeteners and a 450 percent increase of low-calorie sweeteners since 1971. Americans consume more sweeteners than chicken (68.4 pounds), red meat (112 pounds), and wheat flour (135.9 pounds). Though rarely thought of as an important area of agrarian activity, sweeteners have long played an important role in our diet, economy, and foreign policy.

The first recorded acquisition of additive sweeteners is a Paleolithic cave painting in Spain of a man robbing a store of wild honey. Bees and their honey were such an obvious source of sweetener that apiculture had developed in Egypt before 2500 B.C. Honey, however, was difficult to produce in large quantities with the technology then available. A variety of vegetables were used as sweeteners around the world, including carrots in parts of Europe, cornstalks and other vegetables in the New World, and sugarcane throughout much of Asia and the Pacific even in early times. The primary caloric sweeteners in use today include granulated sugar made from cane (primarily *Saccharum officinarum*) and beets (*Beta vulgaris*), starch-based sweeteners such as corn syrup, honey, and syrups from maple and sorghum cane. Artificial sweeteners such as saccharin, cyclamate, and aspartame are of increasing importance in the diet, but are not agricultural products and thus are not discussed here.

Sugarcane

Sugarcane is not a single species, nor are the most common commercial strains able to be propagated except vegetatively through planting cane sections. The crop originated in the Pacific and diffused westward to Southeast Asia and eventually India where it was widely cultivated by 400 B.C. The technology of cane production diffused slowly westward to be introduced to the western Mediterranean and Iberia by the fifteenth century. Columbus brought sugarcane from the Canary Islands for cultivation on Hispaniola on his second voyage, but unfortunately his sugar cultivators died on the voyage and the experiment failed. The Portuguese were more successful in Brazil, but were soon competing with the Spanish and Dutch and later the British and French. The history of sugarcane production, the wars and politics associated with its production and sale, and the problems that this plantation life brought are interesting and too lengthy to explore here.

Experimental cane production was

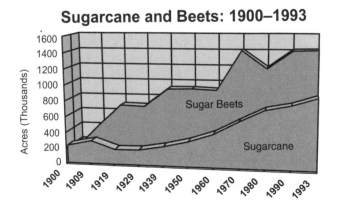

Sugarcane and Beets: 1900–1993

Acres (Thousands)

Sugar Beets

Sugarcane

1900 1909 1919 1929 1939 1950 1960 1970 1980 1990 1993

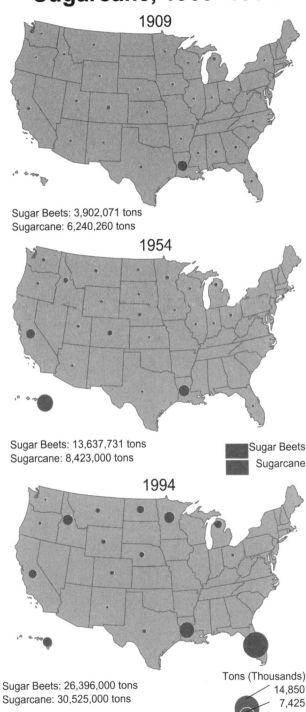

Sugar Beets and Sugarcane, 1909–1994

1909

Sugar Beets: 3,902,071 tons
Sugarcane: 6,240,260 tons

1954

Sugar Beets: 13,637,731 tons
Sugarcane: 8,423,000 tons

Sugar Beets
Sugarcane

1994

Sugar Beets: 26,396,000 tons
Sugarcane: 30,525,000 tons

Tons (Thousands)
14,850
7,425
.5

Sugarcane, 1992

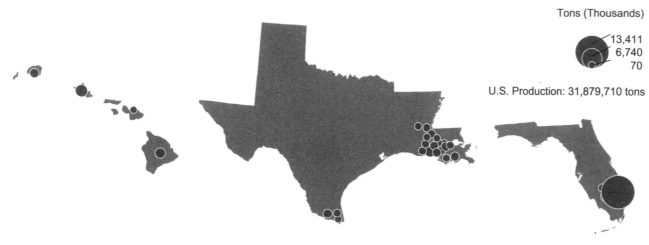

Tons (Thousands)

13,411
6,740
70

U.S. Production: 31,879,710 tons

attempted throughout the southern American colonies, but failed because of climatic limitations. The acquisition of Louisiana in 1803 brought the first successful cane production to the United States. Louisiana remained the center of U.S. cane production until the aftermath of the Civil War devastated the region's economy. Funds were not available to hire wage labor initially, and many plantations turned to the less labor intensive rice cultivation for a few years after the war. A treaty signed in 1876 with the Kingdom of Hawaii allowed the duty-free importation of sugar in the United States, while the acquisition of Puerto Rico and the Philippines brought sugar from those sources in the twentieth century. Florida production began in the late nineteenth century, but did not mature until the twentieth. American cane production is confined to Florida, Louisiana, and Texas (after 1973) on the mainland, in Hawaii in the Pacific, and in the commonwealth of Puerto Rico.

Sugar Beets

Andreas Marggraf, a German, proved that beet sugar was identical to cane in 1747, but the first factory was not built until 1799. Napoleon Bonaparte sponsored research on sugar production from a variety of local materials during the continental blockade and was largely responsible for the realization of the potential of this crop in Europe. The first beet sugar mill in America was constructed in 1838, but failed in the first fifteen years. Successful production did not begin until the construction of a mill at Alvarado, California, in 1870, which continued to operate until 1967. Twenty-nine mills were operating in 1899 in the United States.

The geography of sugar beets has been altered dramatically over the past century, although the need for refineries has meant that it has always been highly concentrated in a few locations. In 1909, for example, the largest concentration was found in the Colorado high plains near Greeley, followed by lesser centers along the Arkansas River

eastward from Rocky Ford, Colorado, into Kansas, the thumb of Michigan centering on Saginaw, along the front edge of the Wasatch south from the Idaho border in Utah, the central coast of California, the Sacramento River delta, and along the Snake River north of Pocatello, Idaho. Almost all of these production areas were associated with German and Polish emigrants bringing the requisite technology to production and refining.

The distribution of beet production underwent a restructuring after World War II as increasing mechanization lowered unit costs but increased total investments. The number of sugar beet producers dropped by two thirds between 1948 and 1974, while the average number of acres per farm increased by four times. Similarly, processing plants peaked in 1972. Virtually all beets are produced under production contracts for a handful of companies today.

Beet production has continued to rise rapidly as the demand for sweeteners increases. Minnesota was the largest pro-

Table III.31
Leading Sugar Beet Producing Counties, 1992

Counties	Tons	Percent of Total
Polk, MN	1,680,630	5.8
Imperial, CA	1,077,979	3.7
Minidoka, ID	1,010,080	3.5
Clay, MN	970,909	3.3
Canyon, ID	900,320	3.1
Renville, MN	896,694	3.1
Cassia, ID	718,338	2.5
Pembina, ND	703,845	2.4
Tuscola, MI	684,003	2.3
Huron, MI	659,827	2.3
Ten county total	9,302,625	31.9
U.S. production	29,124,488	

ducer (6,845,000 tons) in 1992 with production concentrated in the Red River Valley with an additional 3,388,000 tons produced across the river in North Dakota. The sugar beet is an integral part of the potato rotation cycle in Idaho, the second largest beet producing state (4.9 million tons). California is the third largest producer, primarily in the Imperial Valley, where it is a summer crop, and the Sacramento Valley, where it is rotated with wheat and beans. Michigan has slid to fifth, while important production areas in the Llano Estacado of the southern Grassland region,

the Colorado high plains, and Kansas have declined with increasing concerns about the long-term use of the Oglala aquifer as a water source.

Corn Syrup

Gottlieb Kirchhof of the Academy of Science, St. Petersburg, Russia, accidentally discovered during the Napoleonic Wars that sweetener could be effectively produced from dry vegetable starch. Though dextrose could be produced from any starch product, corn became the most popular

source in the United States. The first factory to produce dextrose from cornstarch in this country was constructed in Buffalo, New York, in 1873. Early factories continued to be concentrated in the Northeast in the early years, although today the largest are concentrated in the Corn Belt, especially Iowa and Illinois. Much of the history of this industry is tied to the attempts of investors and corporations to control the industry.

Corn sweeteners are used primarily in food processing. Confectionery companies traditionally have been the largest users of dextrose from corn, followed by the canning, dairy, and baking industries. The decision to utilize dextrose as a major sweetener in soft drinks during the 1980s, based on its significantly lower cost, revolutionized both the diet beverage and corn processing industries. The corn syrup refiners contend that their sales would more than quadruple if government subsidies of traditional sources of sugar were halted.

The Sugar and Corn Sweetener Processing Industries

The evolution of the sugar refining and corn sweetener industries during the late nineteenth and early twentieth centuries was the scene of one of the fiercest contests between big industry and the federal government during the Industrial Revolution (Ballinger, 1978). The Sugar Refineries Company, commonly called the Sugar Trust, was created from eight sugar refining corporations in 1887. Additional firms soon joined the cartel and by 1892 it controlled about 90 percent of the nation's sugar refining capacity. Though periodically

Sugar Beets, 1992

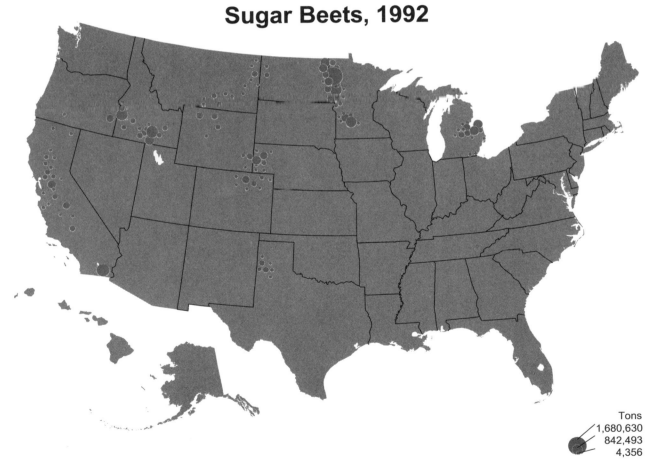

Tons
1,680,630
842,493
4,356

U.S. Production: 29,124,868 tons

Sorghum Syrup, 1992

Gallons (Thousands)

States with less than 200,000
gallons production not indicated

55,305
27,755
205

challenged in court, the renamed American Sugar Company continued its stranglehold on the sugar refining industry through both continual expansion and predatory business practices. By 1905 the Trust still controlled about 75 percent of the refined sugar capacity of the nation, having purchased interests in the newly emerging beet sugar refining industry. The federal government launched an attack in 1910 that culminated in a 1921 consent decree stating that the direct collusive control of the sugar refining industry by American Sugar had been reduced to only 24 percent and that the Company would not attempt to regain its former position in the market.

The Corn Products Company, the first attempt to consolidate the corn sweetener industry, was formed in 1902. CPC, as it is now known, apparently was formed largely by investors in the Standard Oil Company who hoped to break the Sugar Trust for their own gain. Not only did they engage in predatory practices of their own, but attempted to purchase large blocks of American Sugar stock to gain control of that

corporation as well. In 1916 the courts found this corporation "to be an unlawful business combination" and the parent company was trimmed to only three units. CPC survives today as a comprehensive food company with interests in many fields.

While these early sweetener cartels and their competitors no longer exist in their original forms, their efforts established a pattern of domination still felt in the industry today. American Sugar continues as one of the nation's largest sugar refiners, while the corn sweetener industry is dominated today by Archer-Daniels-Midland, calling itself the "Supermarket to the World." The battle continues between the sugar and corn sweetener corporations even today, the focus being primarily on continuing government subsidies.

Sorghum Cane

Sorghum cane was introduced into the United States during the 1850s and became

Honey, 1993

U.S. Production: 18,630,000 gallons

Gallons (Thousands)

33,831
16,929
26

widely grown in the Midwest and South as a sweetener. Chinese sugarcane was introduced to the Midwest through France from China. Another strain from Natal, Brazil, was introduced to the South a few years later where it was widely grown primarily as a noncommercial "home" crop and refined in crude boiling pans on farms. Sorghum syrup production reached levels of millions of gallons during the 1880s when cane sugar remained expensive, but was never important far beyond its local production areas. The strong flavor of sorghum syrup is unappetizing to many and has been largely replaced by the inexpensive table (pancake) syrups based on flavored cane and corn syrups. Sorghum syrup today is almost exclusively an Upland South folk industry and rarely sold out of that area.

Honey

Apiculture is at least 4,500 years old and has long been a source for sweeteners. Modern beekeeping in America is largely associated with the pollination process with honey production representing a valuable byproduct. Per capita consumption of honey has remained largely stable since World War II, although the organic food revolution of the 1980s caused a temporary rise in consumption for a few years before declining to more typical levels. Production today is about 8.2 million gallons per year and is concentrated in the agricultural areas where pollination is crucial for agricultural production, especially California, Florida, and the Great Plains.

Maple Syrup, 1993

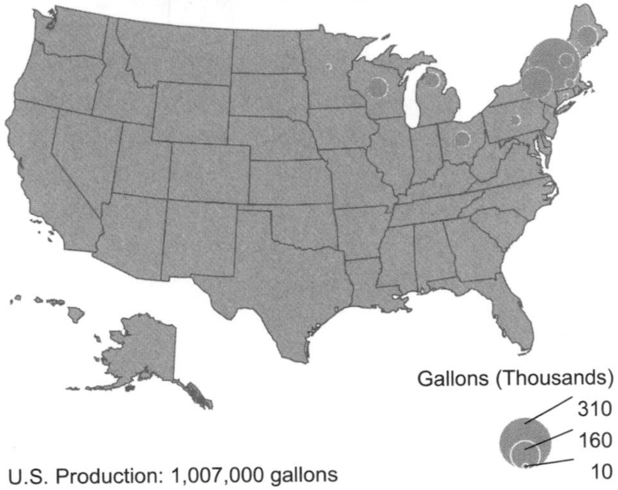

Gallons (Thousands)

310
160
10

U.S. Production: 1,007,000 gallons

Maple Syrup

Maple-flavored syrup has become one of the nation's favorite toppings for pancakes, waffles, and other breakfast favorites, but its high cost has kept consumption comparatively low. Approximately 2.2 million gallons of maple syrup were produced in the United States in 1991, primarily from Vermont, New Hampshire, and New York. Almost another 2 million gallons were imported from Canada.

Only the rapid cadence of the auctioneer's spiel cuts the heavy late summer air in this Carolina tobacco warehouse during sales season. Several bales a minute are sold as the warehouse owner and auctioneer walk down the rows of bales. Tobacco buyers indicate their bids with finger signals across the bales while prices are recorded on weight tickets that are placed on the bales when the farmers deliver their tobacco to the warehouse. (RP/PAP)

Tobacco

The early European exploration and settlement of the Americas was essentially part of vast scramble for individual and national wealth. The rise of the Spanish Empire was substantially fueled by the flow of gold and silver out of its Middle and South American colonies. Europeans settling in areas where minerals riches were not to be found, sought other opportunities. The plantation concept had been introduced to Iberia by the Moors in the thirteenth century and was seized on by the colonizing European nations as the perfect vehicle to exploit the New World. The market and profit potential for known crops such as sugar, indigo, rice, and citrus was great. It was soon determined that the same potential existed for tobacco.

U.S. Tobacco Product Consumption, 1992

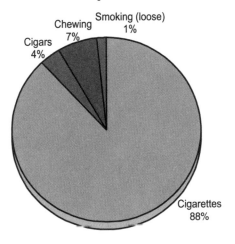

Cigars 4%
Chewing 7%
Smoking (loose) 1%
Cigarettes 88%

Tobacco, 1992

Pounds (Thousands)
32,979
16,490
1

Counties with less than 1,000 pounds harvested not indicated.
U.S. Production: 1,697,831,562 pounds

Table III.32
Leading Tobacco Producing Counties, 1992

Counties	Pounds	Percent of Total
Johnston, NC	32,979,238	1.9
Pitt, NC	31,847,136	1.9
Robeson, NC	29,821,830	1.8
Horry, SC	28,077,424	1.7
Wilson, NC	26,312,016	1.6
Nash, NC	24,675,696	1.5
Pittsylvania, NC	24,663,930	1.5
Wake, NC	22,379,685	1.3
Columbia, NC	21,059,646	1.2
Harnett, NC	20,723,259	1.2
Ten county total	262,539,860	15.5
U.S. production	1,697,831,562	

When Christopher Columbus arrived in the New World in 1492, he saw natives using tobacco in ways we do today—for smoking, chewing, and as snuff. The Indians' cultivation of the "golden leaf" incorporated most of the steps deemed essential today, including topping the plant to ensure vigorous leaf growth, "suckering" (the removal of smaller intermediate leaf growth to encourage growth of larger leaves), and fire curing of the harvested leaf. The common name, tobacco, derives from the European translation of the word *tobago* applied by the natives to the tube used for inhaling smoke and the cylinder of leaf prepared for smoking.

Tobacco is a plant of the genus *Nicotiana,* which contains more than sixty species, and is a member of the nightshade family (*Solanaceae*) grown as an annual crop for its leaves. It is unique among its relatives such as the tomato and potato in that the nonedible leaf is the commercial part of the plant. The leaves are also the source of the insecticidal alkaloid nicotine. It is nicotine and related tobacco alkaloids that furnish the habit-forming and narcotic effects that arguably account for its common use worldwide. Nearly all species are native to the Americas, although several were found by the early explorers in Australia. Most of the globe's agricultural tobacco is *N. tobacum*, with many varieties, which was domesticated before European arrival by Amerindians from coastal Brazil northward to Mexico and the West Indies. *N. rustica* was domesticated in eastern North America and a portion of the southwestern United States and northern Mexico. *N. rustica* is now found commercially in the mahorka tobaccos of Russia, with production found as far north as the Arctic Circle.

Tobacco is grown successfully under a wide range of climatic and soil conditions, although the commercial value of the product depends largely on the environment in which it is produced. Soils do need to be well drained and have sufficient friability for good aeration. Tobacco seeds are exceptionally small—300,000 seeds may weigh no more than an ounce —and thus the crop is usually planted in a sterilized seedbed and transplanted to the field after about 60 days. The plant typically takes another three months to mature.

Harvested tobacco was exported from Santo Domingo to Europe as early as 1531. Tobacco and smoking were known in Spain, France, and Portugal in the 1550s, and in England by the mid-1560s. By the first decade of the 1600s tobacco production had been introduced into Bavaria, Russia, Turkey, Persia, the west coast of Africa, the Philippines, Japan, and China. Sir Walter Raleigh had planted the seed on his estates in Ireland by the turn of the seventeenth century. Thus, the commercial production of tobacco spread to much of the known world before its adoption in North America.

Evolution of Tobacco Production in the United States

The first British settlers in Jamestown, in 1607, encountered a local tobacco, *Nicotiana rustica*, that they described as "poore and weake, and of a byting tast." In 1612 John Rolfe (later famous for his marriage to Pocahontas) acquired a milder seed, *Nicotiana tabacum*, from the Spanish colonies of Trinidad and Caracas. Europeans began producing a tobacco crop using this milder leaf in the Caribbean by the 1530s.

This was the variety that had earlier been introduced to Europe, probably arriving in Britain across the English Channel from the continent.[1] Rolfe sent his first export crop, a few hundred pounds, to England in 1613. London merchants saw the possibilities and demanded more. Settlers in Jamestown saw exporting tobacco as the way to ensure their economic survival—they could find nothing else to grow that could generate sufficient market demand. The North American tobacco industry was born.

Virginia exported 20,000 pounds of tobacco in 1618; 40,000 in 1620. Production expanded so rapidly that Delaware was established expressly for the export of tobacco. George Calvert, the first Lord Baltimore, also asked for a land grant to serve his king and country by planting tobacco. Chesapeake planters began amassing larger and larger plantations to facilitate the more efficient production of the leaf by the middle of the seventeenth century. Recognizing the economic potential of tobacco, the Crown declared tobacco a royal revenue, ordered an end to the growth of tobacco in England, and cut off exports from Spain to further profit from the North American tobacco trade. Tobacco spread northward into New York and

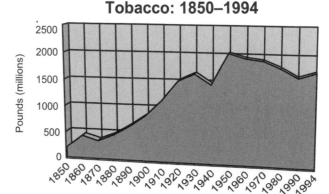

Tobacco: 1850–1994

Connecticut, and south into North Carolina, although Virginia and Maryland produced the bulk of the North American product throughout the colonial period.

These accumulated efforts brought a new problem that continues to plague the tobacco industry—repeated periods of overproduction and great cyclic swings in sale price. The Virginia government attempted to control the problem in the seventeenth century by limiting acreages and banning the importation of "foreign"—North Carolinian—tobacco. Planters became so enraged by these price swings in 1690 that they burned almost a million pounds of leaf in exasperation over low prices. The establishment of inspection warehouses in the early eighteenth century ameliorated the problem somewhat by reducing the export of the poorest quality tobacco. "Tobacco notes" were later introduced to certify a leaf suitable for export.

Tobacco production moved westward across the Appalachians with the expansion of settlement because of increasing demand and soil exhaustion. By the middle of the nineteenth century nearly every state in the Union grew the crop. Total national production topped 200 million pounds by 1839, 400 million by 1859, 900 million by 1898, and 1.5 billion by 1920. In that year twenty-nine states, all of them east of the Rocky Mountains, produced at least some tobacco. The high point in production for the nation was reached in 1951 when 2.33 billion pounds of the leaf were harvested. Harvest levels stabilized around 2 billion pounds annually until the early 1980s when a decade long decline in demand and production lowered output to its current national level of under 1.5 billion pounds annually. A variety of converging factors underlay this decline, including lessening consumption in the face of growing health concerns, increasing imports, and a decrease in the amount of tobacco in each cigarette (the major user of tobacco) with the rise of filtered cigarettes and changing production standards. It took 2.71 pounds of tobacco to manufacture 1,000 cigarettes in 1950; by 1990 that weight had fallen to 1.79 pounds. Only the increasing export of manufactured tobacco products, especially cigarettes, has supported continued production at this level. One third of all American tobacco is now exported.

The Geography of Tobacco

Production of tobacco is geographically localized, and distinctive characteristics set each type apart from the other types. The various forms of tobacco products require leaf of different characteristics. The U.S.

Tobacco, 1859–1994

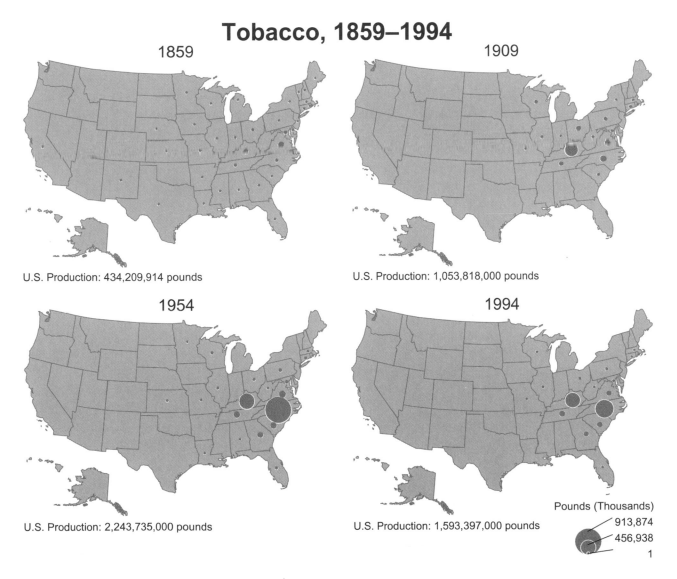

1859
U.S. Production: 434,209,914 pounds

1909
U.S. Production: 1,053,818,000 pounds

1954
U.S. Production: 2,243,735,000 pounds

1994
U.S. Production: 1,593,397,000 pounds

Pounds (Thousands)
913,874
456,938
1

Department of Agriculture designates six major classes of tobacco, each with a distinctive geography of production. The first three classes (flue-, air-, and fire-cured) are named on the basis of the method used in curing. The last three (filler, binder, and wrapper) are named in accordance with their traditional use in cigars. Each class is comprised of two or more different types. Each class is further divided into grades related to stalk position, quality, color, and other important leaf characteristics at the market.

The relative importance of each tobacco type and its region has varied with changing consumer tobacco tastes. Per capita peaks in the use of chewing tobacco, cigars, and smoking tobacco came between 1890 and 1910. Cigarette smoking boomed after 1910 with changing tastes encouraged by technology changes that greatly reduced their costs. Cigarette per capita production reached a peak in the early 1960s and production of cigarette tobaccos peaked in the early 1980s. Only snuff, a very small component of total use, has experienced a recent expansion in use.

Flue-cured tobacco now makes up just over half the total American crop. Nearly all flue-cured tobacco is grown along the Piedmont and Inner Coastal Plain from Virginia through Alabama, with North Carolina the leading producer. The region is divided into four production "belts"—the Old and Middle Belt astride the Virginia/ North Carolina border, the Eastern Belt in eastern North Carolina, the Border Belt on the North Carolina/South Carolina border, and the Georgia and Florida Belt from Georgia through Florida and a bit of Alabama.

Burley, an air-cured tobacco, is grown mostly in and around the Kentucky Blue-grass and Tennessee's Nashville Basin regions, but also is significant in Indiana, Ohio, and Missouri. The mountains of West Virginia, Virginia, and North Carolina provide an additional 35 percent of the total crop. Most burley is blended with flue-cured tobacco to create milder cigarettes. The recent trend toward milder cigarettes has increased the use of burley and supported a continued high level of total production of this type. A related air-cured leaf, Maryland tobacco, is mainly grown in the peninsula between the Potomac River and the Chesapeake Bay in southern Maryland.

Fire-cured tobacco, grown along the western Tennessee/Kentucky border and in Virginia north of the flue-cured area, is used for making snuff, chewing tobacco, strong cigars, and heavy smoking tobacco. Following the substantial national decline in consumption of most of these products, the Tennessee and Kentucky harvest has declined from 30 million pounds in 1950 to fewer than 10 million pounds today. Little more than 100 acres of the crop are now found in its former Virginia district.

Lancaster County, Pennsylvania, the center of cigar filler tobacco cultivation, was at one time the nation's leader in total tobacco production. Declining cigar use has brought a 75 percent decline (from 60 million pounds) in the county's yield since the early 1960s. A narrow band of counties in western Ohio supplies about 15 percent of our filler tobacco, with nearly all the rest grown in Lancaster and several neighboring counties. Many Pennsylvania growers have turned to producing Maryland air-cured tobacco as a more marketable alternative.

Two small groups of counties in south-ern Wisconsin grow nearly all of the country's cigar binder tobacco. This leaf was originally used for binding bunched filler tobacco into the shape of a cigar. Today, however, nearly all American-made cigars use a reconstituted tobacco sheet for this inner binder. Most Wisconsin tobacco is now used for loose-leaf chewing tobacco, a product with expanding consumption until recently. Thus, state production did not drop from its decades-long average of around 20 million pounds until the early 1980s. Total production is now under 10 million pounds and still falling.

Connecticut shade cigar wrapper tobacco, produced in the Connecticut River Valley of Connecticut and Massachusetts, long provided one of the most dramatic agricultural landscapes in America. Grown primarily to be the cigar wrapper, this product must be free of injury, uniform in color, and of generally fine quality. The leaves must be protected from sun and weather extremes with a covering of cheese-cloth over the entire field. As recently as 1965, 15,000 acres of valley land produced a crop of over 20 million pounds, providing a dramatic landscape of checkered white to the airline passenger flying overhead. Connecticut River Valley farmers produced fewer than 3.5 million pounds of tobacco in 1994. Declining consumption, the use of reconstituted tobacco sheets as wrapper, and increasing interest in converting farm-land in the area into exurban estate farms, however, points to the ultimate elimination of the industry in this area. Georgia and Florida also provided this type of tobacco until 1978.

This production regionalization has been maintained since the 1930s at least partly by the federal government's system of growth allotments. The Agricultural Adjustment Act of 1938 authorized the first marketing quotas in which growers are guaranteed a specific parity price for their crop. The aim was, and is, to maintain a balance between yield and demand. Raw tobacco can be kept in storage for years so that a series of surplus production years potentially had disastrous economic consequences for the farmer. The system has been in place for air-cured, burley, and fire-cured tobaccos from the beginning. Cigar binder and Ohio filler tobaccos came under quota in 1951. Pennsylvania filler tobacco was never under quota, and the system no longer covers Maryland and shade binder types.

The quota system sets individual farm production quotas based on overall production goals and the farm's own history. Thus, the system is tied directly to that set of farms that produced tobacco when the first quotas were set in 1938. The system at first used acreage quotas, but rapidly increasing yields between 1950 and 1964 (from 1,269 to 2,067 pounds per acre) forced an adjustment to production quotas instead. One result is that yields have been relatively stable since the mid-1960s. Legislation in the early 1960s further adjusted the system by allowing for the lease and transfer of allotments within counties. That was replaced for flue-cured tobacco in the late 1980s by a system that allowed for the sale of quotas to other growers within the same county. The end result remains a federal system that nearly demands that all tobacco cultivation must be confined to the same set of counties that were important fifty years earlier.

The different demands of tobacco curing have resulted in barns that mirror their use. Traditionally flue-cured tobacco was heat dried in square, airtight barns which were loaded with the tobacco leaf tied into groups of leaves called hands. The hands were hung on removable poles the width of the barn which could be set on wall brackets to allow the tobacco to hang loose without touching other hands or the walls. The barn was then heated for several days to cure the tobacco. This labor intensive system has largely been replaced by a system utilizing more labor-efficient, metal "bulk" units today. Tobacco from the fields is loaded into bins that roll into these highly efficient curing units on conveyors. The entire job of loading the new bulk units can be accomplished by two workers in a matter of minutes. The old flue-cured barns have become a derelict, decaying, and rapidly disappearing part of the landscape. Air-cured burleys, Maryland, and wrapper tobaccos by comparison need to hang for most of a year as they slowly dry. Their barn is much larger and very open, often lined with slats that are opened during periods of nice weather. These barns will be filled from almost floor to rafter after a successful harvest.

Notes

1. While much superior to the local leaf, today it would probably be considered heavy, strong, coarse, and dark.

The demand for large quantities of repetitive hand labor in the vegetable industry has continued even after the development of automated planters and harvesters. This crew is operating a celery planter near Castroville, California. (RP).

Vegetables

The evolution of the American vegetable industry from local market gardening to global behemoth has occurred in five phases. The process began in the late eighteenth century as farmers along the eastern seaboard began increasing production of meat, eggs, vegetables, and other perishables to meet growing local urban demand. Vegetable acreages increased with market demand to allow the beginnings of specialized cultivation a few decades later. The second phase began after the Civil War as market demand supported the creation of distant early- and late-season specialty areas. Tomato production jumped southward to Norfolk, Charleston, Beaufort (South Carolina), and Savannah where fast ships could rush the produce to market. The elaboration of the rail system around Chicago fostered a similar process along its south and westward tentacles into Iowa, southern Illinois, and Mississippi. As late as 1910, however, the primary theme of vegetable cultivation was local production with only minor production clusters around the cities and along the transportation corridors.

Nicolas Appert, a French confectioner, in 1809 developed a process of preserving food in sealed containers after eight years of experimenting with spoiled food and exploding jars. While effective, the Appert process was both slow and expensive. canning could not become an important force in American agriculture until an inexpensive method was developed. The retort, a pressured cooking vessel, began to be used in 1870 to increase the reliability of the cooking process. A variety of other processing innovations rapidly followed, including an automated corn kernel remover (1875), a pea sheller (1883), and a pea viner (1889). The pea sheller, for example, replaced 600 hand shellers in Owasco, New York, the day it was introduced. The creation of an economical container was also a major problem. A tinsmith could make about ten cans per day in 1840. Continuous soldering machines were introduced in the 1870s, which raised productivity to 1,500 cans per worker day. Can costs plummeted with the introduction of the double crimped open-top can after 1910 as production increased to about 35,000 cans per worker day.

This combination of automated preparation equipment, continuous processors, and inexpensive containers allowed for the rapid expansion of the canning industry after 1890. The industry, however, continued to be so labor intensive that few economies of scale could be achieved through plant expansion. Small producers and large had about the same unit costs, and the number of competitors proliferated. The National Canners' Association tabulated 2,412 canneries in 1915, overwhelmingly small operations serving local markets.

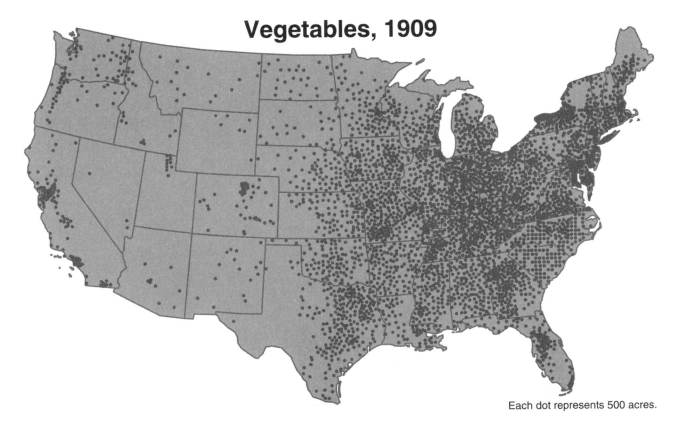

Vegetables, 1909

Each dot represents 500 acres.

Vegetables, 1992

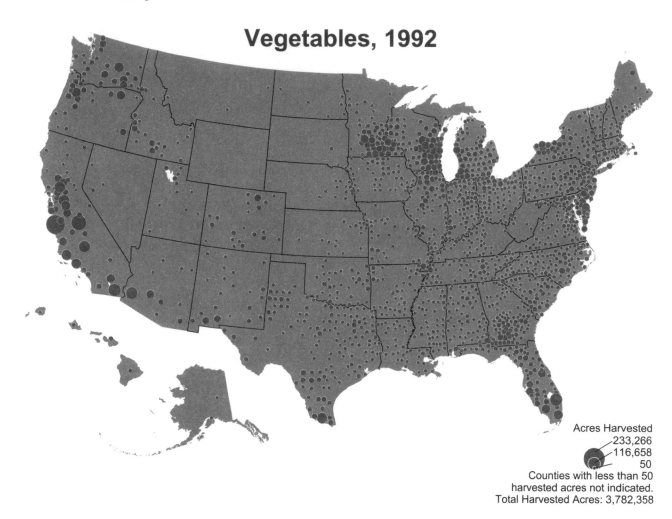

Acres Harvested
233,266
116,658
50
Counties with less than 50
harvested acres not indicated.
Total Harvested Acres: 3,782,358

took place in a remarkably short period. The F. & J. Heinz Company was formed in Sharpsburg (near Johnstown), Pennsylvania, in 1876 and became a household name for pickles, condiments, and other prepared foods by the turn of the century.[1] Dr. John T. Dorrance, a chemist, created the first canned condensed soup at the Joseph Campbell Preserve Food Company of Camden, New Jersey, and ultimately built that company into the world's largest soup producer. The Biardot family (New Jersey) formed the Franco-American Food Company in the 1880s utilizing local produce for their signature spaghetti and specialty products. Green Giant (1903), based in Minneapolis, concentrated on peas, sweet corn, and asparagus grown in southern Minnesota and Wisconsin in the firm's early years. Libby, McNeil and Libby began as a subsidiary of Swift, but was spun off as a separate company in 1918 with offices and production concentrated in the Midwest. The Stokley Brothers started packing in Delaware, but purchased Van Camp

Virginia and Maryland alone had 405 canneries.

The third production elaboration phase began when processor demands outstripped the traditional sources of product for canning—traditionally excess production and visually imperfect fruit and vegetables. This occurred as the emerging industry giants were able to establish market demand for their branded products through advertising and national product distribution. Henry J. Heinz erected the first electrically lit sign in Manhattan in 1900, a forty-foot pickle and message, at 23rd Street and 5th Avenue. Heinz was a mar-

keting genius, who realized very early that his "57 varieties" trademark could become a catchphrase, even though there weren't fifty-seven varieties when he coined it and there were more than 200 by 1900. California Packing Corporation (Del Monte brand) was formed in 1916 and launched the first national advertising campaign by a fruit and vegetable canner in 1917. They also linked with the Great Atlantic and Pacific Tea Company for a cooperative marketing effort that included special wholesale discounts to increase the California company's exposure on the East Coast.

The emergence of the dominant canners

Table III.33
Leading Vegetable Producing Counties, 1992

Counties	Acres Harvested	Percent of Total
Monterey, CA	233,266	6.2
Fresno, CA	143,521	3.8
Palm Beach, FL	84,624	2.2
Imperial, CA	84,569	2.2
Yuma, AZ	75,892	2.0
Kern, CA	68,40	1.8
San Joaquin, CA	59,068	1.6
Yolo, CA	58,695	1.6
Hidalgo, TX	53,855	1.4
Santa Barbara, CA	50,673	1.3
Ten county total	912,570	24.1
Total U.S. acreage	3,782,358	

Table III.34
Principal Vegetables Harvested, 1993

	Acres Harvested (Fresh)	$Thousands (Fresh)	$Thousands (Processed)
Artichokes[a]	8900	45,499	
Asparagus	83,550	115,194	47,872
Beans (green)	90,400	173,261	116,302
Broccoli	107,200	250,826	28,622
Cabbage	71,790	237,404	6270
Cantaloupes	106,250	282,684	
Carrots	101,700	265,251	35,388
Celery	32,320	201,840	
Corn (sweet)	199,030	292,492	196,750
Cucumbers	61,600	292,492	126,255
Lettuce (all types)	274,680	1,478,604	
Onions	150,980	785,079	
Peas (green)[a]	248,700	88,146	
Peppers (sweet)[a]	63,180	402,373	
Tomatoes	132,680	1,113,922	581,893
Watermelon	196,180	251,734	

[a]Fresh and processed

Vegetables: 1925, 1993

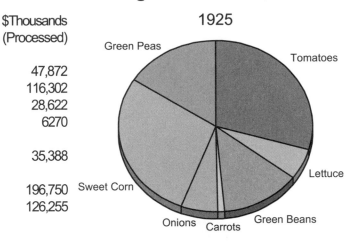

1925

1993

Vegetables for Processing, 1993

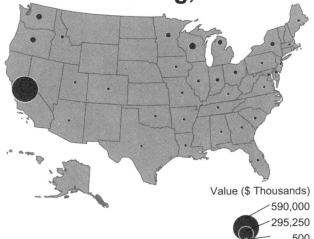

Value ($ Thousands)
590,000
295,250
500

Packing Company in Indiana to allow expansion into the Midwest. The addition of Santa Cruz (California) Fruit Company two years later established the Stokley Brothers as a national company with processing plants in each of the three main processed vegetable areas. California Packing Corporation dominated the California scene, though not without competition. Del Monte's flagship peaches were quickly joined by canned asparagus and a variety of other products primarily grown on their two California ranches. The Hunt Brothers

Fruit Packing Company (1890) was one of the largest competitors in the region.

Each of today's dominant companies began in the late nineteenth century based on local markets and production. The absence of economies of scale through expansion of existing plants, coupled with the desire to enter new markets in other regions, encouraged them to acquire weak competitors in distant markets, rather than expand existing facilities. Ultimately, the distribution of processed vegetable production became a repetition of Earle's 1900 comment that "the particular location of the present shipping centers seems often to have been due to the fact that some pioneer in the business chanced to settle there and to succeed...." The early success of Dorrance, Heinz, the Hunt and Stokley Brothers, and others similarly determined that the premiere processed vegetable centers were to be upstate New York, the Delmarva shore, the central Corn Belt, southwestern Wisconsin and Minnesota, eastern Washington, and the Central Valley of California, instead of just as well-suited locations elsewhere.

The fourth phase of the industry's evolution had its roots in 1915 when Clarence Birdseye went to Labrador on a government expedition for two years. Fascinated with the frozen food phenomenon, he returned to the United States and set about developing a process for freezing fish as an employee of the U.S. Bureau of Fisheries. He formed the General Seafoods Company in 1925, developed a continuous freezing process in 1927, and sold the company to the Postum Cereal Company (now General Foods) in 1929. The new General Foods Company, under the leadership of Marjorie Post, introduced sixteen frozen poultry, meat, fruit, and vegetable products in 1930. Again ahead of its time, the company primarily sold to institutional purchasers until after World War II when frozen food consumption dramatically increased.[2]

The expansion of the frozen food industry, coupled with flattening demand for canned vegetables, laid the foundations for a new geography of vegetable production in the 1960s and 1970s. New freeze-processing plants were needed to meet increasing demand. While most of General Food's frozen food plants were in New York through the 1950s, high land and labor costs made them and other processors reluctant to commit large sums to traditional eastern production areas. Reduced profits from canned produce also brought a round of corporate mergers as packers were swallowed up by the food conglomerates such as Nestlé and Pillsbury. Corporate financing by the new owners made expansion feasible. Inexpensive immigrant labor, few climatic risks, and comparatively low land costs beckoned from the West. Oregon, Washington, and adjacent Idaho became important for green peas, green beans, and sweet corn, while California's production dominance of frozen broccoli and cauliflower was solidified.

The current phase has its roots in the changing demographics of postwar America and the accompanying unparalleled restructuring of traditional American family life. The American diet changed rapidly after 1950 as Americans consumed a vast array of new foods, became more health conscious, and dual worker households created more disposable income for dining away from home more than ever. The perfection of the salad bar in the 1960s in California, the explosion of Chinese, Mexican, and other ethnic cuisines, and the arrival of millions of emigrants from Asia, Africa, and Latin America created a dramatically changing marketplace. Demand for crops consumed raw, such as lettuce, fresh cucumbers, and sweet peppers, exploded; markets for little-known crops associated with the new lifestyle—bok choy, fresh mushrooms, and exotic herbs—were created almost overnight.

This restructuring of demand in an environment dominated by a handful of distributors brought about an unparalleled concentration of production. Almost half of the nation's vegetables are grown in California, more than three-quarters in just four states (California, Florida, Arizona, Texas), and 90 percent in only ten states. Part and parcel of this trend is the dominance of wholesale distribution by a handful of packing houses and retail sales by a declining number of large grocery store chains. All commercial lima beans and spinach,

Most California head vegetables for the fresh market are harvested by hand with the assistance of mobile packing platforms that can interchangeably be used for broccoli (shown here), lettuce, and similar crops. (JF/RP)

Cabbage, 1992

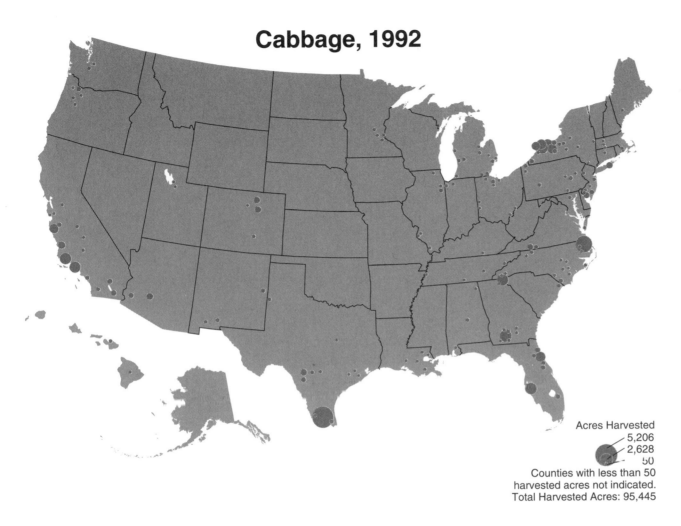

Acres Harvested
5,206
2,628
50
Counties with less than 50
harvested acres not indicated.
Total Harvested Acres: 95,445

popular during the age of discovery because it could be pickled into sauerkraut and used as a method for warding off scurvy. It was one of the most common winter vegetables throughout the nineteenth and early twentieth centuries, as is seen in the production levels even in 1924. It has not fared well in the modern era, however, and its relative importance has slipped below even broccoli's today. Commercial sauerkraut is still predominantly produced in Wisconsin and upstate New York, while the commercial fresh crop is usually grown as a winter fresh vegetable in the Sunbelt.

Carrots

The lowly carrot has undergone a renaissance of interest with the rise of the restaurant salad bar. The vast majority of all carrots (88 percent) are sold fresh; most of the remainder are frozen. About two thirds of the $400-million carrot crop comes from California ($195 million), followed by Washington ($23 million), Florida ($18 million), and Michigan ($17 million). Almost all of the California crop is grown in Kern (Bakersfield) and Imperial counties, while the Washington crop is largely grown under irrigation in central Washington. Washington leads the nation in the harvest of carrots for processing, its production more than doubling between 1991 and 1993.

99.5 percent of the sweet corn, 99 percent of the processed tomato crop, 95 percent of the green peas, and similar amounts of most other vegetables were grown under contract in 1993. Contracts tend to be written with only the largest growers, often in units of a 1,000 or more acres. This process has guaranteed a concentration of production in fewer and fewer growing areas. The continuation of multiple production areas for fresh produce generally has more to do with market timing than competition.

Imports account for about one-quarter of the nation's produce consumption, primarily as off-season fresh vegetables from Latin America. Mexico has long been the most important source of off-season produce for American distributors, but Chile and Costa Rica are of increasing importance. Much of the produce from these nations is grown under contract to American or multinational distributors.

Cabbage

The Brassicaceae family includes a wide range of European vegetables including the cabbage, rutabaga, turnip, mustard, broccoli, cauliflower, and kohlrabi. Cabbage was

Cucumbers

The cucumber was domesticated in India and widely known in Europe during the Middle Ages. It was introduced to the New World in 1498 and to North America by de Soto and Cartier in the sixteenth century. It remained relatively unimportant as a commercial crop in the fresh market until after World War II and the rising importance of salads as a dietary item in the 1950s and 1960s. Florida produces about 40 percent of the crop by weight and half by value. California, the second producer, grows about one third the Florida crop. Cucumber pickles are still grown in twenty-nine states with no single state producing more than 16 percent of the crop. North Carolina's Cates and Mt. Olive pickle packers have made it the largest producer of cucumbers for pickles, while Michigan is second.

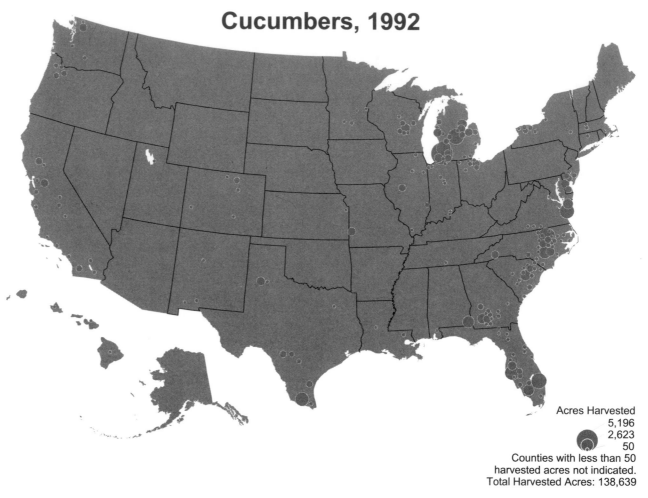

Cucumbers, 1992

Acres Harvested
5,196
2,623
50
Counties with less than 50
harvested acres not indicated.
Total Harvested Acres: 138,639

Carrots

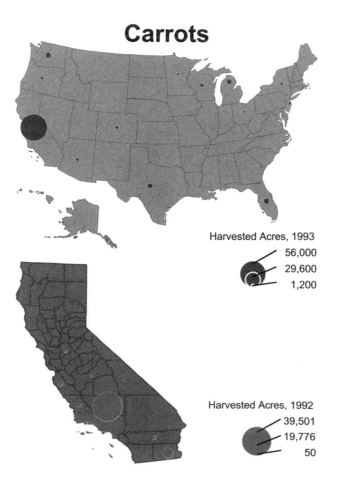

Harvested Acres, 1993
56,000
29,600
1,200

Harvested Acres, 1992
39,501
19,776
50

Green Beans

The green (snap) bean is one of the many American bean varieties that were found in Mexico during the sixteenth century. While the green bean has long been a home garden favorite, it has not shipped or stored well for commercial production. Genetic research focused on creating a plant that was easy to harvest mechanically and ripened all at one time. The vast majority of all green beans consumed today are processed. Early canned production was concentrated in New York and the Del-

marva peninsula with smaller production in the southeastern Wisconsin and Minnesota vegetable zones. Fresh production came primarily from New Jersey and Florida. The decline of vegetable production in New York and New Jersey, along with the changing fortunes of the processors, has brought a major shift in processed production as Wisconsin has risen to dominance. Oregon, Michigan, and New York have about equal production, though New York has been declining in importance in recent years. Fresh production is largely in the South, especially Florida.

Green Beans, 1992

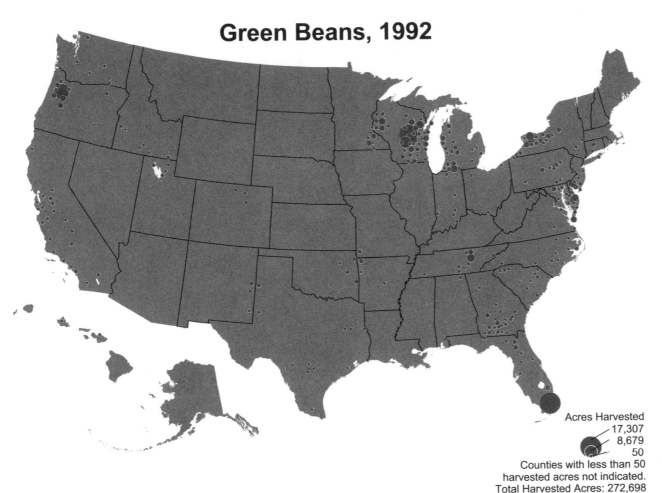

Acres Harvested
17,307
8,679
50
Counties with less than 50
harvested acres not indicated.
Total Harvested Acres: 272,698

Lettuce: 1909–1993

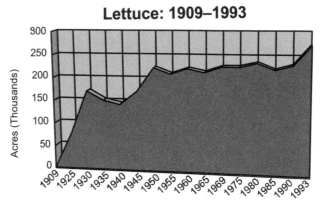

Lettuce

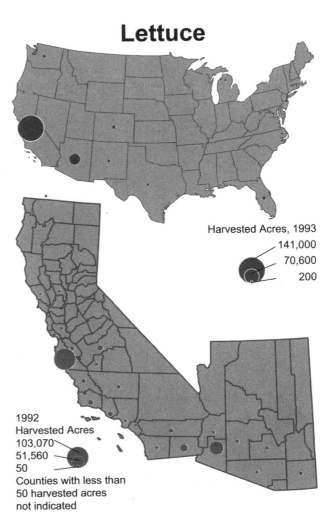

Harvested Acres, 1993
141,000
70,600
200

1992
Harvested Acres
103,070
51,560
50
Counties with less than
50 harvested acres
not indicated

Table III.35
Leading Lettuce Producing Counties, 1992

Counties	Acres Harvested	Percent of Total
Monterey, CA	103,070	35.9
Yuma, AZ	59,429	20.7
Imperial, CA	23,498	8.2
Fresno, CA	14,357	5.0
Santa Barbara, CA	11,676	4.1
Ventura, CA	10,343	3.6
San Luis Obispo, CA	7,684	2.7
Santa Cruz, CA	6,544	2.3
San Benito, CA	6,432	2.2
Palm Beach, FL	5,960	2.1
Riverside, CA	5,787	2.0
Kern, CA	4,562	1.6
Ten county total	259,342	90.2
Total U.S. acreage	287,468	

Lettuce

The growing importance of restaurant salad bars has pushed the value of lettuce production to almost equal the value of fresh tomatoes in recent years. While lettuce was introduced during the early colonial period to the United States, it remained comparatively unimportant until salad bars became a common restaurant feature during the health-conscious 1970s. The 1924 lettuce crop, for example, was valued at only 4 percent of the seven-vegetable crop in the nation, about 8 percent of the 1932 crop, six percent of the 1950 crop, and 22 percent in 1993. The concentration of the crop

This lettuce harvester and crew migrate from their Salinas Valley base to the Imperial Valley during the winter, and to the southern San Joaquin Valley in the spring. (RP/JF)

intensified with its growing market. While only 61 percent of lettuce was produced in California in 1950, four-fifths of the leaf lettuce and three-quarters of the $1.3-billion head lettuce crop came from California in 1993. Arizona and California now account for 94 percent of the nation's head lettuce and 98 percent of the leaf lettuce.

Virtually all lettuce production within California/Arizona is grown on contract with production following the seasons. Harvesting begins the year in the lower Colorado (near Yuma, Arizona) and Imperial Valleys. Production moves to the south-ern San Joaquin in the early spring and to the Salinas Valley for the summer and fall. It returns to the Colorado and Imperial Valleys in early winter and the cycle begins anew. Most equipment, including the harvesting machinery, continuous coolers, and loading equipment, moves from site to site, as do many of the personnel.

Onions

The onion was one of the earliest domesti-cated crops of western Asia and had be-come an integral part of the European diet prior to New World colonization. This ubiquitous crop could be found in the home truck gardens of the early colonists and was one of the first vegetables commer-cially produced in the United States. Its ability to thrive under a wide range of environmental conditions meant that onions were grown both for home and sale throughout almost the entire country. More recently, however, need for concentrated production within an industrialized agricul-ture has tended to bring a high degree of concentration to onion cultivation. Califor-nia is the leading state with $155 million in 1993, primarily because both summer and spring crops are produced. Eastern Oregon's $118-million crop is followed by that of Colorado ($88 million), Texas ($80 mil-lion), New York ($70 million), and Idaho ($62 million). Some areas with small production have successfully created brand recognition for their product, bringing premium prices to growers of Vidalias

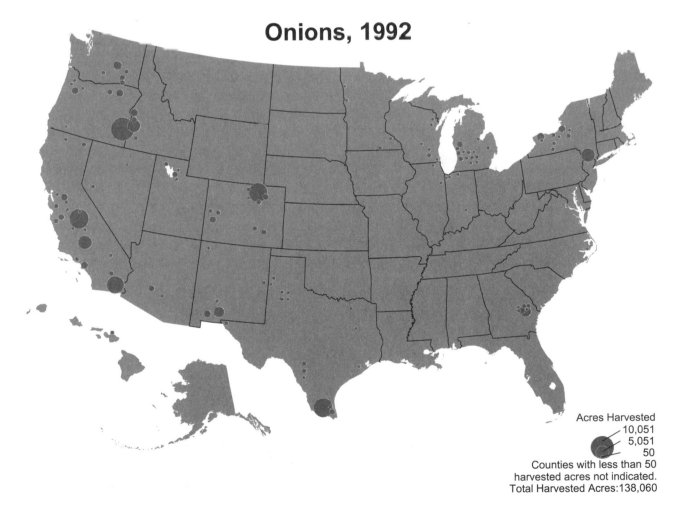

Onions, 1992

Acres Harvested
10,051
5,051
50
Counties with less than 50 harvested acres not indicated.
Total Harvested Acres: 138,060

Table III.36
Leading Onion Producing Counties, 1992

Counties	Acres Harvested	Percent of Total
Malheur, OR	10,051	7.3
Fresno, CA	8,387	6.1
Hidalgo, TX	8,156	5.9
Weld, CO	7,960	5.8
Imperial, CA	7,115	5.2
Orange, NY	5,274	3.8
Kern, CA	5,266	3.8
Canyon, ID	4,966	3.6
Dona Ana, NM	3,806	2.8
Tattnall, GA	3,668	2.7
Ten county total	64,649	46.8
Total U.S. acreage	138,060	

(Georgia) and Walla Wallas (Washington). Production in both of these areas doubled between 1991 and 1993 because of their growing demand.

Sweet Corn

Modern sweet corn is based on a recessive genetic trait and did not become widely consumed until after 1900 when two new varieties were introduced by seed companies. A second group of hybrids was introduced in the 1940s that quickly dominated

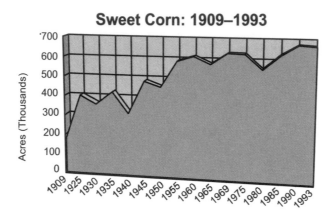

Sweet Corn: 1909–1993

Acres (Thousands)

Table III.37		
Leading Sweet Corn Producing Counties, 1992		
Counties	Acres Harvested	Percent of Total
Palm Beach, FL	33,776	4.4
Grant, WA	27,586	3.6
Fond du Lac, WI	21,846	2.9
Dodge, WI	19,006	2.5
Marion, OR	16,620	2.2
Renville, MN	15,927	2.1
Faribault, MN	13,318	1.8
Portage, WI	11,760	1.6
Goodhue, MN	10,323	1.4
Sheboygan, WI	9,893	1.3
Ten county total	180,055	23.6
Total U.S. acreage	762,132	

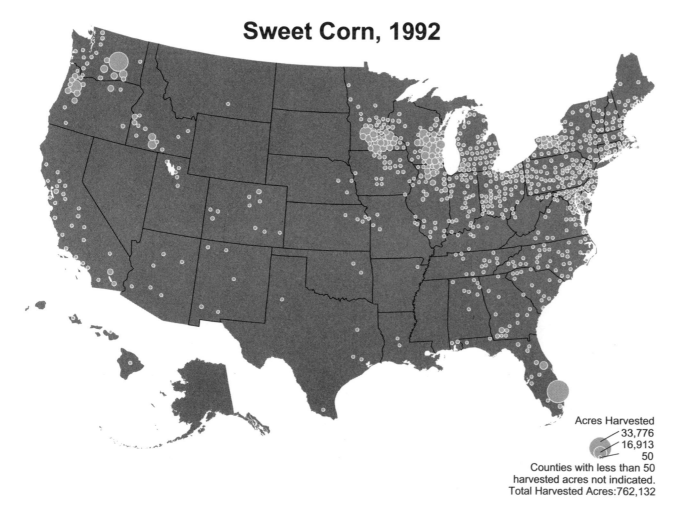

Sweet Corn, 1992

Acres Harvested
33,776
16,913
50
Counties with less than 50 harvested acres not indicated.
Total Harvested Acres: 762,132

the canning and freezing markets. Today's commercial sweet corns are largely based on varieties that keep their field-fresh flavor much longer after harvest than those developed in the 1970s.

While sweet corn can be grown in home gardens in almost every region of the nation, Florida and California are the most important areas of fresh production, though with only 42 percent of the nation's total. No other state produces even 10 percent of the total fresh commercial crop. Minnesota, Washington, Oregon, and Wisconsin are the largest producers of processed sweet corn with about 72 percent of the total, although individual harvests within each state vary significantly from year to year.

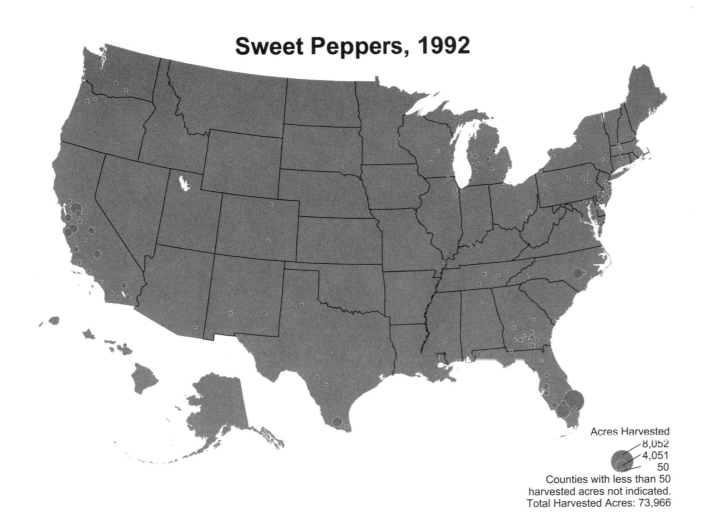

Sweet Peppers, 1992

Acres Harvested
8,052
4,051
50
Counties with less than 50
harvested acres not indicated.
Total Harvested Acres: 73,966

Sweet Peppers

The 1993 sweet pepper crop, valued at more than $400 million, made this unlikely vegetable the fifth most important truck vegetable in the United States. A favorite ingredient of both salad bars and Mexican dishes, the sweet green pepper has had one of the most phenomenal rises in American agricultural history. Florida is the largest producing state with half the harvest. California follows with an additional $131-million crop (33 percent), or a total of 84 percent for the two states. While small amounts are frozen in the increasingly popular "vegetable medleys," virtually the entire crop is sold fresh.

Tomatoes, 1992

Artichokes

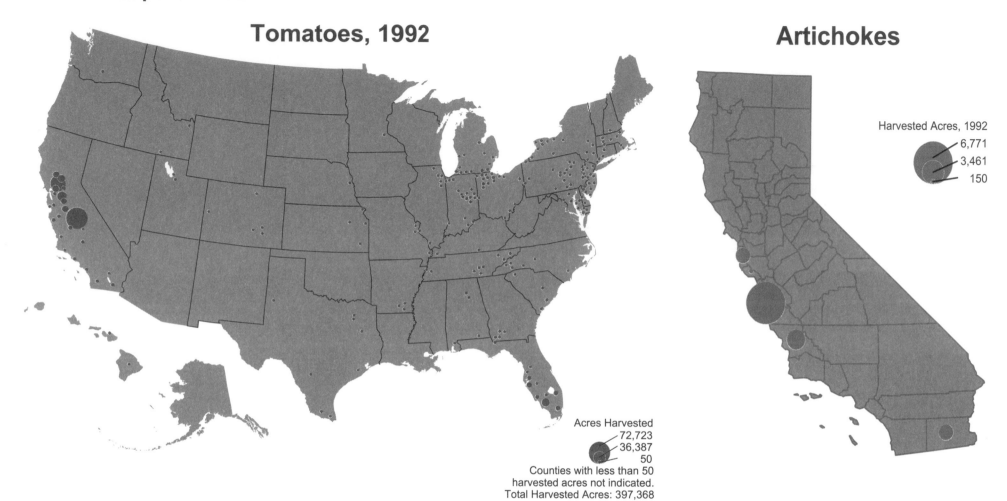

Harvested Acres, 1992
6,771
3,461
150

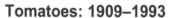

Acres Harvested
72,723
36,387
50
Counties with less than 50
harvested acres not indicated.
Total Harvested Acres: 397,368

Table III.38
Leading Tomato Producing Counties, 1992

Counties	Acres Harvested	Percent of Total
Fresno, CA	72,723	18.3
Yolo, CA	50,531	12.7
San Joaquin, CA	25,560	6.4
Collier, FL	20,289	5.1
Sutter, CA	17,266	4.0
Solano, CA	14,309	3.6
Colusa, CA	13,640	3.4
Merced, CA	13,618	3.4
Stanislaus, CA	11,851	3.0
Manatee, FL	10,533	2.7
Ten county total	250,320	63.0
Total U.S. acreage	397,368	

Tomatoes: 1909–1993

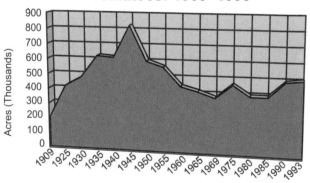

Asparagus

Celery

Cauliflower

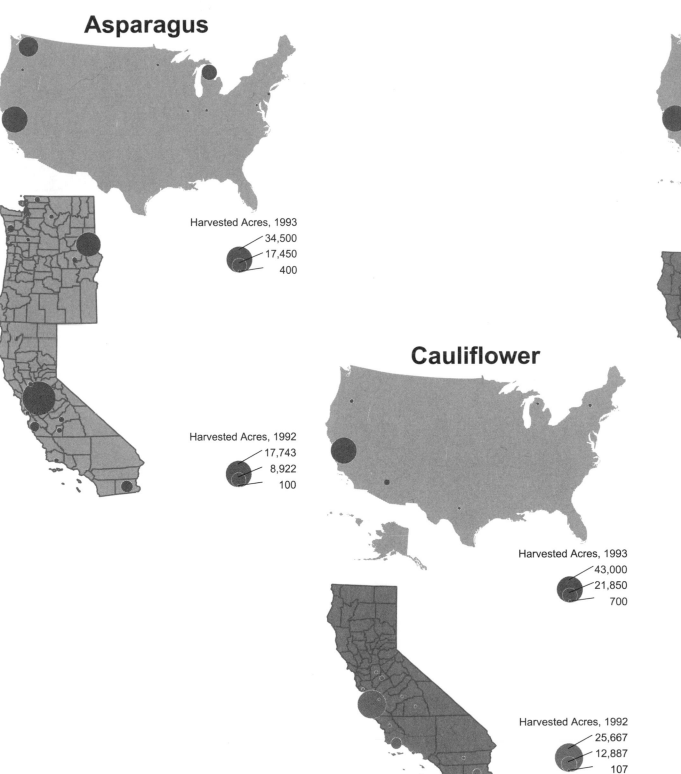

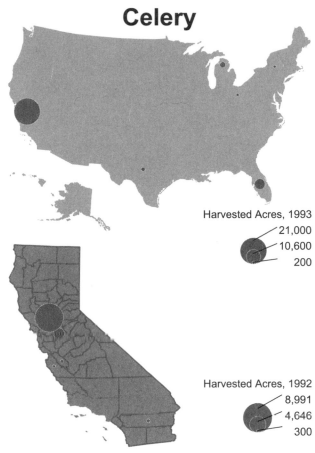

Harvested Acres, 1993
34,500
17,450
400

Harvested Acres, 1992
17,743
8,922
100

Harvested Acres, 1993
21,000
10,600
200

Harvested Acres, 1992
8,991
4,646
300

Harvested Acres, 1993
43,000
21,850
700

Harvested Acres, 1992
25,667
12,887
107

Broccoli

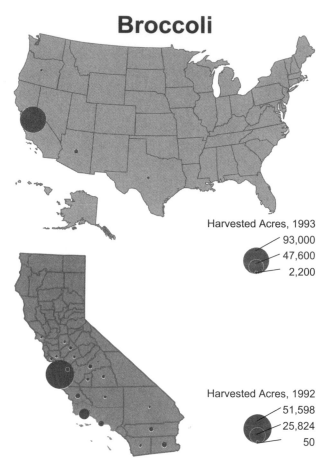

Harvested Acres, 1993
- 93,000
- 47,600
- 2,200

Harvested Acres, 1992
- 51,598
- 25,824
- 50

Green Peas: 1909–1993

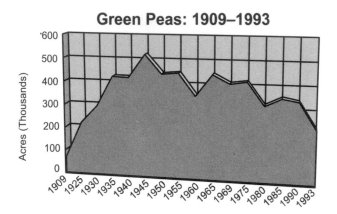

Green Peas, 1992

Acres Harvested
- 27,495
- 13,773
- 50

Counties with less than 50
harvested acres not indicated.
Total Harvested Acres: 328,,287

Tomatoes

The $1.7 billion 1993 tomato harvest made it America's most important vegetable crop, although fresh lettuce production surpassed fresh tomato sales for the first time in history in 1992. The tomato is indigenous to the Peruvian desert coast, but apparently was not domesticated there. The plant was initially brought to Europe from Mexico as an ornamental and a curiosity. Europeans were apparently slow to accept it as a food because of its membership in the nightshade family. It did not become a common part of the American diet until the late nineteenth century, possibly in association with the arrival of large numbers of southern and eastern Europeans.

Commercial tomato production continued in a large number of production areas throughout 1920s. The Delmarva/southern New Jersey district was the largest producing area with more than one-third of the crop, followed by California, with about 10 percent, Indiana, and the remainder of Virginia. Florida supplied most of the winter crop. Some specialty areas, such as Copiah County, Mississippi, and Cherokee County, Texas, developed as rail shipment centers have continued as important production anomalies since that time.

California now produces almost one half of the total crop and 91 percent of processed tomatoes in the United States. The 352,000 acre (1993) California harvest is grown primarily in the Central Valley in a broad swath more than 200 miles long from Colusa to Bakersfield. Small amounts also are produced in the Salinas and Imperial Valleys. Tomatoes cultivated for processing continue to be harvested in Ohio, Indiana, and Michigan as relics of the early origins of Heinz and other packers.[3] Florida's $593-million 1993 fresh market tomato crop was more than twice the fresh production of California. South Carolina, Virginia, Ohio, and Georgia follow California in fresh crop production. Two of the earliest centers, New Jersey and Delaware, produce insignificant amounts of fresh product today.

Other Vegetables

The above-mentioned vegetables were the only crops to exceed $300 million in production in 1993, but six others exceeded $100 million in production that year. The geography of these crops, however, is much like those already described, with the largest production centers located in south Florida and California. Artichokes are possibly the most geographically restricted crop in the nation with only sixty-three farms, located primarily along a fifty-mile section of the central California coast. Increasing demand in recent years has brought the development of more distant growers in the Imperial Valley and near Eureka to the north, but Castroville, California, still correctly touts itself as the artichoke capital of the world. The growth in production of minor vegetables such as Chinese cabbage, herbs, snow peas, and others is quite strong and often associated with smaller producers.

Cantaloupes, 1992

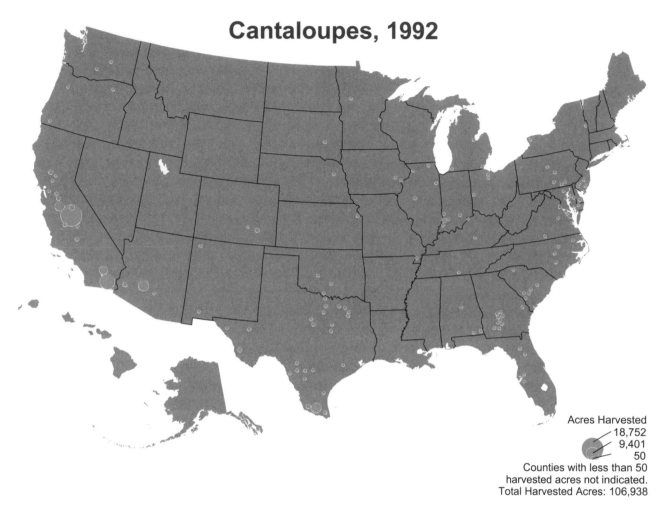

Acres Harvested
— 18,752
— 9,401
— 50
Counties with less than 50
harvested acres not indicated.
Total Harvested Acres: 106,938

family originated in Africa and was cultivated in classical Greece. Once introduced to the New World, the muskmelon quickly spread through the Amerindian population after its introduction in the sixteenth century. The cantaloupe as we know it was created in the Central Valley of California in the 1920s and perfected in the 1930s. Two thirds of the nation's production still comes from there and 92 percent from California, Arizona, and Texas combined.

Notes

1. Reorganized as the H. J. Heinz Company in 1888.
2. The demand for frozen convenience foods during this period is mirrored by the introduction of the first TV dinner in 1954.
3. Vegetables for processing are usually cultivated at a very large scale under contract with processors. In 1993, for example, tomatoes harvested for fresh market had an average value of $36.30 per hundredweight. Tomatoes harvested for processing brought an average $58 per ton. Farmers specializing in processed vegetables believe that their ability to cultivate and harvest vast quantities of produce without concern about market demand, competition from other regions, or shipment to distant markets more than outweighs the disadvantages of lower prices for their output.

Melons

The USDA includes melon production with its vegetable statistics. The watermelon is the most important melon with 196,000 acres planted, valued at approximately $250 million. The watermelon is an African crop that was introduced to Europe (Cordoba, Spain) by the Moorish invaders in 961. The Spanish introduced watermelon to Florida before 1576, and it was common in most parts of colonial Spanish America by 1650. Amerindians quickly adopted this new melon and, like many Old World

crops, its cultivation raced ahead of exploration. Amerindian farmers cultivated it throughout its environmental range in the eastern United States at the time of first local European contact. Texas harvested the most acreage in 1993 (42,000), followed by Florida (37,000), and Georgia (30,000), although the early season production in Florida ($67 million) and California ($58 million) gave these states the highest crop values.

Cantaloupes are the most widely grown of the muskmelons, which in turn are related to the cucumber. This muskmelon

Watermelons, 1992

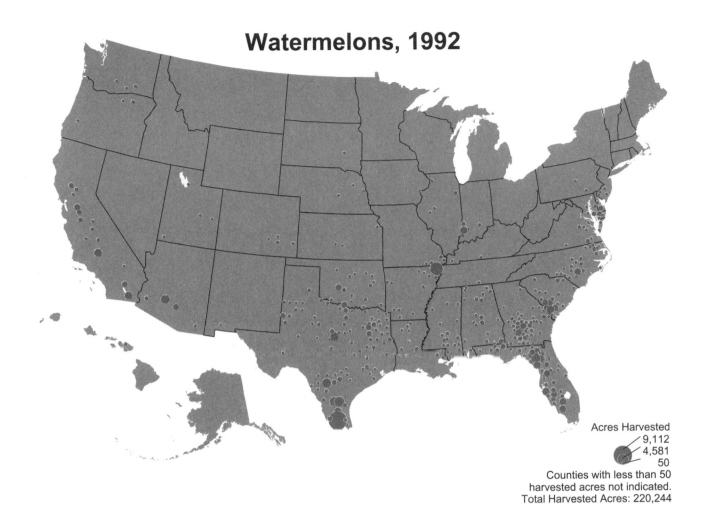

Acres Harvested
9,112
4,581
50
Counties with less than 50
harvested acres not indicated.
Total Harvested Acres: 220,244

Wheat dominates the rolling Palouse Hills of eastern Washington. (RP)

Wheat

Wheat (*Triticum—*) is one of the earliest domesticated plants of the Neolithic Revolution in southwestern Asia. It is believed to have evolved from spelt, which had, in turn, earlier evolved from emmer. Emmer has been cultivated in the region for at least 9,000 years and all three grains continue to be cultivated today. Wheat's higher yields and adaptability soon made it the favored grain of the region and the single most important element in the Neolithic diet. Wheat production spread from this core area into northern India, across Asia to northern China, and westward to northern Africa and Europe. At the beginning of the Age of Discovery, wheat bread (with or without additions of other grains such as rye, barley, and oats) was the most important dietary staple throughout Europe.

Most early European colonists to the United States were initially forced to alter their diet while they established a traditional western European agrarian society. The first recorded wheat crop was planted in 1602 on Elizabeth Island in Buzzards Bay, Massachusetts, followed by plantings at Jamestown (1611) and Plymouth Colony (1621). A dietary dichotomy evolved in the American colonies with most people consuming cornbread while the economic elite consumed the more expensive wheat bread.

Wheat was attempted in all areas with varying success. New England farmers were plagued with wheat rust, small fields, and competition from comparatively inexpensive imported foreign grain. Southern planters concentrated on more profitable export crops with little interest in food production. Middle Atlantic farmers were the most successful, and Pennsylvania soon became the nation's breadbasket.

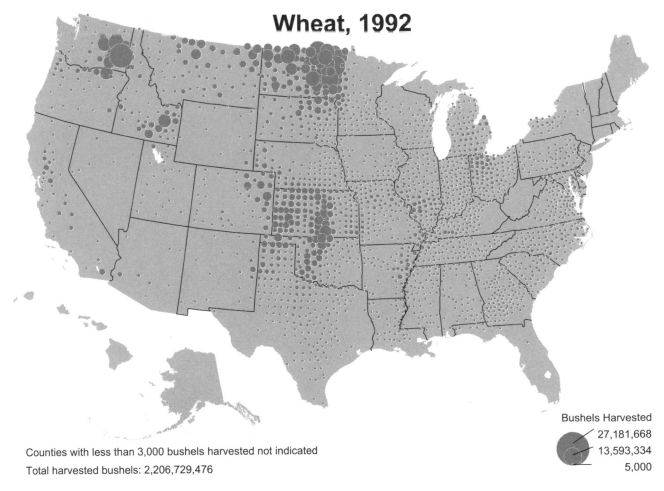

Wheat, 1992

Counties with less than 3,000 bushels harvested not indicated

Total harvested bushels: 2,206,729,476

Bushels Harvested
27,181,668
13,593,334
5,000

Table III.39
Leading Wheat Producing Counties, 1992

Counties	Bushels	Percent of Total
Whitman, WA	27,181,668	1.2
Polk, MN	21,799,164	1.0
Cavalier, ND	21,714,592	1.0
Cass, ND	19,853,537	0.9
Marshall, MN	18,723,403	0.8
Lincoln, WA	17,092,595	0.8
Walsh, ND	16,283,920	0.7
Barnes, ND	16,129,139	0.7
Stutsman, ND	16,026,065	0.7
Ward, ND	15,768,090	0.7
Ten county total	190,572,173	8.6
U.S. production	2,206,729,476	

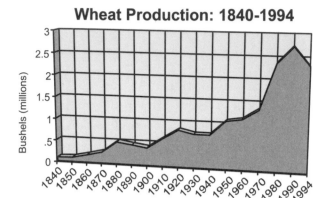

Wheat Production: 1840-1994

Wheat, 1859–1994

1859

U.S. Production: 838,793,000 bushels

1909

U.S. Production: 2,552,190,000 bushels

1954

U.S. Production: 908,928,000 bushels

1994

U.S. Production: 2,438,000,000 bushels

Bushels (Thousands)
388,500
194,307
114

The center of wheat production migrated westward in the 1820s after the completion of the Erie Canal. The Midwest became the center of wheat production throughout the midnineteenth century, until it was displaced by even more efficient farming operations in the Great Plains. Wheat continued to be grown in the Corn Belt until the corn-clover-wheat crop rotation evolved into the corn-soybean rotation and periodic wheat rotations became unnecessary.

Production in the Far West began to be important in the late nineteenth century in California and later in eastern Washington and southern Idaho. Creating new equipment (e.g., the crawling tractor) and strategies (e.g., alternate year dry fallow) specifically adapted for their environments, western farmers continue to be competitive with Great Plains farmers today. While alternative crops have lured many California wheat farmers into other endeavors in recent years, eastern Washington and southern Idaho farmers continue to be major producers with some of the highest yields per acre in the nation. Common white wheat, sown as both winter wheat and spring wheat, continues to be the most important variety for both of these areas.

Wheat Production Cycle

Wheat typically is planted in either the fall (winter wheat) or spring (spring wheat) depending on the winter temperatures of the growing area. Winter wheat is planted from early to mid-September in most of the eastern United States. Spring wheat is planted between early March (Kansas) and mid-April (North Dakota). Freezing temperatures on plant tissue are only one of several hazards of winter wheat production that include heaving, smothering, and physiological drought (inaccessibility to moisture because it is frozen). Actual hardiness and the degree of tolerance to cold vary from variety to variety, even within the major types of wheat.

Wheat has been left largely unattended between planting and harvest for thousands of years. Some contemporary farmers have begun to spray for insects and disease since World War II, but more striking is the rise of irrigated wheat in the Far West. The invention and spread of center pivot and other motorized sprinkler systems for alfalfa and potatoes made irrigation of other crops in the rotation cycle easy and inexpensive. The highest wheat yields ever recorded occurred in the sprinkler irrigated fields of eastern Washington.

Harvest in the United States typically begins in May in central Texas and marches inevitably northward. Most Great Plains wheat was once harvested by contract harvesters who began their seasons with their work crews, combines, trucks, and trailers in central Texas in late May to finish in the northern Canadian prairies in August and September. Great Plains farmers are increasingly harvesting their own crop today, though hundreds of migrant harvesting crews still roam the plains each summer.

Wheat Varieties

There are four main types and hundreds of varieties of wheat grown in the United States. Hard wheats are used primarily for bread. Production of hard red winter wheat, the most common type, is concentrated in Kansas and surrounding states, though new varieties have allowed it to expand northward in recent years. Hard red spring wheat, concentrated in the Dakotas, is considered the standard for bread dough, partially explaining the importance of Minneapolis as the nation's first center of large-scale bread flour milling. Soft red winter wheat, used primarily for pastry and cakes, traditionally was most common in the Corn Belt, but has been disappearing from the scene in recent years as those farmers have favored a corn-soybean crop rotation. White wheat, used primarily for pastry and breakfast foods, is almost entirely grown in eastern Washington and adjacent Oregon. Durum, used primarily as the basis for pasta dough, is the least important of the major wheats. The growth in popularity of Italian cuisine in this country, as well as a massive increase in restaurants featuring bagels (which also are made with durum), has caused significant shortages and production is beginning to increase.

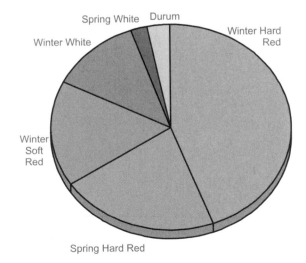

Wheat Varieties, 1993

Spring White
Durum
Winter Hard Red
Winter White
Winter Soft Red
Spring Hard Red

Contemporary Production

Wheat acreages have been stable or declining slightly for many years, though total production has been increasing through higher yields. Approximately a third of the entire annual crop is grown in North Dakota and Kansas each year. Combined production from only five states produces more than half the total national crop; twelve states produce more than three-quarters of the total. This is in sharp contrast to the turn of the century when the production of twelve states had to be combined to account for half the national annual production. Temperature-tolerant varieties and changing tastes have also altered distribution with a general expansion of the traditional Winter Wheat Belt northward and a commensurate shrinking of the Spring Wheat Belt. Increases in durum acreages (for the rapid expansion of pasta and bagel consumption in the United States) are just beginning to become evident.

Production in the Far West is also shifting, especially in the general decline of acreages in California as irrigation expanded into traditional dry farming areas of the Central Valley. Acreages continue to increase in the irrigated sections of the Ranching and Oasis region, especially in Idaho. Canola production is also beginning to appear, but has had little overall impact. Columbia Plain wheat producers continue to favor white wheat used primarily in pastry and breakfast cereals. Irrigated acreages continue to expand in the Ranching and Oasis region where yields can approach 170 bushels per acre, more than four times greater than typical yields in many areas of the East.

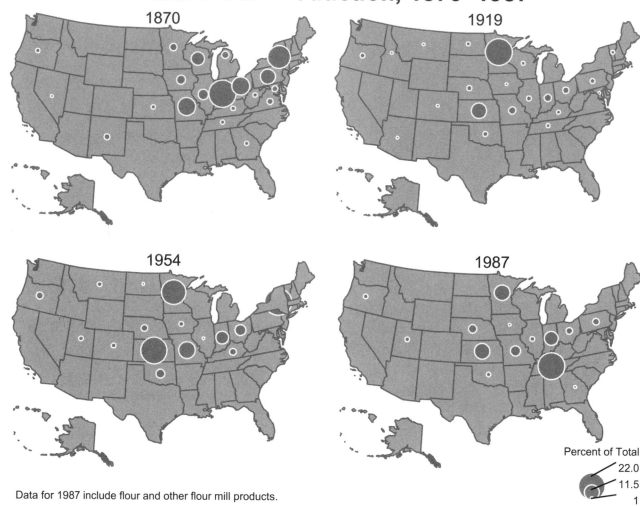

Wheat Flour Production, 1870–1987

1870

1919

1954

1987

Data for 1987 include flour and other flour mill products.

Percent of Total
22.0
11.5
1

Wheat generally has declined throughout the eastern United States where it was used primarily used as a crop rotation element. Significant production continues on the southern Atlantic Coastal Plain, where it is used as a winter ground cover and harvested in late spring, but overall even there production has decreased with declining cropland acreages.

Flour Milling

Flour milling was widely distributed throughout the nation until the beginnings of the railroad era when increasing percentages of grain were shipped long distances through selected gateway cities. This funneling effect, coupled with the declining importance of local production in most of the nation, made it possible for the growth of milling companies in Buffalo, Chicago, Cincinnati, and other transshipment cities. The beginnings of this agglomeration of the industry are evident in 1870. New York (14 percent) was the largest center of milling and Pennsylvania (12 percent) the second, though the overall pattern of decentralization is still evident. By 1900 the situation had altered dramatically. The center of wheat production had moved westward. Minnesota, tapping the Burlington and Great Northern routes across North Dakota, accounted for 22 percent of the nation's wheat flour. Kansas, the second largest producing state, was also the second largest center of milling. More interesting is the concentration of mills producing over 100,000 barrels per annum. Minnesota accounted for almost a third of the nation's wheat flour in that year while the aggregate production from only three states (Minnesota, Kansas, and New York) accounted for more than one half the nation's total. The pattern today has altered even further with the concentration of production almost entirely in transshipment cities, bringing the relative decline of Minnesota to 12 percent, and the relative increase of the mills around Buffalo, New York, almost to a par, much as was seen in 1860. Tennessee production, focusing on Memphis, is the third largest center, followed by Kansas City and St. Louis.

Right: Grain elevators are often the largest structures punctuating the landscape in wheat-producing areas. (RP/JF)

Exports

The United States is the world's largest exporter of wheat and wheat flour (28.5 million metric tons in 1990) with about a third more tonnage per year than Canada (17.5 million tons) and France (17.3 million tons). Valued at more than $4.4 billion in 1990, wheat export continues to play an important role in the nation's foreign policy and balance of payments. China was the largest single destination of American wheat and flour in that year followed by Japan. Sub-Saharan Africa and Pakistan were also major importers of American flour.

IV
The State of the American Cornucopia: An Afterword

The American grain harvest routinely exceeds the nation's storage capacity. (JF/RP)

The report of my death was an exaggeration.
Samuel Clemens

Much has been written about the demise of American agriculture in recent years, but like Samuel Clemens's famous 1897 statement to the press, these prognostications are more sensational than true. Gross farm income in 1992 passed $197 billion, while net income was estimated at $48.6 billion. The lean years of the late 1970s and early 1980s, when farm debt stretched beyond hope, have passed for most farmers and net income has been rising for more than a decade. Certainly many farmers and some farm regions remain in economic jeopardy, and even larger numbers of farmers receive less than bank interest on their farm investment, but increasingly these represent the exception, rather than the rule. Most farm regions are undergoing at least some growth and some are experiencing unparalleled expansion. Hundreds of bright new chicken houses and hog growout barns are appearing across the South and the Midwest, while many western dairy feedlot operators are barely able to keep expanding to meet market demand. Our image of contemporary American agriculture is thus a matter of perspective, perspective that is often missing in the Sunday newspaper and on television magazine shows.

A weekend jaunt to the country or a summer vacation across America's rural landscape would make the average American conclude that farming is little changed from his or her childhood, or for that matter from the childhood of their grandparents. The tractors are still green, the corn is still as high as an elephant's eye, and the farmyards are still neatly kept. This seeming continuity, however, masks one of the most far-reaching economic transformations taking place in contemporary America. The breadth of these changes reaches into virtually every aspect of farm life.

Escalating farm size and declining rural population are the most well-documented aspects of these changes, but are only small parts of the total picture. The relatively high cost of labor, though still well below prices paid in industry, has brought about an almost complete mechanization of the farming cycle. The typical farm equipment storage building is filled with machinery designed to lower labor costs and make farm tasks more efficient, safer, and more comfortable. While much of this equipment is easily recognizable to the nonfarmer, even more is not. For example, green silage has virtually replaced hay in many areas. Traditional rectangular balers have been replaced by silage cutters or round balers. The millions of reusable wooden lug boxes, long stored as familiar grey mountains next to the packing sheds in processed fruit areas, have become collectors items in antique stores, replaced by half-ton field bins, and bulk trailers, fork lifts, and other bulk-handling equipment. The once venerable cow pony is increasingly becoming an ornament on western ranches where all-terrain vehicles and helicopters do the same tasks faster and more reliably. Even the ten-gallon hat (for work) is disappearing in the West and being replaced by the ubiquitous baseball (CAT Power, etc.) cap because of the problems of constantly entering and exiting vehicles with the larger headgear.

Other changes are also taking place that have revolutionized farming even more with little visible indication of their presence. Contract farming has become so important in the production of some crops that virtually no market exists for them when grown on speculation. The computer has ceased being a toy for thousands of farmers who are able to project crop production trends based on information from the Department of Agriculture, receive direct weather data from the National Oceanic and Atmosphere Agency, and watch commodity trading on the floor of the Chicago Board of Trade. Aggressive farmers buy and sell crop futures to protect themselves from the vagaries of the weather and market. Even the crops themselves have been altered through bioengineering and genetic splicing to produce faster maturing broilers, corns for every purpose, and the perfect catfish.

It would be attractive to paint a portrait of the typical American farm, but that is impossible. While the new agriculture is creating an atmosphere that promotes concentration and repetition, it is simultaneously allowing more diversity than ever. Few continue to be isolated from the mainstream of agricultural innovation as farm agents, agricultural cooperatives, equipment and seed manufacturers, and the farm media inundate the farmer with information about new agricultural products and methods. The result has been an increasing standardization of production paralleled by greater independence and a willingness to try alternative approaches and crops. A recent visit to what appeared to be a typical Wisconsin dairy farm found dad working the last alfalfa cut, while the eldest son was shredding recycled newspaper bedding in the dairy barn. Younger brother on the farm across the road, on the other hand, had left the dairy business altogether. His barns had been modified into duck houses for his flocks of 10,000 birds in each cycle. Such diversity is endemic to the current agricultural environment, which concurrently encourages large-scale production and niche marketing. Thus Wisconsin may still be perceived to be the "Nation's Dairyland," but the sale of dairy products accounts for more than half of all agricultural sales in only 56 percent of the state's counties.

Even ignoring hobby farms, estate farms, and part-time operations, today's "typical" farm is taking on multiple personalities. The most well-documented farms are the large, heavily mechanized operations utilizing economies of scale to achieve high net returns. Once concentrated primarily in the West, today they are found in every farming region. The livestock producers—cattle and dairy feedlots, chickens, and hogs—are the most frequently documented of these agricultural giants, but actually the phenomenon has spread across the entire spectrum of farming from the Kern County Land Company, specializing in fruit, nut, and field crops on more than a hundred thousand acres of California's San Joaquin Valley, to the King Ranch raising sugarcane, cattle, and other crops on ranches in Texas and Florida. While seemingly unrelated except in size, these operations are alike in their achievement of vast profits through economies of scale, the utilization of state of the art technology, and a shrewd sense of business.

Simultaneously, there has been an explosion of small farms producing crops for markets too small for the behemoths to negotiate. Growing both highly specialized and common crops, these new "small" farms have adopted strategies that allow

them to thrive. Ginseng, for example, has become the single most important export crop (by value) produced in the state of Wisconsin. Only a handful of farmers are able to occupy this niche, but those that do, do very well. A farmer with only a few acres near Raleigh, North Carolina, has become the nation's largest producer of truffles in another example of the role of the exotic in small farming.

Successful small farmers specializing in common crops tend to focus on local markets where they are able to eliminate most distribution costs and receive higher prices from retailers for fresher, more flavorful produce. Urban foraging has become a major vacation and idle weekend activity in almost every section of the country as city dwellers ostensibly drive into the countryside to watch the change of the seasons, search out cute antique stores, or seek recreation. Millions of these travelers stop at roadside stands to purchase everything from boiled peanuts in the South to raspberries in the Northwest. Astute marketers among these roadside operators have discovered that they can increase their profits even further by processing part of their produce into jams, jellies, pies, apple cider, and a host of other products. Lancaster County, Pennsylvania, for example, attracts millions of tourists each year searching for the elusive Amish carriage. Its farmers earned more $4.7 million from roadside sales in 1992. Similarly, farmers with roadside sales in nearby Delaware County, Pennsylvania, utilizing their closer proximity to Philadelphia and Wilmington, Delaware, *averaged* more than $25,000 per farm from direct sales. Considering that an unknown portion of direct sales goes unrecorded, it is clear that road-

side sales are no longer only the farm wife's source of egg money on thousands of the nation's farms.

It would be easy to assume that all farms are large, that all crops are concentrated, and that all farming is industrial in intensity. However, nothing could be further from the truth. New Jersey may no longer be the Garden State, but more than $500 million of nursery crops, vegetables, peaches, and other fruits, berries, and field crops were produced there in 1992. While most of this book has focused on the zones of highest product concentration, the small-scale, commercial cultivation of crops for human consumption continues in almost every state of the Union where production is environmentally possible. Wisconsin, for example, had commercial apple production in all but eight counties in 1992, almost the same number of counties with no commercial dairying. Large-scale, commercial agriculture is the dominant theme of today's agriculture, but the pockets, individual quirks, and nuances of its farm landscape still remain as important elements of the total fabric.

Bibliography

Nearly all research is based firmly on preceding work. This atlas is no exception. It would not have been possible without the diverse literature in American agriculture and most especially without the publications of the United States Department of Agriculture and the Bureau of the Census.

Sources of Special Interest

We are especially indebted to the following six works which played a crucial role in developing our understanding of contemporary American agriculture. They are the best currently available and should be on everyone's reading list.

Bowler, Ian (ed.). *The Geography of Agriculture in the Developed Market Economies*. London: Longman Scientific & Technical, 1992.

An excellent conceptual introduction to the sweeping changes that are rapidly modifying the geography of agriculture in the developed world.

Bryant, Christopher, and Thomas Johnston. *Agriculture in the City's Countryside*. Toronto: University of Toronto Press, 1992.

The urban fringe is the most rapidly changing agricultural environment in America today. An insightful exploration of this dynamic landscape.

Hart, John Fraser. *The Land That Feeds Us*. New York: Norton, 1991.

Hart has been a keen student of the geography of American agriculture for decades. His book offers a tour of agriculture in the eastern United States, with a continuing look at how the agricultural economy and landscape have changed over the past 20 years.

Heilman, Grant. *Farm*. New York: Abbeville, 1988.

Heilman is the country's leading photographer of the farm scene. This beautifully illustrated book offers both a glance at the country's major agricultural regions and a survey of several dozen of the farm economy's key products.

Hudson, John C. *Making the Corn Belt: A Geographical History of Middle-Western Agriculture*. Bloomington: Indiana University Press, 1994.

Our best look at the historical geography of an American agricultural region. Rich in detail yet broad conceptually. We need a dozen others like it!

Sauer, Jonathan D. *Historical Geography of Crop Plants*. Boca Raton, FL: CRC, 1993.

Sauer provides a carefully documented review of the history and geography of some seventy crop plants, from alfalfa to walnuts, from their wild progenitor to the present. Enormously valuable.

State Agriculture Statistics

Most states publish an annual report of agricultural statistics along with the USDA's National Agricultural Statistics Service. While they vary somewhat in quality and format, most provide a wealth of data, maps, and other graphics plus some generally limited commentary.

United States Bureau of the Census. *Census of Agriculture*. Washington: Bureau of the Census, 1950– .

The census first began to ask agricultural enumeration questions in 1840. The agricultural census was included in the national decennial census until 1950, with special agricultural censuses in 1925, 1935, and 1945. The agricultural census was separated from the decennial census after 1950. Post-1950 agricultural censuses were conducted in 1954, 1959, 1964, 1969, 1974, 1978, 1982, 1987, and 1992. The plan is to continue a five-year census cycle in years ending in 2 and 7.

Most of the maps in this atlas are derived from census data.

United States Department of Agriculture. National Agricultural Statistics Service. Various titles. Various dates

In existence since 1961, but known as the USDA Statistical Reporting Service prior to 1986. The service publishes both annual surveys of production and more lengthy extended reviews of selected crops. Of special use to us was their *Crop Production*, an annual survey (released in January) of production by state for nearly fifty crops.

———. *Agricultural Statistics*. Washington, DC: Superintendent of Documents, 1936– .

Published annually since 1936. A lengthy statistical review of American (and foreign) agriculture. Includes sections on income and expenses, price support programs, consumption, taxes and credit programs, and import and exports in addition to production data. Much of the statistical data had previously been published annually in the *Agricultural Yearbook*. A necessary resource for anyone interested in the geography of American agriculture.

———. *Yearbook of Agriculture*. Washington, DC: United States Department of Agriculture, 1894– .

Published annually since 1894 under slight variations in title. Prior to 1936 agricultural statistics (see above) were published as a statistical section of the *Yearbook*. Each yearbook has focused on a single theme since 1938's *Soils and Men*. Earlier yearbooks contained often lengthy reviews of a wide variety of agricultural issues.

Other Sources of Interest

Adams, Leon D. *The Wines of America* (3d ed.). New York: McGraw-Hill, 1985.

Albrecht, Don E., and Steve H. Murdock. *The Sociology of U.S. Agriculture: An Ecological Approach*. Ames: Iowa State University Press, 1990.

Ankli, Robert, and Alan Olmstead. "The Adoption of the Gasoline Tractor in California." *Agricultural History* 55 (1982): 213–30.

Armstrong, R. Warwick (ed.). *Atlas of Hawaii* (2d ed.). Honolulu: University of Hawaii Press, 1983.

Atack, Jeremy, and Fred Bateman. *To Their Soil. Agriculture in the Antebellum North*. Ames: Iowa State University Press, 1987.

Axton, W. F. *Tobacco and Kentucky*. Lexington: University of Kentucky Press, 1975.

Baker, Donald G., et al. "Agriculture and the Recent 'Benign Climate' in Minnesota." *Bulletin of the American Meteorological Society* 74 (1993): 1035–40.

Baker, Oliver E. *A Graphic Summary of American Agriculture Based Largely on the Census*. USDA Miscellaneous Publication 105. Washington, DC: Superintendent of Documents, 1931.

———. "Agricultural Regions of North America." *Economic Geographer*, 1926–32. (An 11-part work that appeared serially.)

Ball, Carleton. "The Grain Sorghums: Immigrant Crops That Have Made Good." *Yearbook of the United States Department of Agriculture, 1913*. Washington, DC: Government Printing Office, 1914, pp. 221–38.

Ballinger, Roy A., *A History of Sugar Marketing through 1974*. Washington, DC: United States Department of Agriculture, 1978.

Baltensperger, Bradley H. *Nebraska: A Geography*. Boulder, CO: Westview, 1985.

Beal, George M., and Henry Bakken. *Fluid Milk Marketing*. Madison, WI: Mimir, 1956.

Bidwell, Percy, and John Falconer. *History of Agriculture in the Northern United States, 1620–1860*. Washington, DC: Carnegie Institution of Washington, 1925.

Birdsall, Stephen S., and John W. Florin. *Regional Landscapes of the United States and Canada* (4th ed.). New York: Wiley, 1992.

Bixby, Donald E., et al. *Taking Stock: The North American Livestock Census*. Blacksburg, VA: McDonald and Woodward, 1994.

Blouet, Brian, and Frederick Luebke (eds.). *The Great Plains: Environment and Culture*. Lincoln: University of Nebraska Press, 1979.

Bogue, Allan G. *From Prairie to Cornbelt: Farming on the Illinois and Iowa Prairies in the Nineteenth Century*. Chicago: University of Chicago Press, 1963.

Bogue, Margaret B. "The Lake and the Fruit: The Making of Three Farm-Type Areas." *Agricultural History* 59 (1985): 493–522.

Borchert, John R. *America's Northern Heartland: An Economic and Historical Geography of the Upper Middle West*. Minneapolis: University of Minnesota Press, 1986.

Bowler, Ian, et al. (eds.). *Contemporary Rural Systems in Transition*. Vol. 1: *Agriculture and Environment*. Wallingford, UK: CAB International, 1992.

Broadway, Michael J., and Terry Ward. "Recent Changes in the Structure and Localization of the U.S. Meat Packing Industry." *Geography*, 75 (1990): 76–79.

Brown, F. Lee, and Helen M. Ingram. *Water and Poverty in the Southwest*. Tucson: University of Arizona Press, 1990.

Bruchey, Stuart (comp. and ed.). *Cotton and the Growth of the American Economy: 1790–1860: Sources and Readings*. New York: Harcourt, Brace, 1967.

Burbach, Roger, and Patricia Flynn. *Agribusiness in the Americas*. New York: Monthly Review Press, 1980.

Carlson, Alvar W. *The Spanish-American Homeland: Four Centuries in New Mexico's Rio Arriba*. Baltimore: Johns Hopkins University Press, 1990.

Carlson, Paul. "Indian Agriculture: Subsistence Patterns and the Environment on the Southern Great Plains. *Agricultural History* 66 (1992): 52–60.

Carney, Judith, and Richard Porcher. "Geographies of the Past: Rice, Slaves and Technological Transfer in South Carolina." *Southeastern Geographer* 33 (1993): 127–47.

Cochrane, Willard W. *The Development of American Agriculture: A Historical Analysis* (2d ed.). Minneapolis: University of Minnesota Press, 1993.

Cohen, David. *The Dutch-American Farm*. New York: New York University Press, 1992.

Comeaux, Malcolm. *Arizona: A Geography*. Boulder, CO: Westview, 1990.

Conrat, Maisie, and Richard Conrat. *The American Farm: A Photographic History*. San Francisco: California Historical Society, 1977.

Conzen, Michael P., et al. *A Scholar's Guide to Geographical Writing on the American and Canadian Past*. Chicago: University of Chicago, Department of Geography Research Paper # 235, 1993.

Corbett, L.C., et al. "Horticultural Manufactures." *Agriculture Yearbook, 1925*. Washington, DC: U.S. Government Printing Office, 1926.

Cuff, Timothy. "A Weighty Issue Revisited: New Evidence on Commercial Swine Weights and Pork Production in Mid-Nineteenth Century America." *Agricultural History* 66 (1992): 55–74.

Daniel, Pete. *Breaking the Land: The Transformation of Cotton, Tobacco, and Rice Cultures Since 1880*. Urbana: University of Illinois Press, 1985.

Dent, Borden. *Cartography: Thematic Map Design* (3d ed.). Dubuque, IA: William C. Brown, 1993.

Dethloff, Henry. *A History of the American Rice Industry, 1685–1985*. College Station: Texas A&M University Press, 1988.

Dippel, Joseph. *Connecticut's Farmland Preservation Program 15th Anniversary (1978–1993) Report*. Hartford: Department of Agriculture, 1994.

Donnell, E.J. *Chronological and Statistical History of Cotton*. New York: by the author, 1872. Reprinted. Wilmington, DE: Scholarly Resources, 1973.

Doolittle, William. "Agriculture in North America on the Eve of Contact: A Reassessment." *Annals*, Association of American Geographers 82 (1992): 386–401.

Durand, Loyal. "The Migration of Cheese Manufacture in the United States." *Annals*, Association of American Geographers 42 (1952): 263–82.

Earle, F. S., "Development of Trucking Interest." *Yearbook of the United States Department of Agriculture, 1900*. Washington, DC: U.S. Government Printing Office, 1901, 437–52.

Eck, Paul. *The American Cranberry*. New Brunswick, NJ: Rutgers University Press, 1990.

Fernald, Edward (ed.). *Atlas of Florida*. Tallahassee: Florida State University Foundation, 1981.

Folger, J.C. "The Commercial Apple Industry in the United States." *Yearbook of American Agriculture*. Washington, DC: U.S. Government Printing Office, 1918.

Francaviglia, Richard. *The Mormon Landscape*. New York: AMS, 1978.

Garner, Wightman. *The Production of Tobacco* (1st rev. ed.). New York: Blakiston, 1951.

Garrison, J. Ritchie. *Landscape and Material Life in Franklin County, Massachusetts, 1770-1860*. Knoxville: University of Tennessee Press, 1991.

Gates, Paul W. *The Farmer's Age: Agriculture, 1815–1860*. New York: Harper and Row, 1968.

Golze, Alfred R. *Reclamation in the United States*. New York: McGraw-Hill, 1952.

Gregor, Howard F. *Industrialization of U.S. Agriculture: An Interpretive Atlas*. Boulder, CO: Westview, 1982.

Greig, W. Smith. *Economics and Management of Food Processing*. Westport, CT: AVI, 1984.

Griffiths, Mel, and Lynnell Rubnight. *Colorado: A Geography*. Boulder, CO: Westview, 1983.

Grotewold, Andreas. *Regional Changes in Crop Production in the United States from 1909 to 1949*. Chicago: University of Chicago, Department of Geography Research Paper # 40, 1955.

Hampe, Edward, Jr., and Merle Wittenberg. *The Lifeline of America: Development of the Food Industry*. New York: McGraw-Hill, 1964.

Hanfman, Deborah, et al. *The Potentials of Aquaculture: An Overview and Bibliography*. Beltsville, MD: National Agricultural Library, 1989.

Hart, John Fraser. "Change in the Corn Belt." *Geographical Review* 76 (1986): 51–72.

———. "The Character of Tobacco Barns and Their Role in the Tobacco Economy of the United States." *Annals*, Association of American Geographers 51 (1961): 274–93.

———. "The Middle West." *Annals*, Association of American Geographers 62 (1972): 258–82.

Haystead, Ladd, and Gilbert C. Fite. *The Agricultural Regions of the United States*. Norman: University of Oklahoma Press, 1955.

Hayter, Earl W. *The Troubled Frontier, 1850–1900: Rural Adjustment to Industrialization*. DeKalb: Northern Illinois University Press, 1968.

Hewes, Leslie. *The Suitcase Farming Frontier: A Study in the Historical Geography of the Central Great Plains*. Lincoln: University of Nebraska Press, 1983.

Higbee, Edward. *American Agriculture: Geography, Resources, Conservation*. New York: Wiley, 1958.

Hilliard, Sam B. *Atlas of Antebellum Southern Agriculture*. Baton Rouge: Louisiana State University Press, 1984.

———. *Hog Meat and Hoecake: Food Supply in the Old South, 1840–1860*. Carbondale: Southern Illinois University Press, 1972.

Hudson, John C. *Plains Country Towns*. Minneapolis: University of Minnesota Press, 1982.

Isaacson, Dorris (ed.) *Maine: A Guide 'Down East'* (2d ed.). Rockland, ME: Courier-Gazette, 1970.

Jordan, Terry G. "The Origins and Distribution of Open-Range Ranching." *Social Science Quarterly* 53 (1972): 105–21.

———. *Trails to Texas: Southern Roots of Western Cattle Ranching*. Lincoln: University of Nebraska Press, 1981.

Jull, M.S. et al. "The Poultry Industry." *Agriculture Yearbook, 1924*. Washington, DC: U.S. Government Printing Office, 1925, 377–455.

Kelso, Maurice M., William E. Martin, and Lawrence E. Mack. *Water Supplies and Economic Growth in an Arid Environment: An Arizona Case Study*. Tucson: University of Arizona Press, 1973.

Kennedy, Joseph C.G. *Agriculture of the United States in 1860; Compiled from the Original Returns of the Eighth Census*. Washington, DC: U.S. Government Printing Office, 1864.

Kolmos, John. "The Height and Weight of West Point Cadets: Dietary Change in Antebellum America." *Journal of Economic History* 47 (1987): 897–927.

Krapovickas, A. "The origins, variability, and spread of the groundnut (Arachis hypgaea)." In P. A. Xoc and S. A. Banack (eds.), *The Domestication and Exploitation of Plants and Animals*. Chicago: Aldine, 1964.

Lantis, David W., with Rodney Steiner and Arthur Karinen. *California: Land of Contrast* (2d ed.). Belmont, CA: Wadsworth, 1970.

Larson, C.W. "The Dairy Industry." *Yearbook of the Department of Agriculture, 1922*. Washington, DC: U.S. Government Printing Office, 1923, 281–394.

Lemon, James T. *The Best Poor Man's Country*. Baltimore: Johns Hopkins University Press, 1972.

Lewis, Thomas R. "Declining Cigar Tobacco Production in Southern New England." *Journal of Geography* 79 (1980): 108–11.

Lewthwaite, Gordon R. "Wisconsin Cheese and Farm Type: A Locational Hypothesis." *Economic Geography* 40 (1964): 95–112.

Maass, Arthur, and Raymond L. Anderson. *. . . and the Desert Shall Rejoice: Conflict, Growth, and Justice in Arid Environments*. Cambridge, MA: MIT Press, 1978.

Malin, James C. *Winter Wheat in the Golden Belt of Kansas: A Study in Adaptation of Subhumid Geographical Environment*. Lawrence: University of Kansas Press, 1944.

Martin, Harold. *A Good Man. . . A Great Dream: D. W. Brooks of Gold Kist*. Atlanta: Gold Kist, 1982.

Mather, E. Cotton. "The American Great Plains." *Annals, Association of American Geographers* 62 (1972): 237–57.

Mayo, Nathan. *Possibilities of the Everglades* (rev.). Tallahassee: Department of Agriculture, State of Florida, 1940.

McCorkle, James Jr. "Moving Perishables to Market: Southern Railroads and the Nineteenth-Century Origins of Southern Truck Farming." *Agricultural History* 66 (1966): 42–62.

McKnight, Tom L. *Regional Landscapes of the United States and Canada*. Englewood Cliffs, NJ: Prentice-Hall, 1992.

McMichael, Philip (ed.). *The Global Restructuring of Agro-Food Systems*. Ithaca, NY: Cornell University Press, 1994.

McPhee, John. *Oranges*. New York: Farrar, Straus, 1967.

Meinig, Donald W. *The Great Columbia Plain; A Historical Geography, 1805–1910*. Seattle: University of Washington Press, 1968.

———. "The Growth of Agricultural Regions in the Far West, 1850–1910." *Journal of Geography* 54 (1955): 211–32.

Molnar, Joseph J. (ed.) *Agricultural Change: Consequences for Southern Farms and Rural Communities*. Boulder, CO: Westview, 1986.

Monmonier, Mark. *How to Lie with Maps*. Chicago: University of Chicago Press, 1991.

Morgan, Joseph R. *Hawaii: A Geography*. Boulder, CO: Westview, 1983.

Morgan, Dan. *Merchants of Grain*. New York: Viking, 1979.

Norris, Frank. *The Octopus: A Story of California*. Garden City, NY: Doubleday, 1901.

Oemler, A. *Truck-Farming at the South: A Guide to the Raising of Vegetables for Northern Markets*. New York: Orange Judd Company, 1883.

Parker, N. William. "Two Hidden Sources of Productivity Growth in American Agriculture, 1860–1890." *Agricultural History* 56 (1982): 648–61.

Parman, Donald L. "New Deal Indian Agricultural Policy and the Environment: The Papagos as a Case Study." *Agricultural History* 66 (1992): 23–33.

Peet, J. Richard. "The Spatial Expansion of Commercial Agriculture in the Nineteenth Century: A Von Thunen Analysis." *Economic Geography* 45 (1969): 283–301.

Percy, David O. "Ax or Plow? Significant Colonial Landscape Alteration Rates in the Maryland and Virginia Tidewater." *Agricultural History* 66 (1992): 66–75.

Pfister, A.J. *The Salt River Project: Keeping the Spirit Strong*. New York: Newcomen Society of the United States, 1987.

Pierce, John. "Towards the Reconstruction of Agriculture: Paths of Change and Adjustment." *Professional Geographer* 46 (1994): 178–90.

Pillsbury, Richard. *From Boarding House to Bistro: The American Restaurant Then and Now*. London: Unwin & Hyman, 1990.

Pinney, Thomas. *A History of Wine in America: From the Beginnings to Prohibition*. Berkeley: University of California Press, 1989.

Pisani, Donald J. *From the Family Farm to Agribusiness: The Irrigation Crusade in California and the West, 1850–1931*. Berkeley: University of California Press, 1984

———. *To Reclaim a Divided West: Water, Law, and Public Policy, 1842–1902*. Albuquerque: University of New Mexico Press, 1992.

PlantFinder. Pembroke Pines, FL: Betrock Information Systems (various dates).

"The Power of Pork." *Raleigh News and Observer* 2/19/95 to 2/26/95.

Punch, Walter T. (ed.). *Keeping Eden: A History of Gardening in America*. Boston: Little, Brown, 1992.

Rafferty, Milton. *The Ozarks: Land and Life*. Norman: University of Oklahoma Press, 1980.

Robbelen, Gerhard, et al. *Oil Crops of the World: Their Breeding and Utilization*. New York: McGraw Hill, 1989.

Russell, Howard S. *A Long, Deep Furrow: Three Centuries of Farming in New England*. Hanover, NH: University Press of New England, 1976.

Sayer, Robert F. (ed.). *Take this Exit; Rediscovering the Iowa Landscape*. Ames: Iowa State University Press, 1989.

Scheuring, Ann (ed.). *A Guidebook to California Agriculture*. Berkeley: University of California Press, 1983.

Schwartz, Marvin. *Tyson: From Farm to Market*. Fayetteville: University of Arkansas Press, 1991.

Shortridge, James R. *The Middle West: Its Meaning in American Culture*. Lawrence: University of Kansas Press, 1989.

Skaggs, Jimmy M. *Prime Cut: Livestock Raising and Meat Packing in the United States, 1607–1983*. College Station: Texas A&M University Press, 1987.

Spencer, Joseph E., and Ronald J. Horvath. "How Does an Agricultural Region Originate." *Annals*, Association of American Geographers 53 (1963): 74–92.

Sommers, Lawrence M. *Michigan: A Geography*. Boulder, CO: Westview, 1984.

Soule, Judith D., and Jon K. Piper. *Farming in Nature's Image: An Ecological Approach to Agriculture*. Washington, DC: Island, 1992.

Spillman, W.J. "Types of Farming in the United States." *Yearbook of the United States Department of Agriculture, 1908*. Washington, DC: U.S. Government Printing Office, 1909, 351–66.

Street, James H. *The New Revolution in the Cotton Economy: Mechanization and its Consequences*. Chapel Hill: University of North Carolina Press, 1957.

Swanson, Louis E. (ed.). *Agriculture and Community Change in the U.S.* Boulder, CO: Westview, 1988.

Taylor, William. "The Influence of Refrigeration on the Fruit Industry," *Yearbook of the United States Department of Agriculture, 1900*. Washington, DC: U.S. Government Printing Office, 1901, 561–80.

Thompson, John, (ed.). *Geography of New York State*. Syracuse, NY: Syracuse University Press, 1966.

Tucker, David M. *Kitchen Gardening in America; A History*. Ames: Iowa State University, 1993.

U.S. Bureau of the Census. *Historical Statistics of the United States, Colonial Times to 1957*. Washington, DC: U.S. Government Printing Office, 1960.

United States Department of Agriculture. *A Chronology of American Agriculture, 1776–1976*. Washington, DC: USDA Economic Research Service, 1977.

Vogeler, Ingolf, et al. *Wisconsin: A Geography*. Boulder, CO: Westview, 1986.

Wahl, Richard W. *Markets for Federal Water: Subsidies, Property Rights and the Bureau of Reclamation*. Washington, DC: Resources for the Future, 1989.

Walsh, Margaret A. *The Rise of the Midwestern Meat Packing Industry*. Lexington: University Press of Kentucky, 1982.

———. "Pork Packing as a Leading Edge Midwestern Industry, 1835–1875." *Agricultural History* 51 (1977): 704.

Ward, Ronald, and Richard Kilmer. *The Citrus Industry: A Domestic and International Economic Perspective*. Ames: Iowa State University Press, 1989.

Webb, Walter Prescott. *The Great Plains*. New York: Ginn, 1931.

Whayne, Jeannie. "Creation of a Plantation System in the Arkansas Delta in the Twentieth Century." *Agricultural History* 66 (1992): 63–84.

Wiest, Edward. *The Butter Industry in the United States: An Economic Study of Butter and Oleomargarine, Studies in History, Economics and Public Law, Whole Number 165*. New York: Columbia University Press, 1916.

Wik, Renold. "The Radio in Rural America During the 1920s." *Agricultural History* 55 (1981): 339–50.

Wilhelm, Eugene J. "Animal Drives: A Case Study in Historical Geography." *Journal of Geography* 66 (1967): 327–34.

Williams, Michael. *Americans and Their Forests: A Historical Geography*. Cambridge: Cambridge University Press, 1989.

Winsberg, Morton D. "Agricultural Specialization in the United States Since World War II." *Agricultural History* 56 (1982): 692–701.

———. "South Florida Agriculture." In Thomas Boswell (ed.), *South Florida: The Winds of Change*. Washington, DC: Association of American Geographers, 1991, 17–30.

Yamajuchi, Mas. *World Vegetables: Principles, Production and Nutritional Values*. Westport, CT: AVI, 1983.

General Index

Geographic Index